15th Meeting on the Mathematics of Language (MoL 2017)

London, United Kingdom
13 – 14 July 2017

Editors:

Makoto Kanazawa
Philippe de Groote
Mehrnoosh Sadrzadeh

ISBN: 978-1-5108-4360-8

MOL 2017

The 15th Meeting on the Mathematics of Language

Proceedings of the Conference

July 13–14, 2017
London, UK

Sponsored by

Introduction

This is the proceedings of the 15th Meeting on the Mathematics of Language (MOL 2017), held at Queen Mary University of London, on July 13–14, 2017.

The volume contains eleven regular papers and two invited papers. It also includes an abstract of a third invited talk. The regular papers were selected from a total of 23 submissions, using the EasyChair conference management system.

The conference benefited from the financial support of the British Logic Colloquium (`http://www.blc-logic.org`) and of AYLIEN (`http://aylien.com`), which we gratefully acknowledge.

Last but not least, we would like to express our sincere gratitude to all the reviewers for MOL 2017 and to all the people who helped with the local organization.

Makoto Kanazawa, Philippe de Groote, and Mehrnoosh Sadrzadeh (editors)

Program Chairs:

Makoto Kanazawa, National Institute of Informatics (Japan)
Philippe de Groote, INRIA Nancy (France)

Local Chair:

Mehrnoosh Sadrzadeh, Queen Mary University of London (UK)

Local Organization:

Sophie Chesney, Queen Mary University of London (UK)
Maximilian Droog-Hayes, Queen Mary University of London (UK)
Dimitri Kartsaklis, University of Cambridge (UK)
Shalom Lappin, University of Gothenburg (Sweden)
Stephen McGregor, Queen Mary University of London (UK)
Graham White, Queen Mary University of London (UK)
Sue White, Queen Mary University of London (UK)
Gijs Jasper Wijnholds, Queen Mary University of London (UK)

Program Committee:

Henrik Björklund, Umeå University (Sweden)
David Chiang, University of Notre Dame (USA)
Alexander Clark, King's College London (UK)
Carlos Gómez-Rodríguez, University of A Coruña (Spain)
Jeffrey Heinz, University of Delaware (USA)
Gerhard Jäger, University of Tübingen (Germany)
Greg Kobele, University of Chicago (USA)
Marco Kuhlmann, Linköping University (Sweden)
Giorgio Magri, CNRS (France)
Andreas Maletti, Universität Leipzig (Germany)
Jens Michaelis, Bielefeld University (Germany)
Larry Moss, Indiana University, Bloomington (USA)
Valeria de Paiva, Nuance Communications (USA)
Gerald Penn, University of Toronto (Canada)
Carl Pollard, The Ohio State University (USA)
Jim Rogers, Earlham College (USA)
Ed Stabler, Nuance Communications (USA)
Mark Steedman, Edinburgh University (UK)
Anssi Yli-Jyrä, University of Helsinki (Finland)

Additional Reviewer:

Justin Debenedetto, University of Notre Dame (USA)

Invited Speakers:

Stephen Clark, University of Cambridge (UK)
Shay Cohen, University of Edinburgh (UK)
Frank Drewes, Umeå University (Sweden)

Table of Contents

Conference Program

Thursday, July 13

09:00-09:30 *Registration and Opening*

Session 1

09:30-10:10 *BE Is Not the Unique Homomorphism That Makes the Partee Triangle Commute*
Junri Shimada

10:10-10:50 *How Many Stemmata with Root Degree k?*
Armin Hoenen, Steffen Eger and Ralf Gehrke

10:50-11:05 *Coffee Break*

Session 2

11:05-11:45 *On the Logical Complexity of Autosegmental Representations*
Adam Jardine

11:45-12:25 *Extracting Forbidden Factors from Regular Stringsets*
James Rogers and Dakotah Lambert

12:25-14:10 *Lunch*

14:10-14:30 *S.-Y. Kuroda Prize Ceremony*

14:30-14:35 *Break*

Thursday, July 13 (continued)

Session 3: Invited Talk

14:35-15:35   *Latent-Variable PCFGs: Background and Applications*
Shay Cohen

15:35-15:50   *Coffee Break*

Session 4

15:50-16:30   *A Proof-Theoretic Semantics for Transitive Verbs with an Implicit Object*
Nissim Francez

16:30-17:10   *Why We Speak*
Rohit Parikh

17:10-17:50   *A Monotonicity Calculus and Its Completeness*
Thomas Icard, Lawrence Moss and William Tune

19:00-   *Conference Dinner at QMUL Octagon*

Friday, July 14

Friday, July 14 (continued)

Session 5: Invited Talk

09:30-10:30   *DAG Automata for Meaning Representation*
Frank Drewes

10:30-10:45   *Coffee Break*

Session 6

10:45-11:25   *(Re)introducing Regular Graph Languages*
Sorcha Gilroy, Adam Lopez, Sebastian Maneth and Pijus Simonaitis

11:25-12:05   *Graph Transductions and Typological Gaps in Morphological Paradigms*
Thomas Graf

12:05-13:50   *Lunch*

13:50-14:30   *Business Meeting*

14:30-14:35   *Break*

BE Is Not the Unique Homomorphism
That Makes the Partee Triangle Commute

Junri Shimada

Tokyo Keizai Univeristy, Tokyo, Japan
Meiji Gakuin University, Tokyo, Japan
Keio University, Tokyo, Japan
junrishimada@gmail.com

Abstract

Partee (1986) claimed without proof that the function BE is the only homomorphism that makes the Partee triangle commute. This paper shows that this claim is incorrect unless "homomorphism" is understood as "complete homomorphism." It also shows that BE and A are *the* inverses of each other on certain natural assumptions.

1 Introduction

In a famous and influential paper, Partee (1986) discussed type-shifting operators for NP interpretations, including lift, ident and BE:

$$\text{lift} = \lambda x \lambda P.\, P(x),$$
$$\text{ident} = \lambda x \lambda y.\, [y = x],$$
$$\text{BE} = \lambda \mathscr{P} \lambda x.\, \mathscr{P}(\lambda y.\, [y = x]).$$

She pointed out that these operators satisfy the equality BE ∘ lift = ident, so the following diagram, now often referred to as the Partee triangle, commutes.

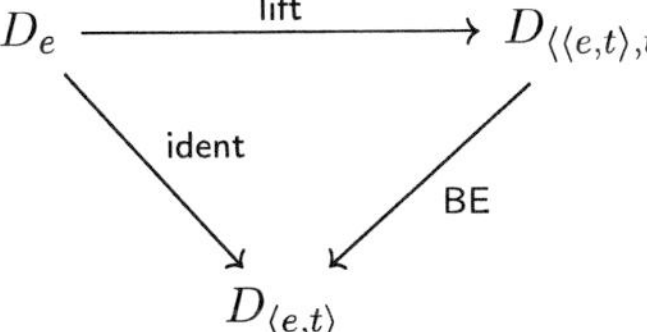

Diagram 1: The Partee triangle

Partee declared that BE is "natural" because of the following two "facts."

Fact 1. BE *is a homomorphism from* $\langle \langle e, t \rangle, t \rangle$ *to* $\langle e, t \rangle$ *viewed as Boolean structures, i.e.,*

$$\text{BE}(\mathscr{P}_1 \sqcap \mathscr{P}_2) = \text{BE}(\mathscr{P}_1) \sqcap \text{BE}(\mathscr{P}_2),$$
$$\text{BE}(\mathscr{P}_1 \sqcup \mathscr{P}_2) = \text{BE}(\mathscr{P}_1) \sqcup \text{BE}(\mathscr{P}_2),$$
$$\text{BE}(\neg \mathscr{P}_1) = \neg \text{BE}(\mathscr{P}_1).$$

Fact 2. BE *is the unique homomorphism that makes the diagram commute.*

While Fact 1 is immediate, Fact 2 is not obvious. Partee (1986) nevertheless did not give a proof of Fact 2, but only a note saying, "Thanks to Johan van Benthem for the fact, which he knows how to prove but I don't." Meanwhile, van Benthem (1986) referred to Partee's work and stated Fact 2 on p. 68, but gave no proof either. Despite this quite obscure exposition, because of the classic status of Partee's and van Benthem's work, I suspect that many linguists take Fact 2 for granted while unable to explain it. Not only is this unfortunate, but it is actually as expected, because Fact 2 turns out to be not quite correct unless "homomorphism" is read as "complete homomorphism." The main purpose of this paper is to rectify this detrimental situation.

Van Benthem (1986) took the domain of entities to be finite, writing, "Our general feeling is that natural language requires the use of *finite models* only" (p. 7). Fact 2 is indeed correct on this assumption. However, natural language has predicates like *natural number* whose extensions are obviously infinite. Also, if we take the domain of portions of matter in the sense of Link (1983) to be a nonatomic join-semilattice, then the domain of entities will surely be infinite, whether countable or uncountable. It is a fact that a single sentence of natural language, albeit only finitely long, can talk about an infinite number of entities, as exemplified in (1).

(1)   a.   Every natural number is odd or even.
         b.   All water is wet. (Link, 1983)

Given this, it is linguistically unjustified to assume the domain of entities to be finite. Since Partee (1986) herself discussed Link (1983), she was certainly aware that the domain of entities might very

Proceedings of the 15th Meeting on the Mathematics of Language, pages 1–10,
London, UK, July 13–14, 2017. ©2017 Association for Computational Linguistics

well be infinite, so it is unlikely that Partee followed van Benthem about the size of the domain of entities.

What difference does it make if the domain D_e of entities is infinite, then? Fact 2 would be correct if "homomorphism" were read as "complete homomorphism." A complete Boolean homomorphism is a Boolean homomorphism that in addition preserves infinite joins and meets. It is clear from the equalities given in Partee's Fact 1 that she did not mean complete homomorphism by the word "homomorphism." When D_e is finite, this does not matter since in that case, $D_{\langle\langle e,t\rangle,t\rangle}$ is also finite, and consequently, every Boolean homomorphism from $D_{\langle\langle e,t\rangle,t\rangle}$ is necessarily complete. However, when D_e is infinite, so is $D_{\langle\langle e,t\rangle,t\rangle}$, and in that case, a Boolean homomorphism from $D_{\langle\langle e,t\rangle,t\rangle}$ can be incomplete, and "Fact" 2 turns out to be false.

This paper essentially consists of extended notes on Partee (1986). Section 2 shows that BE is the unique complete homomorphism that makes the Partee triangle commute and also that it is not the unique homomorphism that does so if D_e is infinite. Section 3 discusses why it is important that BE is complete by examining its interaction with A. Finally, Section 4 shows that A is special among the many inverses of BE. The paper assumes the reader's basic familiarity with Boolean algebras and does not provide definitions or explanations of the technical terms that are used. I would suggest Givant and Halmos (2009) as a good general reference.

2   Uniqueness and Nonuniqueness Proofs

Since it is cumbersome to work with functions, let's adopt set talk. The operators lift, ident and BE and the Partee triangle can be rendered as follows, where $D = D_e$ is a nonempty set of entities and $\wp$ denotes power set.[1]

$$\text{lift} = \lambda x.\,\{P \in \wp(D) \mid x \in P\},$$
$$\text{ident} = \lambda x.\,\{x\},$$
$$\text{BE} = \lambda \mathscr{P}.\,\{x \in D \mid \{x\} \in \mathscr{P}\}.$$

---

[1] Here and below, λ's are used merely to describe functions; they are not meant to be symbols in a logical language that are to be interpreted.

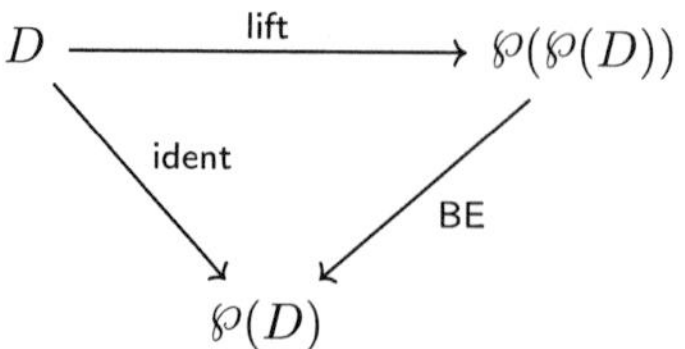

Diagram 2: The Partee triangle
(set talk rendition)

Theorem 1. BE *is a complete homomorphism from* $\wp(\wp(D))$ *to* $\wp(D)$.

Proof. It suffices to show that BE preserves arbitrary unions and complements (denoted by ^{c}). If $\{\mathscr{P}_i\}_{i \in I}$ is an arbitrary family in $\wp(\wp(D))$,

$$\text{BE}\Big(\bigcup_{i \in I} \mathscr{P}_i\Big) = \Big\{x \in D \mid \{x\} \in \bigcup_{i \in I} \mathscr{P}_i\Big\}$$
$$= \bigcup_{i \in I}\{x \in D \mid \{x\} \in \mathscr{P}_i\}$$
$$= \bigcup_{i \in I} \text{BE}(\mathscr{P}_i).$$

For all $\mathscr{P} \in \wp(\wp(D))$,

$$\text{BE}(\mathscr{P}^{\text{c}}) = \{x \in D \mid \{x\} \in \mathscr{P}^{\text{c}}\}$$
$$= \{x \in D \mid \{x\} \notin \mathscr{P}\}$$
$$= \{x \in D \mid \{x\} \in \mathscr{P}\}^{\text{c}}$$
$$= \text{BE}(\mathscr{P})^{\text{c}}. \qquad \square$$

Lemma 2. *Let* h *be a homomorphism from* $\wp(\wp(D))$ *to* $\wp(D)$. *The following conditions are equivalent.*

(i) $h = \text{BE}$.

(ii) *For all* $x \in D$ *and all* $\mathscr{P} \in \wp(\wp(D))$, *if* $x \in h(\mathscr{P})$ *then* $\{x\} \in \mathscr{P}$.

Proof. (i) $\Rightarrow$ (ii). Obvious since $x \in \text{BE}(\mathscr{P})$ iff $\{x\} \in \mathscr{P}$.

(ii) $\Rightarrow$ (i). We show the contrapositive. Suppose $h \neq \text{BE}$, so there exists some $\mathscr{P} \in \wp(\wp(D))$ such that

$$h(\mathscr{P}) \neq \text{BE}(\mathscr{P}) = \{x \in D \mid \{x\} \in \mathscr{P}\}.$$

Then, either there is some $a \in D$ such that $a \in h(\mathscr{P})$ and $\{a\} \notin \mathscr{P}$ or there is some $a \in D$ such that $a \notin h(\mathscr{P})$ and $\{a\} \in \mathscr{P}$. In the latter case, we have $a \in h(\mathscr{P})^{\text{c}} = h(\mathscr{P}^{\text{c}})$ and $\{a\} \notin \mathscr{P}^{\text{c}}$. Thus, in either case, (ii) does not hold. $\qquad \square$

Lemma 3. *Let* h *be a homomorphism from* $\wp(\wp(D))$ *to* $\wp(D)$ *such that* $h \circ \text{lift} = \text{ident}$. *For any nonsingleton set* $P \in \wp(D)$, $h(\{P\}) = \varnothing$.

Proof. We first show that $h(\{\varnothing\}) = \varnothing$. Assume that there exists some $a \in h(\{\varnothing\})$. Since $\varnothing \notin \mathsf{lift}(a)$, we have $\{\varnothing\} \subseteq \mathsf{lift}(a)^{\mathsf{c}}$. Since h is a homomorphism and hence preserves order,

$$
\begin{aligned}
h(\{\varnothing\}) &\subseteq h(\mathsf{lift}(a)^{\mathsf{c}}) \\
&= h(\mathsf{lift}(a))^{\mathsf{c}} \\
&= (h \circ \mathsf{lift})(a)^{\mathsf{c}} \\
&= \mathsf{ident}(a)^{\mathsf{c}} \\
&= \{a\}^{\mathsf{c}},
\end{aligned}
$$

so $a \notin h(\{\varnothing\})$, a contradiction.

Next, we show that $h(\{P\}) = \varnothing$ if $|P| \geq 2$. Assume that for some P with $|P| \geq 2$, there exists some $a \in h(\{P\})$. Since $|P| \geq 2$, there is some $b \in P$ with $b \neq a$. Since $P \in \mathsf{lift}(b)$ and hence $\{P\} \subseteq \mathsf{lift}(b)$, we have

$$
\begin{aligned}
h(\{P\}) &\subseteq h(\mathsf{lift}(b)) \\
&= (h \circ \mathsf{lift})(b) \\
&= \mathsf{ident}(b) \\
&= \{b\}.
\end{aligned}
$$

Since $a \in h(\{P\})$, we obtain $b = a$, a contradiction. $\qquad\square$

Theorem 4. BE *is the unique complete homomorphism that makes the Partee triangle commute.*

Proof. To see that BE makes the Partee triangle commute, observe that for any $a \in D$,

$$
\begin{aligned}
&\mathsf{BE}(\mathsf{lift}(a)) \\
&= \{x \in D \mid \{x\} \in \mathsf{lift}(a)\} \\
&= \{x \in D \mid \{x\} \in \{P \in \wp(D) \mid a \in P\}\} \\
&= \{x \in D \mid a \in \{x\}\} \\
&= \{x \in D \mid x = a\} \\
&= \{a\} \\
&= \mathsf{ident}(a).
\end{aligned}
$$

Now, let h be a complete homomorphism from $\wp(\wp(D))$ to $\wp(D)$ such that $h \circ \mathsf{lift} = \mathsf{ident}$. We show that $h = \mathsf{BE}$. Let $a \in D$ and $\mathscr{P} \in \wp(\wp(D))$ satisfy $a \in h(\mathscr{P})$. By Lemma 2, it is sufficient to show that $\{a\} \in \mathscr{P}$. Since h is a complete homomorphism,

$$
\begin{aligned}
\bigcup_{P \in \mathscr{P} \cap \mathsf{lift}(a)} h(\{P\}) &= h\left(\bigcup_{P \in \mathscr{P} \cap \mathsf{lift}(a)} \{P\}\right) \\
&= h(\mathscr{P} \cap \mathsf{lift}(a)) \\
&= h(\mathscr{P}) \cap h(\mathsf{lift}(a)) \\
&= h(\mathscr{P}) \cap \mathsf{ident}(a) \\
&= h(\mathscr{P}) \cap \{a\} \\
&= \{a\}.
\end{aligned}
$$

It follows that for some $P \in \mathscr{P} \cap \mathsf{lift}(a)$,

$$
h(\{P\}) = \{a\}.
$$

By Lemma 3, P must be a singleton set. Since the only singleton set contained in $\mathsf{lift}(a)$ is $\{a\}$, we have $\{a\} = P \in \mathscr{P} \cap \mathsf{lift}(a)$, so $\{a\} \in \mathscr{P}$. $\qquad\square$

Note that Theorem 4 immediately follows from Keenan and Faltz's (1985) Justification Theorem (p. 92) as well. Individuals in Keenan and Faltz's theory can be identified with the elements of the set $\{I_x \mid x \in D\}$, where $I_x = \mathsf{lift}(x)$. Given a function f from the set of individuals into $\wp(D)$ such that $f(I_x) = \mathsf{ident}(x)$ for all $x \in D$, the Justification Theorem says that there exists a unique complete homomorphism from $\wp(\wp(D))$ to $\wp(D)$ that extends f.

When D is finite, a homomorphism from $\wp(\wp(D))$ to $\wp(D)$ is necessarily a complete homomorphism, so by Theorem 4, BE is automatically the unique homomorphism that makes the Partee triangle commute. This is not the case, however, when D is infinite. To consider such cases, the following lemma plays an important role of giving (unique) representations of homomorphisms that make the Partee triangle commute.

Lemma 5. *Let h be a function from $\wp(\wp(D))$ into $\wp(D)$. The following conditions are equivalent.*

(i) *h is a homomorphism from $\wp(\wp(D))$ to $\wp(D)$ and $h \circ \mathsf{lift} = \mathsf{ident}$.*

(ii) *There is a family $\{\mathfrak{U}_x\}_{x \in D}$ of subsets of $\wp(\wp(D))$ such that each $\mathfrak{U}_x$ is an ultrafilter in the Boolean algebra $\wp(\mathsf{lift}(x))$ satisfying $\mathsf{lift}(x) \cap \mathsf{lift}(y) \notin \mathfrak{U}_x$ for all $y \neq x$, and*

$$
h = \lambda\mathscr{P}.\,\{x \in D \mid \mathscr{P} \cap \mathsf{lift}(x) \in \mathfrak{U}_x\}.
$$

Proof. (i) $\Rightarrow$ (ii). Assume (i). For each $x \in D$, let

$$
\mathfrak{U}_x = \{\mathscr{P} \in \wp(\mathsf{lift}(x)) \mid x \in h(\mathscr{P})\}.
$$

To begin with, we show that $\mathfrak{U}_x$ is an ultrafilter in the Boolean algebra $\wp(\mathrm{lift}(x))$. First, since $x \in \{x\} = \mathrm{ident}(x) = h(\mathrm{lift}(x))$, the top element $\mathrm{lift}(x)$ of $\wp(\mathrm{lift}(x))$ belongs to $\mathfrak{U}_x$. Second, if $\mathscr{P}, \mathscr{Q} \in \mathfrak{U}_x$, then $x \in h(\mathscr{P})$ and $x \in h(\mathscr{Q})$ and hence $x \in h(\mathscr{P}) \cap h(\mathscr{Q}) = h(\mathscr{P} \cap \mathscr{Q})$, so $\mathscr{P} \cap \mathscr{Q} \in \mathfrak{U}_x$. Third, if $\mathscr{P} \in \mathfrak{U}_x$ and $\mathscr{Q} \in \wp(\mathrm{lift}(x))$ satisfy $\mathscr{P} \subseteq \mathscr{Q}$, then $x \in h(\mathscr{P}) \subseteq h(\mathscr{Q})$, so $\mathscr{Q} \in \mathfrak{U}_x$. This establishes that $\mathfrak{U}_x$ is a filter in $\wp(\mathrm{lift}(x))$. To see that $\mathfrak{U}_x$ is an ultrafilter, suppose $\mathscr{P} \in \wp(\mathrm{lift}(x))$ and $\mathscr{P} \notin \mathfrak{U}_x$. Since $x \in h(\mathrm{lift}(x)) = h(\mathscr{P} \cup (\mathscr{P}^{\mathrm{c}} \cap \mathrm{lift}(x))) = h(\mathscr{P}) \cup h(\mathscr{P}^{\mathrm{c}} \cap \mathrm{lift}(x))$ and $x \notin h(\mathscr{P})$, we have $x \in h(\mathscr{P}^{\mathrm{c}} \cap \mathrm{lift}(x))$, so $\mathscr{P}^{\mathrm{c}} \cap \mathrm{lift}(x) \in \mathfrak{U}_x$. Thus, the complement of $\mathscr{P}$ in $\wp(\mathrm{lift}(x))$ belongs to $\mathfrak{U}_x$.

Next, observe that if $\mathrm{lift}(x) \cap \mathrm{lift}(y) \in \mathfrak{U}_x$, then $x \in h(\mathrm{lift}(x) \cap \mathrm{lift}(y)) = h(\mathrm{lift}(x)) \cap h(\mathrm{lift}(y))$, so $x \in h(\mathrm{lift}(y)) = \mathrm{ident}(y) = \{y\}$ and therefore $x = y$. It follows that $\mathrm{lift}(x) \cap \mathrm{lift}(y) \notin \mathfrak{U}_x$ for all $y \neq x$. This establishes that the family $\{\mathfrak{U}_x\}_{x \in D}$ has the desired properties.

Now, for all $\mathscr{P} \in \wp(\wp(D))$, we have

$$h(\mathscr{P} \cap \mathrm{lift}(x)^{\mathrm{c}}) \subseteq h(\mathrm{lift}(x)^{\mathrm{c}})$$
$$= h(\mathrm{lift}(x))^{\mathrm{c}}$$
$$= \mathrm{ident}(x)^{\mathrm{c}}$$
$$= \{x\}^{\mathrm{c}},$$

so $x \notin h(\mathscr{P} \cap \mathrm{lift}(x)^{\mathrm{c}})$. It follows that

$$x \in h(\mathscr{P})$$

iff $\quad x \in h((\mathscr{P} \cap \mathrm{lift}(x)) \cup (\mathscr{P} \cap \mathrm{lift}(x)^{\mathrm{c}}))$

iff $\quad x \in h(\mathscr{P} \cap \mathrm{lift}(x)) \cup h(\mathscr{P} \cap \mathrm{lift}(x)^{\mathrm{c}})$

iff $\quad x \in h(\mathscr{P} \cap \mathrm{lift}(x))$

iff $\quad \mathscr{P} \cap \mathrm{lift}(x) \in \mathfrak{U}_x$.

Thus $h(\mathscr{P}) = \{x \in D \mid \mathscr{P} \cap \mathrm{lift}(x) \in \mathfrak{U}_x\}$.

(ii) $\Rightarrow$ (i). Assume (ii). To show that h is a homomorphism, it suffices to check that it preserves finite union and complement. Being an ultrafilter, $\mathfrak{U}_x$ is a prime filter. Therefore, for all $\mathscr{P}, \mathscr{Q} \in \wp(\wp(D))$,

$$x \in h(\mathscr{P} \cup \mathscr{Q})$$

iff $\quad (\mathscr{P} \cup \mathscr{Q}) \cap \mathrm{lift}(x) \in \mathfrak{U}_x$

iff $\quad (\mathscr{P} \cap \mathrm{lift}(x)) \cup (\mathscr{Q} \cap \mathrm{lift}(x)) \in \mathfrak{U}_x$

iff $\quad \mathscr{P} \cap \mathrm{lift}(x) \in \mathfrak{U}_x$ or $\mathscr{Q} \cap \mathrm{lift}(x) \in \mathfrak{U}_x$

iff $\quad x \in h(\mathscr{P})$ or $x \in h(\mathscr{Q})$

iff $\quad x \in h(\mathscr{P}) \cup h(\mathscr{Q})$,

so $h(\mathscr{P} \cup \mathscr{Q}) = h(\mathscr{P}) \cup h(\mathscr{Q})$. Also, for all $\mathscr{P} \in \wp(\wp(D))$, since $\mathscr{P} \cap \mathrm{lift}(x)$ and $\mathscr{P}^{\mathrm{c}} \cap \mathrm{lift}(x)$ are complements of each other in $\wp(\mathrm{lift}(x))$ and since $\mathfrak{U}_x$ is an ultrafilter in $\wp(\mathrm{lift}(x))$,

$$x \in h(\mathscr{P}^{\mathrm{c}}) \quad \text{iff} \quad \mathscr{P}^{\mathrm{c}} \cap \mathrm{lift}(x) \in \mathfrak{U}_x$$
$$\text{iff} \quad \mathscr{P} \cap \mathrm{lift}(x) \notin \mathfrak{U}_x$$
$$\text{iff} \quad x \notin h(\mathscr{P})$$
$$\text{iff} \quad x \in h(\mathscr{P})^{\mathrm{c}},$$

so $h(\mathscr{P}^{\mathrm{c}}) = h(\mathscr{P})^{\mathrm{c}}$.

It remains to show that $h \circ \mathrm{lift} = \mathrm{ident}$. While $\mathrm{lift}(x) \cap \mathrm{lift}(y) \notin \mathfrak{U}_x$ for every $y \neq x$, we have $\mathrm{lift}(x) \cap \mathrm{lift}(x) = \mathrm{lift}(x) \in \mathfrak{U}_x$ since an ultrafilter in $\wp(\mathrm{lift}(x))$ contains the top element of $\wp(\mathrm{lift}(x))$. Consequently,

$$h(\mathrm{lift}(x)) = \{y \in D \mid \mathrm{lift}(x) \cap \mathrm{lift}(y) \in \mathfrak{U}_y\}$$
$$= \{x\}$$
$$= \mathrm{ident}(x). \qquad \square$$

Lemma 6. *Let $\mathfrak{U}_x$ be a principal ultrafilter in $\wp(\mathrm{lift}(x))$. The following conditions are equivalent.*

(i) $\mathrm{lift}(x) \cap \mathrm{lift}(y) \notin \mathfrak{U}_x$ for all $y \neq x$.

(ii) $\mathfrak{U}_x$ is generated by $\{\{x\}\}$.

Proof. (i) $\Rightarrow$ (ii). Assume (i). Since $\mathfrak{U}_x$ is a principal filter in $\wp(\mathrm{lift}(x))$, there is some $\mathscr{Q} \in \wp(\mathrm{lift}(x))$ that generates it, i.e.,

$$\mathfrak{U}_x = {\uparrow}\mathscr{Q} = \{\mathscr{P} \in \wp(\mathrm{lift}(x)) \mid \mathscr{Q} \subseteq \mathscr{P}\}.$$

We show that $\mathscr{Q} = \{\{x\}\}$. For every $y \neq x$, we have $\mathrm{lift}(x) \cap \mathrm{lift}(y) \notin \mathfrak{U}_x$, and because $\mathfrak{U}_x$ is an ultrafilter, this implies that its complement $\mathrm{lift}(x) \cap \mathrm{lift}(y)^{\mathrm{c}}$ in $\wp(\mathrm{lift}(x))$ belongs to $\mathfrak{U}_x$, so $\mathscr{Q} \subseteq \mathrm{lift}(x) \cap \mathrm{lift}(y)^{\mathrm{c}}$. It follows that

$$\mathscr{Q} \subseteq \bigcap_{y \neq x} (\mathrm{lift}(x) \cap \mathrm{lift}(y)^{\mathrm{c}})$$
$$= \mathrm{lift}(x) \cap \left(\bigcap_{y \neq x} \mathrm{lift}(y)^{\mathrm{c}} \right)$$
$$= \{P \in \wp(D) \mid x \in P\}$$
$$\quad \cap \left(\bigcap_{y \neq x} \{P \in \wp(D) \mid y \in P\}^{\mathrm{c}} \right)$$
$$= \{P \in \wp(D) \mid x \in P\}$$
$$\quad \cap \left(\bigcap_{y \neq x} \{P \in \wp(D) \mid y \notin P\} \right)$$
$$= \{P \in \wp(D) \mid x \in P \text{ and } y \notin P \text{ for all } y \neq x\}$$
$$= \{\{x\}\}.$$

Since $\mathfrak{U}_x = {\uparrow}\mathscr{Q}$ is an ultrafilter, $\mathscr{Q} \neq \varnothing$. Hence $\mathscr{Q} = \{\{x\}\}$.

(ii) $\Rightarrow$ (i). Assume $\mathfrak{U}_x = \uparrow\{\{x\}\}$, i.e.,

$$\mathfrak{U}_x = \{\mathscr{P} \in \wp(\mathrm{lift}(x)) \mid \{\{x\}\} \subseteq \mathscr{P}\}$$
$$= \{\mathscr{P} \in \wp(\mathrm{lift}(x)) \mid \{x\} \in \mathscr{P}\}.$$

Since

$$\mathrm{lift}(x) \cap \mathrm{lift}(y)$$
$$= \{P \in \wp(D) \mid x \in P\} \cap \{P \in \wp(D) \mid y \in P\}$$
$$= \{P \in \wp(D) \mid \{x, y\} \subseteq P\},$$

if $y \neq x$, then $\{x\} \notin \mathrm{lift}(x) \cap \mathrm{lift}(y)$, so $\mathrm{lift}(x) \cap \mathrm{lift}(y) \notin \mathfrak{U}_x$. $\square$

Lemma 7. *Let D be infinite. For every $x \in D$, there exists a nonprincipal ultrafilter $\mathfrak{U}_x$ in $\wp(\mathrm{lift}(x))$ such that $\mathrm{lift}(x) \cap \mathrm{lift}(y) \notin \mathfrak{U}_x$ for all $y \neq x$.*

Proof. For $x \in D$, let $\mathfrak{C}_x$ be the following subset of $\wp(\mathrm{lift}(x))$:

$$\mathfrak{C}_x = \{\{\{x\}\}\} \cup$$
$$\{\mathrm{lift}(x) \cap \mathrm{lift}(y) \mid y \in D \text{ and } y \neq x\}.$$

If $\mathfrak{F}$ is a finite subset of $\mathfrak{C}_x$, then since D is infinite, there is some $y \in D$ such that $\mathrm{lift}(x) \cap \mathrm{lift}(y) \notin \mathfrak{F}$, so in particular $\{x, y\} \notin \bigcup \mathfrak{F}$ and thus $\bigcup \mathfrak{F}$ cannot equal $\mathrm{lift}(x)$, the top element of $\wp(\mathrm{lift}(x))$. $\mathfrak{C}_x$ thus has the finite join property, so the ideal $\mathfrak{I}_x$ generated by $\mathfrak{C}_x$ in $\wp(\mathrm{lift}(x))$ is proper. By the Boolean prime ideal theorem, $\mathfrak{I}_x$ can be extended to a prime ideal, i.e., a maximal ideal $\mathfrak{M}_x$ in $\wp(\mathrm{lift}(x))$.[2] Let $\mathfrak{U}_x$ be the dual ultrafilter of $\mathfrak{M}_x$ in $\wp(\mathrm{lift}(x))$:

$$\mathfrak{U}_x = \{\mathscr{P}^{\mathsf{c}} \cap \mathrm{lift}(x) \mid \mathscr{P} \in \mathfrak{M}_x\}.$$

For all $y \neq x$, since $\mathrm{lift}(x) \cap \mathrm{lift}(y) \in \mathfrak{C}_x \subseteq \mathfrak{M}_x$, we have $\mathrm{lift}(x) \cap \mathrm{lift}(y) \notin \mathfrak{U}_x$. Since $\{\{x\}\} \in \mathfrak{C}_x \subseteq \mathfrak{M}_x$, we also have $\{\{x\}\} \notin \mathfrak{U}_x$. Lemma 6 then implies that $\mathfrak{U}_x$ is not a principal filter. $\square$

Theorem 8. *If D is infinite, there are uncountably many homomorphisms h from $\wp(\wp(D))$ to $\wp(D)$ such that $h \circ \mathrm{lift} = \mathrm{ident}$.*

Proof. Let D be infinite. By Lemma 5, a homomorphism h from $\wp(\wp(D))$ to $\wp(D)$ such that $h \circ \mathrm{lift} = \mathrm{ident}$ is written

$$h = \lambda \mathscr{P}. \{x \in D \mid \mathscr{P} \cap \mathrm{lift}(x) \in \mathfrak{U}_x\},$$

---

[2] Thus Lemma 7 (and hence also Theorem 8) uses the Boolean prime ideal theorem, a weaker form of the axiom of choice.

where $\mathfrak{U}_x$ is an ultrafilter in $\wp(\mathrm{lift}(x))$ such that $\mathrm{lift}(x) \cap \mathrm{lift}(y) \notin \mathfrak{U}_x$ for all $y \neq x$. By Lemmata 6 and 7, there are at least two such ultrafilters $\mathfrak{U}_x$ for each $x \in D$: a principal one and a nonprincipal one. For each x, different choices for $\mathfrak{U}_x$ clearly give rise to different homomorphisms. It follows that the cardinality of the set of homomorphisms h such that $h \circ \mathrm{lift} = \mathrm{ident}$ is at least $2^{|D|}$. $\square$

Observe that since

$$\{x\} \in \mathscr{P} \quad \text{iff} \quad \{\{x\}\} \subseteq \mathscr{P}$$
$$\text{iff} \quad \{\{x\}\} \subseteq \mathscr{P} \cap \mathrm{lift}(x)$$
$$\text{iff} \quad \mathscr{P} \cap \mathrm{lift}(x) \in \uparrow\{\{x\}\},$$

we can write

$$\mathsf{BE} = \lambda \mathscr{P}. \{x \in D \mid \{x\} \in \mathscr{P}\}$$
$$= \lambda \mathscr{P}. \{x \in D \mid \mathscr{P} \cap \mathrm{lift}(x) \in \uparrow\{\{x\}\}\}.$$

Thus, in Lemma 5's representation of BE, each $\mathfrak{U}_x$ is a principal filter in $\wp(\mathrm{lift}(x))$. This also explains why BE has to be the unique homomorphism that makes the Partee triangle commute when D is finite, because in that case, each $\wp(\mathrm{lift}(x))$ is finite, and every filter in a finite Boolean algebra is necessarily principal.

Now suppose D is infinite. By Theorem 8, there is a homomorphism $h \neq \mathsf{BE}$ that makes the Partee triangle commute. By Theorem 4, we know that h is not a complete homomorphism. It may be illuminating to confirm this fact directly. The observation in the previous paragraph implies that in Lemma 5's representation of h, there is some $a \in D$ such that $\mathfrak{U}_a$ is a nonprincipal ultrafilter in $\wp(\mathrm{lift}(a))$. We have $\{\{a\}\} \cap \mathrm{lift}(a) = \{\{a\}\} \notin \mathfrak{U}_a$ because $\{\{a\}\} \in \mathfrak{U}_a$ would imply $\mathfrak{U}_a = \uparrow\{\{a\}\}$ but $\mathfrak{U}_a$ is nonprincipal. Also, for all $x \neq a$, we have $\{\{a\}\} \cap \mathrm{lift}(x) = \varnothing \notin \mathfrak{U}_x$ because an ultrafilter does not contain the bottom element. Thus

$$h(\{\{a\}\}) = \{x \in D \mid \{\{a\}\} \cap \mathrm{lift}(x) \in \mathfrak{U}_x\} = \varnothing.$$

Since $\{\{a\}\}$ is the only singleton set in $\mathrm{lift}(a)$, for every $P \in \mathrm{lift}(a)$ such that $P \neq \{\{a\}\}$, we have $h(\{P\}) = \varnothing$ by Lemma 3. It follows that

$$\bigcup_{P \in \mathrm{lift}(a)} h(\{P\}) = \bigcup_{P \in \mathrm{lift}(a)} \varnothing = \varnothing.$$

On the other hand,

$$h\left(\bigcup_{P \in \mathrm{lift}(a)} \{P\}\right) = h(\mathrm{lift}(a)) = \mathrm{ident}(a) = \{a\}.$$

Thus

$$h\left(\bigcup_{P \in \mathrm{lift}(a)} \{P\}\right) \neq \bigcup_{P \in \mathrm{lift}(a)} h(\{P\}).$$

So h does not generally preserve an infinite union.

3   Why do we need a complete homomorphism?

Partee (1986) proposes that BE is a type-shifting operator naturally employed in natural language semantics on the grounds that it is a Boolean homomorphism and it makes the Partee triangle commute. As we have seen, however, when D is infinite, there are infinitely many such homomorphisms. Couldn't they then perhaps be employed as type-shifting operators in place of BE? What distinguishes BE from all the rest is the fact that it is the only complete one. So the question boils down to this: how should being a complete homomorphism matter?

To answer this question, let's recall Partee's (1986) discussion of the functions THE and A from $D_{\langle e,t \rangle}$ into $D_{\langle \langle e,t \rangle,t \rangle}$, which in set talk can be rendered as the following functions from $\wp(D)$ into $\wp(\wp(D))$.

$$\mathsf{THE} = \lambda P.\{Q \in \wp(D) \mid |P| = 1 \text{ and } P \subseteq Q\},$$
$$\mathsf{A} = \lambda P.\{Q \in \wp(D) \mid P \cap Q \neq \varnothing\}.$$

Partee argues that THE and A are "natural" since they are inverses of BE in the sense that for all $P \in \wp(D)$,

$$\mathsf{BE}(\mathsf{THE}(P)) = \begin{cases} P & \text{if } P \text{ is a singleton,} \\ \varnothing & \text{otherwise,} \end{cases}$$
$$\mathsf{BE}(\mathsf{A}(P)) = P.$$

One should then wonder whether analogous equalities hold with other homomorphisms that make the Partee triangle commute.

It is immediate that an analogous equality holds with THE.

Theorem 9. *Let h be a homomorphism from $\wp(\wp(D))$ to $\wp(D)$ such that $h \circ \mathsf{lift} = \mathsf{ident}$. For all $P \in \wp(D)$,*

$$h(\mathsf{THE}(P)) = \begin{cases} P & \text{if } P \text{ is a singleton,} \\ \varnothing & \text{otherwise.} \end{cases}$$

Proof. For any $x \in D$, $\mathsf{THE}(\{x\}) = \mathsf{lift}(x)$, so $h(\mathsf{THE}(\{x\})) = h(\mathsf{lift}(x)) = \mathsf{ident}(x) = \{x\}$. If $P \in \wp(D)$ is not a singleton, then $\mathsf{THE}(P) = \varnothing$, so $h(\mathsf{THE}(P)) = h(\varnothing) = \varnothing$. $\square$

With A, by contrast, an analogous equality does not generally hold, and this is where (in)completeness becomes crucial.

Theorem 10. *Let h be a homomorphism from $\wp(\wp(D))$ to $\wp(D)$ such that $h \circ \mathsf{lift} = \mathsf{ident}$. For all $P \in \wp(D)$,*

$$h(\mathsf{A}(P)) \supseteq P.$$

In particular, if P is finite,

$$h(\mathsf{A}(P)) = P.$$

Proof. Let $P \in \wp(D)$. We have

$$\begin{aligned} \mathsf{A}(P) &= \{Q \in \wp(D) \mid P \cap Q \neq \varnothing\} \\ &= \bigcup_{x \in P} \{Q \in \wp(D) \mid x \in Q\} \\ &= \bigcup_{x \in P} \mathsf{lift}(x). \end{aligned}$$

Thus, for every $x \in P$, $\mathsf{lift}(x) \subseteq \mathsf{A}(P)$ and hence $\{x\} = \mathsf{ident}(x) = h(\mathsf{lift}(x)) \subseteq h(\mathsf{A}(P))$. Thus

$$P = \bigcup_{x \in P} \{x\} \subseteq h(\mathsf{A}(P)).$$

If P is finite, then the homomorphism properties of h ensure that

$$\begin{aligned} h(\mathsf{A}(P)) &= h\left(\bigcup_{x \in P} \mathsf{lift}(x)\right) \\ &= \bigcup_{x \in P} h(\mathsf{lift}(x)) \\ &= \bigcup_{x \in P} \mathsf{ident}(x) \\ &= \bigcup_{x \in P} \{x\} \\ &= P. \qquad \square \end{aligned}$$

Theorem 10 suggests that homomorphisms other than BE are undesirable as a type-shifting operator to replace BE, even if they make the Partee triangle commute. To see this point, imagine that some such homomorphism $h \neq$ BE were actually employed as a type-shifter.

First, consider the following example.

(2)   André is a girl.

Following Partee (1986), let's assume that the verb *be* is semantically vacuous and a type-shifter is inserted to convert a quantifier into a predicate. (2) would then be analyzed as

(3)   André $\in h(\llbracket \text{a girl} \rrbracket) = h(\mathsf{A}(\llbracket \text{girl} \rrbracket))$.

Now suppose $\llbracket \text{girl} \rrbracket$ is finite, as would be the case, say $\llbracket \text{girl} \rrbracket = \{\text{Mari}, \text{Meiko}, \text{Hana}\}$. Then by Theorem 10,

$$h(\mathsf{A}(\llbracket \text{girl} \rrbracket)) = \llbracket \text{girl} \rrbracket = \{\text{Mari}, \text{Meiko}, \text{Hana}\},$$

so (3) is equivalent to

(4)   André $\in$ {Mari, Meiko, Hana},

or what amounts to the same thing,

(5)   André $=$ Mari or
        André $=$ Meiko or
        André $=$ Hana.

These are indeed the desired truth conditions for (2), so no problem arises in this case.

Now consider the examples in (6).

(6)   a. π is a natural number.
      b.   This is some water.

These would be analyzed as in (7), assuming that *this* denotes an entity and that $[\![\text{some}]\!] = [\![\text{a}]\!] = \mathsf{A}$.

(7)   a. $\pi \in h(\mathsf{A}([\![\text{natural number}]\!]))$
      b. $[\![\text{this}]\!] \in h(\mathsf{A}([\![\text{water}]\!]))$.

In contrast to the previous case, $[\![\text{natural number}]\!]$ and $[\![\text{water}]\!]$ ought to be infinite sets. According to Theorem 10, what we can know is then only that

$$h(\mathsf{A}([\![\text{natural number}]\!])) \supseteq [\![\text{natural number}]\!]\,,$$
$$h(\mathsf{A}([\![\text{water}]\!])) \supseteq [\![\text{water}]\!]\,.$$

What these inequalities imply is that even though $\pi \notin [\![\text{natural number}]\!]$, (7-a) might hold and hence (6-a) come out true, and similarly, even if $[\![\text{this}]\!] \notin [\![\text{water}]\!]$, (7-b) might hold and so (6-b) come out true. Such states of affairs would be clearly undesirable. This suggests that h should not be used as a type-shifter in natural language semantics.

The above argument does not show, however, that undesirable states of affairs necessarily ensue, as the inequality $h(\mathsf{A}(P)) \supseteq P$ in Theorem 10 is not necessarily a proper inclusion. Then, even in a case where D is infinite, might there perhaps be a homomorphism $h \neq \mathsf{BE}$ such that $h(\mathsf{A}(P)) = P$ for all $P \in \wp(D)$? The following theorem tells us that this possibility never obtains. Note that it also characterizes BE without directly mentioning completeness or the property of making the Partee triangle commute.

Theorem 11. BE *is the unique homomorphism h from $\wp(\wp(D))$ to $\wp(D)$ such that $h \circ \mathsf{A}$ is the identity map on $\wp(D)$.*

Proof. Since BE is a complete homomorphism, by substituting BE for h in the last set of equalities in the proof of Theorem 10, we can see that $\mathsf{BE}(\mathsf{A}(P)) = P$ for all $P \in \wp(D)$.

To show the uniqueness, assume to the contrary that there is a homomorphism $h \neq \mathsf{BE}$ such that $h \circ \mathsf{A}$ is the identity map. By Lemma 2, for some $a \in D$ and some $\mathscr{P} \in \wp(\wp(D))$, we have $a \in h(\mathscr{P})$ and $\{a\} \notin \mathscr{P}$. Since

$$\begin{aligned}
\mathsf{A}(D\backslash\{a\}) &= \{Q \in \wp(D) \mid (D\backslash\{a\}) \cap Q \neq \varnothing\} \\
&= \{Q \in \wp(D) \mid Q\backslash\{a\} \neq \varnothing\} \\
&= \{Q \in \wp(D) \mid Q \neq \{a\}, \varnothing\} \\
&= \wp(D)\backslash\{\{a\}, \varnothing\}
\end{aligned}$$

and since $\{a\} \notin \mathscr{P}$, we have

$$\mathscr{P} \subseteq \wp(D)\backslash\{\{a\}\} = \mathsf{A}(D\backslash\{a\}) \cup \{\varnothing\}.$$

Since $h \circ \mathsf{A}$ is the identity map, it follows that

$$\begin{aligned}
h(\mathscr{P}) &\subseteq h(\mathsf{A}(D\backslash\{a\}) \cup \{\varnothing\}) \\
&= h(\mathsf{A}(D\backslash\{a\})) \cup h(\{\varnothing\}) \\
&= (D\backslash\{a\}) \cup h(\{\varnothing\}) \\
&= (D\backslash\{a\}) \cup \varnothing \qquad \text{(by Lemma 3)} \\
&= D\backslash\{a\}.
\end{aligned}$$

This contradicts $a \in h(\mathscr{P})$. $\qquad\square$

It follows from Theorems 10 and 11 that if $h \neq \mathsf{BE}$ is a homomorphism from $\wp(\wp(D))$ to $\wp(D)$ and $h \circ \text{lift} = \text{ident}$, then there exists some infinite set $P \in \wp(D)$ such that $h(\mathsf{A}(P)) \supsetneq P$. Indeed, we can find a concrete example. Since $h \neq \mathsf{BE}$, in Lemma 5's representation, there is some $a \in D$ such that $\mathfrak{U}_a$ is a nonprincipal ultrafilter in $\wp(\text{lift}(a))$. We have $\{\{a\}\} \notin \mathfrak{U}_a$ since $\mathfrak{U}_a$ is nonprincipal. Since

$$\begin{aligned}
\mathsf{A}(D\backslash\{a\}) \cap \text{lift}(a) &= (\wp(D)\backslash\{\{a\}, \varnothing\}) \cap \text{lift}(a) \\
&= \{\{a\}\}^{\mathsf{c}} \cap \text{lift}(a)
\end{aligned}$$

is the complement of $\{\{a\}\}$ in $\wp(\text{lift}(a))$ and since $\mathfrak{U}_a$ is an ultrafilter in $\wp(\text{lift}(a))$, we have

$$\mathsf{A}(D\backslash\{a\}) \cap \text{lift}(a) \in \mathfrak{U}_a.$$

According to Lemma 5, this implies that

$$a \in h(\mathsf{A}(D\backslash\{a\})).$$

Combining with Theorem 10, we conclude that

$$h(\mathsf{A}(D\backslash\{a\})) = D \supsetneq D\backslash\{a\}.$$

The discussion in this section is just another example demonstrating the significance of the notion of completeness of the Boolean structures and homomorphisms between them that are employed in natural language semantics, which was extensively argued for by Keenan and Faltz (1985).

4 Inverses of BE

Having discussed the naturalness of BE, Partee (1986) asks what possible determiners δ are inverses of BE, i.e., $\mathrm{BE}(\delta(P)) = P$ for all $P \in \wp(D)$. It is immediate that a necessary and sufficient condition for δ to be an inverse of BE is that

(8)   for all $P \in \wp(D)$,
$$\{x \in D \mid \{x\} \in \delta(P)\} = P,$$

so there exist many inverses of BE. A is one but so is $[\![\text{exactly one}]\!]$. Partee suggests that "nice" formal properties such as being increasing (in both arguments) and being symmetric might distinguish A from the others. Contrary to her claim, though, symmetry fails to distinguish A from $[\![\text{exactly one}]\!]$, as both of these are symmetric. On the other hand, the property of being increasing certainly distinguishes A from $[\![\text{exactly one}]\!]$ since A is increasing and $[\![\text{exactly one}]\!]$ is not. Still, there are many inverses of BE other than A that are increasing, as the reader can easily check. Then, how might formal properties single A out?

At this point, we shall recall Keenan and Stavi's (1986) view that all possible determiners are expressible as Boolean combinations of "basic" determiners, which are all increasing and weakly conservative.[3] These two properties are defined as follows, where δ is an arbitrary function from $\wp(D)$ into $\wp(\wp(D))$.

(9) δ is *increasing*
$\Leftrightarrow$ for all $P, Q_1, Q_2 \in \wp(D)$, if $Q_1 \in \delta(P)$ and $Q_1 \subseteq Q_2$, then $Q_2 \in \delta(P)$.

(10) δ is *weakly conservative*
$\Leftrightarrow$ for all $P, Q \in \wp(D)$, if $Q \in \delta(P)$ then $P \cap Q \in \delta(P)$.

Keenan and Stavi (1986) proved that the functions obtained as Boolean combinations of basic determiners are exactly those functions from $\wp(D)$ into $\wp(\wp(D))$ that are conservative:

(11) δ is *conservative*
$\Leftrightarrow$ for all $P, Q \in \wp(D)$, $Q \in \delta(P)$ iff $P \cap Q \in \delta(P)$.

Keenan and Stavi proposed that this accounts for

---

[3] In Keenan and Stavi's (1986) theory, the determiner *no*, which is not increasing, is not a basic determiner. Keenan and Stavi suggest that increasingness may be a universal property of monomorphemic determiners, if negative determiners like *no* are analyzed as bimorphemic, consisting of a negative morpheme *N-* and a stem.

the apparent fact that all determiners are conservative.[4] Now, if we restrict our attention to inverses of BE that are increasing and weakly conservative, it turns out that there remain only two.

Lemma 12. *Let δ be an increasing, weakly conservative function from $\wp(D)$ into $\wp(\wp(D))$ such that $\mathrm{BE} \circ \delta$ is the identity map on $\wp(D)$. Then either*

(i) $\delta = \mathsf{A}$ *or*

(ii) $\delta(P) = \begin{cases} \mathsf{A}(P) & \text{if } P \neq D, \\ \wp(D) & \text{if } P = D. \end{cases}$

Proof. Let $P \in \wp(D)$ and $x \in P$. By (8), $\{x\} \in \delta(P)$. Since δ is increasing, for every $Q \in \wp(D)$ such that $\{x\} \subseteq Q$, we have $Q \in \delta(P)$, so $\mathrm{lift}(x) \subseteq \delta(P)$. Hence

(12) $\mathsf{A}(P) = \bigcup_{x \in P} \mathrm{lift}(x) \subseteq \delta(P)$.

We now show that $\delta(P) = \mathsf{A}(P)$ if $P \neq D$. Suppose $\delta(P) \neq \mathsf{A}(P)$. (12) implies $\delta(P) \not\subseteq \mathsf{A}(P)$, so there exists some $Q \in \delta(P)$ such that $Q \notin \mathsf{A}(P)$, which means $P \cap Q = \varnothing$ by the definition of A. Since δ is weakly conservative and $Q \in \delta(P)$, we have $\varnothing = P \cap Q \in \delta(P)$. Because δ is increasing, for every $x \in D$, we have $\{x\} \in \delta(P)$ since $\varnothing \in \delta(P)$ and $\varnothing \subseteq \{x\}$. (8) then implies $P = D$.

Finally, observe that by (12),

$$\begin{aligned}
\delta(D) &\supseteq \mathsf{A}(D) \\
&= \{Q \in \wp(D) \mid D \cap Q \neq \varnothing\} \\
&= \{Q \in \wp(D) \mid Q \neq \varnothing\} \\
&= \wp(D)/\{\varnothing\},
\end{aligned}$$

so either $\delta(D) = \mathsf{A}(D) = \wp(D)/\{\varnothing\}$ or $\delta(D) = \wp(D)$. It follows that either (i) or (ii) holds. $\square$

How can we distinguish A from the other increasing, weakly conservative inverse of BE described in Case (ii) of the above lemma? One possibility might be to note that while $\mathsf{A}(D)$ is a sieve, $\delta(D)$ in Case (ii) is not a sieve, in the sense of Barwise and Cooper (1981):

(13) $\mathscr{P} \in \wp(\wp(D))$ is a *sieve*
$\Leftrightarrow \mathscr{P} \neq \wp(D)$ and $\mathscr{P} \neq \varnothing$.

---

[4] Conservativity coincides with Barwise and Cooper's (1981) 'lives-on-its-argument' property, which Barwise and Cooper propose to be a universal property of determiners.

A non-sieve is either true of every predicate or false of every predicate, and therefore would be pointless to use in normal conversation. Van Benthem (1986) suggests that a determiner δ is generally expected to be such that $\delta(P)$ is a sieve for all $P \neq \varnothing$ (Variety, p. 9).

Another, presumably more appealing way is to invoke the notion of logicality, for which I refer to Westerståhl (1985). So far, we have fixed a model, whose domain of entities is D, and have not strictly distinguished linguistic expressions and their model-theoretic interpretations. We should now get rigorous about this distinction because logicality is a property of an object language symbol, and not of its interpretation in a particular model. Henceforth, let's take BE and A to be object language symbols such that for every model $\mathcal{M} = \langle D, [\![\ ]\!]^{\mathcal{M}} \rangle$,

$$[\![BE]\!]^{\mathcal{M}} = \lambda \mathscr{P}. \{x \in D \mid \{x\} \in \mathscr{P}\},$$
$$[\![A]\!]^{\mathcal{M}} = \lambda P. \{Q \in \wp(D) \mid P \cap Q \neq \varnothing\}.$$

Now according to Westerståhl (1985), an object language symbol is logical if and only if it has the two properties called constancy and topic-neutrality.[5] It turns out that constancy alone is sufficient to single A out. Here is the relevant definition (Westerståhl, 1985, p. 393, with slight adaptation).

(14)   A determiner δ is *constant*
   $\Leftrightarrow$ for all models $\mathcal{M}_1 = \langle D_1, [\![\ ]\!]^{\mathcal{M}_1} \rangle$
   and $\mathcal{M}_2 = \langle D_2, [\![\ ]\!]^{\mathcal{M}_2} \rangle$, if $D_1 \subseteq D_2$,
   then for all $P, Q \subseteq D_1$, we have
   $Q \in [\![\delta]\!]^{\mathcal{M}_1}(P)$ iff $Q \in [\![\delta]\!]^{\mathcal{M}_2}(P)$.

A can now be characterized as in the theorem below. So long as we assume all determiners to be conservative, this theorem tells us that A is the only increasing, logical determiner that is an inverse of BE.

Theorem 13. A *is the unique increasing, weakly conservative, constant inverse of* BE.[6]

Proof. What this theorem asserts precisely is that

---

[5] Constancy corresponds to (invariance for) Extension (of the context) in van Benthem's (1986) terminology. Topic-neutrality is a generalized notion of permutation invariance (cf. Keenan and Stavi, 1986; van Benthem, 1986).

[6] Being both conservative and constant is equivalent to being conservative* in Westerståhl's (1985) terminology. So we could alternatively say that A is the unique increasing and conservative* inverse of BE.

(i) A is constant and for every model $\mathcal{M} = \langle D, [\![\ ]\!]^{\mathcal{M}} \rangle$, $[\![A]\!]^{\mathcal{M}}$ is increasing and weakly conservative and $[\![BE]\!]^{\mathcal{M}} \circ [\![A]\!]^{\mathcal{M}}$ is the identity map on $\wp(D)$,

and that

(ii) if δ is constant and for every model $\mathcal{M} = \langle D, [\![\ ]\!]^{\mathcal{M}} \rangle$, $[\![\delta]\!]^{\mathcal{M}}$ is increasing and weakly conservative and $[\![BE]\!]^{\mathcal{M}} \circ [\![\delta]\!]^{\mathcal{M}}$ is the identity map on $\wp(D)$, then for every model $\mathcal{M}$, $[\![\delta]\!]^{\mathcal{M}} = [\![A]\!]^{\mathcal{M}}$.

Showing (i) is straightforward. Here, let's just verify the constancy of A. Let $\mathcal{M}_1 = \langle D_1, [\![\ ]\!]^{\mathcal{M}_1} \rangle$ and $\mathcal{M}_2 = \langle D_2, [\![\ ]\!]^{\mathcal{M}_2} \rangle$ be models such that $D_1 \subseteq D_2$. For all $P, Q \subseteq D_1$,

$$Q \in [\![A]\!]^{\mathcal{M}_1}(P)$$
$$\text{iff} \quad Q \in \{R \in \wp(D_1) \mid P \cap R \neq \varnothing\}$$
$$\text{iff} \quad Q \in \{R \in \wp(D_2) \mid P \cap R \neq \varnothing\}$$
$$\text{iff} \quad Q \in [\![A]\!]^{\mathcal{M}_2}(P).$$

Thus A is constant.

To show (ii), assume for a contradiction that δ has the described properties but there exists some model $\mathcal{M}_1 = \langle D_1, [\![\ ]\!]^{\mathcal{M}_1} \rangle$ such that $[\![\delta]\!]^{\mathcal{M}_1} \neq [\![A]\!]^{\mathcal{M}_1}$. By Lemma 12, $[\![\delta]\!]^{\mathcal{M}_1}(D_1) = \wp(D_1)$, so in particular,

$$\varnothing \in [\![\delta]\!]^{\mathcal{M}_1}(D_1).$$

Now let $\mathcal{M}_2 = \langle D_2, [\![\ ]\!]^{\mathcal{M}_2} \rangle$ be a model with $D_2 \supsetneq D_1$. By Lemma 12, $[\![\delta]\!]^{\mathcal{M}_2}(D_1) = [\![A]\!]^{\mathcal{M}_2}(D_1) = \{Q \in \wp(D_2) \mid D_1 \cap Q \neq \varnothing\}$, so

$$\varnothing \notin [\![\delta]\!]^{\mathcal{M}_2}(D_1).$$

This contradicts the constancy of δ. $\square$

5   Conclusion

Given that the domain of entities of a model for natural language semantics should generally be infinite, BE is characterized not as the unique homomorphism that makes the Partee triangle commute (Theorem 8), but as the unique *complete* homomorphism that makes it commute (Theorem 4). In light of Keenan and Faltz (1985), who have shown the importance of considering complete (rather than plain) homomorphisms in natural language semantics, this is a welcome result. BE can alternatively be characterized as the unique homomorphism that is an inverse of A (Theorem 11), while A can be characterized as the unique increasing, (weakly) conservative, constant/logical inverse of BE (Theorem 13). From this viewpoint, the naturalness of BE and that of A complement each other. On the other hand, despite Partee's (1986) conjecture that A and THE are the most "natural" determiners, it is not clear whether THE may be mathematically viewed as equally natural as A is. I hope that this paper has elucidated some finer mathematical points of the Partee triangle that have gone unnoticed and will help rid the linguistic community of any misunderstandings or confusion regarding Partee's (1986) Fact 2. Partee's statement in Fact 2 was not precise, but after all, the results of this paper reinforce her intuition that BE is nice and natural.

Acknowledgments

I would like to thank Makoto Kanazawa, who kindly read a very early draft of this paper and informed me that van Benthem (1986), while assuming the domain of entities to be finite, asserted Partee's (1986) Fact 2 without proof.

References

Jon Barwise and Robin Cooper. 1981. Generalized quantifiers and natural language. *Linguistics and Philosophy* 4:159–219.

Steven Givant and Paul Halmos. 2009. *Introduction to Boolean Algebras*. Springer, New York.

Edward L. Keenan and Leonard M. Faltz. 1985. *Boolean Semantics for Natural Language*. D. Reidel Publishing Company, Dordrecht.

Edward L. Keenan and Jonathan Stavi. 1986. A semantic characterization of natural language determiners. *Linguistics and Philosophy* 9:253–326.

Gödehard Link. 1983. The logical analysis of plurals and mass terms: A lattice-theoretical approach. In Rainer Bäuerle, Christoph Schwarze, and Arnim von Stechow, editors, *Meaning, Use, and Interpretation of Language*, Walter de Gruyter, Berlin, pages 303–323.

Barbara H. Partee. 1986. Noun phrase interpretation and type-shifting principles. In Jeroen Groenendijk, Dick de Jongh, and Martin Stokhof, editors, *Studies in Discourse Representation Theory and the Theory of Generalized Quantifiers*, Foris Publications, Dordrecht, pages 115–143.

Johan van Benthem. 1986. *Essays in Logical Semantics*. D. Reidel Publishing Company, Dordrecht.

Dag Westerståhl. 1985. Logical constants in quantifier languages. *Linguistics and Philosophy* 8:387–413.

How Many Stemmata with Root Degree k?

Armin Hoenen
CEDIFOR
Goethe University Frankfurt
hoenen@em.uni-frankfurt.de

Steffen Eger
UKP
TU Darmstadt
eger@ukp.informatik.tu-darmstadt.de

Ralf Gehrke
CEDIFOR
Goethe University Frankfurt
gehrke@rz.uni-frankfurt.de

Abstract

We are investigating parts of the mathematical foundations of stemmatology, the science reconstructing the copying history of manuscripts. After Joseph Bédier in 1928 got suspicious about large amounts of root bifurcations he found in reconstructed stemmata, Paul Maas replied in 1937 using a mathematical argument that the proportion of root bifurcating stemmata among all possible stemmata is so large that one should not become suspicious to find them abundant. While Maas' argument was based on one example with a tradition of three surviving manuscripts, we show in this paper that for the whole class of trees corresponding to Maasian reconstructed stemmata and likewise for the class of trees corresponding to complete historical manuscript genealogies, root bifurcations are apriori the most expectable root degree type. We do this by providing a combinatorial formula for the numbers of possible so-called *Greg trees* according to their root degree (Flight, 1990). Additionally, for complete historical manuscript trees (regardless of loss), which coincide mathematically with *rooted labeled trees*, we provide formulas for root degrees and derive the asymptotic degree distribution. We find that root bifurcations are extremely numerous in both kinds of trees. Therefore, while previously other studies have shown that root bifurcations are expectable for true stemmata, we enhance this finding to all three philologically relevant types of trees discussed in breadth until today.

1 Introduction

Stemmatology is the science trying to reestablish the copy history of a text surviving in a number of versions. One of the editors' objectives in stemmatology can be approaching the original authorial wording, which itself is most probably lost, given the body of extant text variants (Cameron, 1987).

In order to do so, the philologist may reconstruct the copy history of the manuscripts so as to better understand which variants are most likely original. Usually, the visual reconstruction is a graph or more precisely a tree where the nodes symbolize manuscripts and the copy processes are depicted by the edges. Such a visual reconstruction is then called a stemma. For an example of a stemma, see Figure 1.

Maybe the biggest and surely most famous problem in philology is an observation that the French philologist Joseph Bédier made editing the medieval French text "Le lai de l'ombre" in 1890, 1913 and 1928 (Bédier, 1890, 1913, 1928). Bédier observed that 105 out of 110 stemmata, the vast majority, in a collection he had made without controlling for root degree patterns had a bifurcation immediately below their root, an observation repeated multiple times thereafter on different collections, compare Table 1.

This observation was worrisome. If there are exactly two texts (nodes) directly below the assumed authorial original (root),[1] the implications for text reconstruction of the urtext are the following. An editor may choose one of the two texts as his/her preferred base text at will and reconstruct the ancestral text from this base text eliciting only in special cases the second or yet another variant.

---

[1]More precisely, in most cases, a root of such a tree represents a hypothetical intermediary: the latest common ancestor of all survivors. It corresponds to the oldest objectively reconstructible text and is called archetype.

Proceedings of the 15th Meeting on the Mathematics of Language, pages 11–21,
London, UK, July 13–14, 2017. ©2017 Association for Computational Linguistics

Collection	root bifurcations	root tri- or multifurcations
Bédier (1928)	95.5%	4.5%
Castellani (1957)	82.5%	17.5%
Haugen (2015) Bibliotheca A.	85.5%	14.5%
Haugen (2015) Editiones A.	80.5%	19.5%

Table 1: Percentages of root bifurcative stemmata in four collections, reported in (Haugen, 2015). Note that extending his collection through stemmata which are not yet viewed as conclusive by the composer, Castellani (1957, p.24) reports only $75 - 76\%$ root bifurcating trees.

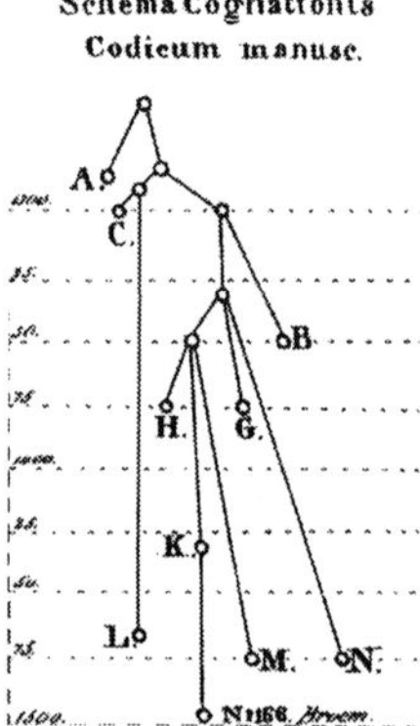

Figure 1: First modern stemma by Schlyter, 1827, from O'Hara (1996).

Bédier was worried about editors consciously or subconsciously choosing a base manuscript for the urtext after their taste and justifying this by postulating root bifurcations in their stemmata. As a second explanation for a large incidence of root bifurcations in reconstructed stemmata he suspected a methodology-inherent tendency for overseparation since editors always look for the *one* authorial in opposition to all other variants (a fallacy of the stemmatic method).

One can easily imagine that the subsequent debate had far-reaching consequences for textual criticism and editing. The community divided into best text editors (or Bédierists) which abandoned stemmatic approaches altogether and based their editions on a good available manuscript and those which continued and continue to produce stemmata (or Lachmannians). More realistically, any modern editor may choose among one of those approaches depending on his/her material and circumstances. Nevertheless, the argument has ever since stimulated much research repeatedly including mathematical argumentation, see for instance, Greg (1931), Maas (1937), Fourquet (1946), Whitehead (1951), Pasquali (1952), Castellani (1957), Hering (1967), Kleinlogel (1968), Weitzman (1982), Weitzman (1987), Grier (1989), Haugen (2002), Timpanaro (2005), Haugen (2010), Haugen (2015), Hoenen (2016). Maas argued that the number of stemmata with a root bifurcation among all possible stemmata which can be reconstructed (thus regarding stemma generation apriori as a random process) would be naturally high. One should thus rather not be too surprised of large proportions in real reconstructed stemmata: those were no good reason to abandon the stemmatic method. Maas numerically based this counter argument on the example of traditions with three surviving manuscripts.[2] Bédierists could have reacted to this and could have tried to seek a generalization of his argument. However, neither Bédierists nor Lachmannians have ever come up with such a generalization. What if Maas' argument would only hold for three surviving manuscripts, but witness completely different proportions for 4, 5, or 60 survivors? Would those numbers reveal justification for being suspicious of the real-world reconstructions?

In fact, Maas himself estimated numbers of possible stemmata for a number of surviving manuscripts of up to 5 according to Flight (1990), who decades later generalized the type of graphs Maas had considered for the modeling of stemmata. Flight (1990) provided a formula to count numbers of these so-called Greg trees, given a certain number of survivors. However, the question of the proportion of root bifurcating stem-

---

[2]Maas distinguishes two kinds of traditions of medieval texts: texts read by many and texts read by few. He assumes that strict stemmatics fails for texts read by many, which should be characterized by a larger number of survivors. Yet, not all philologists follow this distinction. Pasquali and Pieraccioni (1952) distinguish *open* and *closed* traditions, where the latter are such which are largely free of flaws complicating stemmatic assessment. Closed traditions are not straighforwardly connected with the number of survivors, compare also West (1973), which is why there is no reason to limit the range of surviving manuscripts to very small numbers and surely not to just one or two examples.

mata and how this proportion develops—thus ultimately the generalization of Maas' argument—has not yet been answered. In this paper, we fill this gap and provide a formula for the numbers of possible root k-furcating stemmata given m surviving manuscripts and compute the proportion of root bifurcating stemmata among all stemmata given m survivors.

Our work connects to a tradition both in linguistics and biology to count certain subclasses of graphs. In our case these graphs are trees, whereas other works have counted alignments between two or multiple sequences, that is, certain bi- or multi-partite graphs (Griggs et al., 1990; Covington, 2004; Eger, 2015).

2 Counting Manuscript Trees: Prerequisites

The theoretical entity used to model manuscript genealogy is a *tree*. A tree, as a concept from graph theory, is a set of *nodes* V together with a set of (unordered) *edges* E, with $E \subseteq \{\{u, v\} \mid u, v \in V\}$. The two defining properties of trees is that they must be free of cycles (including self-cycles) and connected. General works on counting different types of trees appear early on (Cayley, 1889), and research on trees is comprehensive, compare Moon (1970). The similarity of the three disciplines of historical linguistics, phylogeny and stemmatology has likewise been noticed early and led to various transfers and adaptations between methods of those fields, compare O'Hara (1996). Especially in the domain of phylogeny the understanding of trees is a central issue and consequently much research has focussed on phylogenetic trees, see for instance Felsenstein (1978); Swofford (1990); Huson (1998); Felsenstein (2004). One characteristic of phylogenetic trees is that they are apriori exclusively bifurcating. Thus, the question for a proportion of root bifurcating trees becomes meaningless. Apart from this, the manual reconstruction of a consistent and complete genome or characterome of ancestors is by no means as central an issue as in stemmatics (Platnick and Cameron, 1977; Cameron, 1987).

In the context of manuscript trees, although a number of the above enumerated philological studies count stemmatic trees under certain conditions or elaborate on specific phenomena, Flight (1990) is apparently the first to provide a generalized definition for stemmas. He aims at solving the

question, which he attributes to Maas (1958), how many different stemmas may exist for some given number of surviving manuscripts (Flight, 1990, p.122).

To solve this, he counts so called *Greg trees*.[3] Based on Flight (1990), we define a rooted directed Greg tree (which Flight names after the textual critic W. W. Greg) as a tree with a distinguished root, m labeled nodes standing for surviving manuscripts and n unlabeled nodes symbolizing hypothetical manuscripts. The latter must have an outdegree of at least two. There can be neither chains of hypothetical manuscripts (unlabeled nodes) with indegree one and outdegree one nor unlabeled leafs. This restriction corresponds to philological practice (Maas, 1937). A rooted Greg tree therefore symbolizes a reconstructed stemma. With this definition, Flight (1990) recovers the numbers of possible trees for three surviving manuscripts as postulated by Maas (1937), see Figure 2. Flight (1990) gives a recursive formula for the enumeration of unrooted and rooted Greg trees, building on all (four) generalized conditions on how to add a new labeled node and tabulates all possible Greg trees for up to 12 labeled nodes. Thus, he extends values mentioned by Maas as well as corrects Maas' numbers. From the 22 rooted Greg trees for 3 survivors, there are 12 root bifurcating ones, compare again Figure 2. The recursive formula Flight gives for rooted Greg trees $g(m, n)$ on m labeled and n unlabeled nodes is:[4]

$$
\begin{aligned}
g(m, n) = {} & (m + n - 2) \cdot g(m - 1, n - 1) \\
& + (2m + 2n - 2) \cdot g(m - 1, n) \\
& + (n + 1) \cdot g(m - 1, n + 1).
\end{aligned}
$$

If we fix m, the number of unlabeled nodes n can vary in the range of $\{0, 1, \ldots, m-1\}$ and the sum, over n, of all such (m, n)-trees for a fixed m is the number $g(m)$ of possible rooted Greg trees for m survivors (Flight, 1990). This gives the number of possible stemmata one can reconstruct for m surviving manuscripts adhering to philological principles.[5]

---

[3]According to Josuat-Vergès (2015), a similar problem in phylogeny has been described and tackled by Felsenstein (1978) as recognized by Knuth (2005).

[4]Flight refers to these as g^*, but for brevity and since we do not deal with unrooted Greg trees, we denote them simply as g.

[5]The number sequence $g(m)$ is listed as integer sequence A005264 in the On-Line Encyclopedia of Integer Sequences (OEIS), published electronically at https://oeis.org.

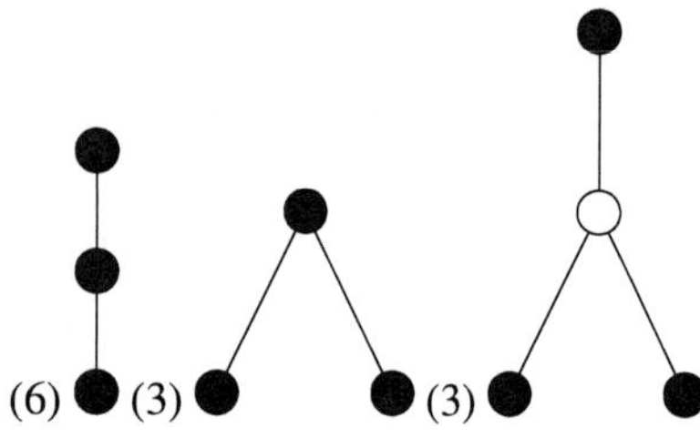

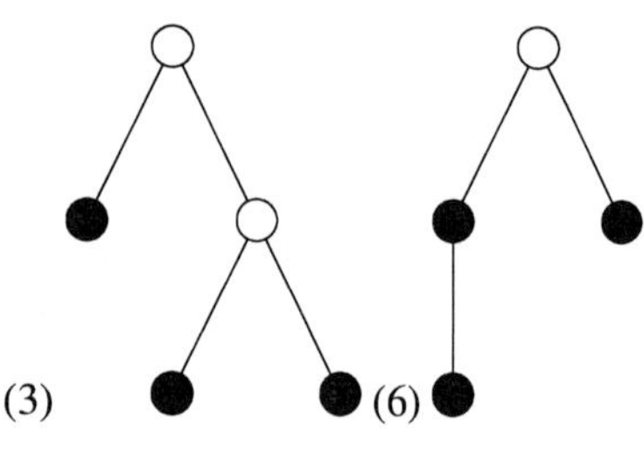

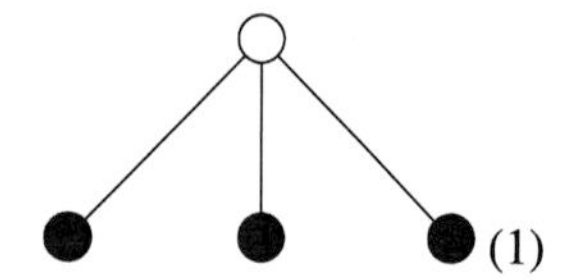

Figure 2: The unlabeled rooted (root topmost node) topologies of possible stemmata for three surviving manuscripts as thought of by Maas (1937). White nodes symbolize reconstructed lost manuscripts (unlabeled) whereas black nodes are survivors (labeled). The number in brackets refers to the number of possible distinct labeled trees (label permutations) for each topology.

While Flight (1990) does not compute numbers of Greg trees according to their root degree, Hering (1967), referring to a colleague of his,[6] tabulates the numbers of root k-furcating Greg trees (and the numbers of rooted Greg trees being the sum over all k) up to $m = 6$. The sums for all k at a fixed m coincides exactly with $g(m)$ calculated by Flight (1990). Alas, there is no formula provided by Hering (1967). Furthermore, he states that a calculation for more than 6 survivors would be difficult. This is demoralizing insofar as surely numbers (much) larger than $m = 6$ are relevant to the philological debate. For instance, according to Weitzman (1987), numbers of survivors in Greek and Latin traditions can range from 1 to "well over 100".

3 Counting Manuscript Trees: New Formulas

3.1 A Meta Formula

First, we present a general formula for counting trees with fixed root degree and two different types of nodes (e.g., black and white), which we use later on to derive our main results. We write $\mathcal{T}$ for a class of trees and T for $|\mathcal{T}|$.

If the root of a rooted tree has degree k and the tree has μ black nodes and ν white nodes, it means that the tree has k subtrees, which we also perceive as rooted. The root node, r, is either black or white. We connect r to the root of each subtree. Each of these subtrees can have some size $s_1 + p_1, \ldots, s_k + p_k$, where s_i is the number of black nodes in branch i and p_i is the number of white nodes in the same branch. The sum of the s_i must equal $\mu - \delta_B$ and the sum of the p_i must equal $\nu - \delta_W$, since there are in total μ black nodes and ν white nodes. Here, δ_B is a binary variable indicating whether r is a black node and analogously for δ_W, where $\delta_B = 1$ iff $\delta_W = 0$. If the black nodes are distinguishable, we can choose the subsets of nodes of sizes $s_1, \ldots, s_k$ from a total of $\mu - \delta_B$ nodes, and analogously for the white nodes. There are $\binom{\mu - \delta_B}{s_1, \ldots, s_k}$ possibilities to do so, where $\binom{m}{k_1, \ldots, k_\ell} = \frac{m!}{k_1! \cdots k_\ell!}$ are the multinomial coefficients.

Now, we specialize. We assume that the black nodes are distinguishable and the white nodes are indistinguishable. Then, for any class of rooted trees $\mathcal{T}_{\mu,\nu}$ with μ such black nodes and ν such

---

[6]Prof. Dr. Wolfgang Engel, a mathematician from Rostock University.

14

white nodes, the number $T_{\mu,\nu,k}$ of rooted labeled trees from $\mathcal{T}_{\mu,\nu}$ in which the root has degree k has the form

$$\mu \sum_{(\mathbf{s},\mathbf{p})\in\mathcal{C}((\mu-1,\nu),k)} \binom{\mu-1}{\mathbf{s}} F(\mathbf{s},\mathbf{p})$$

$$+ \sum_{(\mathbf{s},\mathbf{p})\in\mathcal{C}((\mu,\nu-1),k)} \binom{\mu}{\mathbf{s}} F(\mathbf{s},\mathbf{p}).$$

Here, $\mathcal{C}((a,b),\ell)$ denotes the number of *vector compositions* (Eger, 2017) of the 'vector' $(a,b)\in\mathbb{N}^2$ with ℓ parts; that is,

$$\mathcal{C}((a,b),\ell) = \{(s_1,\dots,s_\ell),(p_1,\dots,p_\ell)\,| $$
$$s_1+\cdots+s_\ell = a,\ p_1+\cdots+p_k = b\}.$$

Moreover, by $\mathbf{s}$ and $\mathbf{p}$, we denote tuples $(s_1,\dots,s_k)$ and $(p_1,\dots,p_k)$, respectively. The above sum formula arises because the root node can either be black or white. If it is black, we have the additional factor μ because the black nodes are distinguishable and each of them can be the root.

Finally, F is a function of the sizes $s_1,\dots,s_k,p_1,\dots,p_k$ which will be specified in any particular case.

Now, we have overcounted $T_{\mu,\nu,k}$ since we have counted subtrees as if they were ordered, while in reality different orders of the subtrees do not constitute a distinct tree $t\in\mathcal{T}_{\mu,\nu,k}$. Thus, we have to divide by $k!$ to finally arrive at:

$$T_{\mu,\nu,k} = \frac{\mu}{k!} \sum_{(\mathbf{s},\mathbf{p})\in\mathcal{C}((\mu-1,\nu),k)} \binom{\mu-1}{\mathbf{s}} F(\mathbf{s},\mathbf{p})$$

$$+ \frac{1}{k!} \sum_{(\mathbf{s},\mathbf{p})\in\mathcal{C}((\mu,\nu-1),k)} \binom{\mu}{\mathbf{s}} F(\mathbf{s},\mathbf{p}).$$

$$(1)$$

It is possible that $T_{\mu,\nu,k}$ can be expressed simpler—e.g., as a linear combination of the terms $T_{\mu+\tau,\nu+\rho,k+\kappa}$ for integers τ,ρ,κ—for specific choices of F.

3.2 *Root k-furcating Greg Trees*

We are now ready to derive the general formula for the number $g_k(m,n)$ of root k-furcating Greg trees for m survivors (labeled nodes) and n hypothetical (unlabeled) nodes.

The only question remaining from above is how we have to specify the function $F(\mathbf{s},\mathbf{p})$ on the k subtrees. This is very simple, however. Since all branches i are independent of each other, $F(\mathbf{s},\mathbf{p})$ takes the form of a product of individual factors:

$$F(\mathbf{s},\mathbf{p}) = \prod_{i=1}^{k} g(s_i,p_i)$$

where g is the function of Flight (1990). The number $g_k(m,n)$ of root k-furcating Greg trees for m survivors and n hypothetical nodes is hence given by (1) with this specification of F.

We make three additional remarks. The s_i satisfy $s_i\geq 1$, since the specification of Greg trees disallows to have only unlabeled nodes (i.e., $s_i = 0$) in a branch. In contrast, the p_i may take on the value zero and therefore satisfy $p_i\geq 0$. Moreover, the p_i actually satisfy $0\leq p_i < s_i$ because of the link restrictions on unlabeled nodes in Greg trees. While the constraint on the p_i's is automatically taken care of by the function g of Flight (1990), explicitly accounting for it can speed up computations.[7] Finally, when $k = 1$, we have to exclude the second term in (1) from consideration because, by definition, the root of a Greg tree cannot have degree one when it is unlabeled.

The numbers $g_k(m)$ of root k-furcating Greg trees for m survivors and an arbitrary number of hypothetical manuscripts n is the sum over n of root k-furcating (m,n)-trees. In other words,

$$g_k(m) = \sum_{n\geq 0} g_k(m,n).$$

Table 2 shows the growth of $g_k(m)$ until $m,k = 15$.

We are now interested in the proportions of root bifurcating Greg trees among all Greg trees since this was alluded to in Bédier (1928). That is, we investigate the ratio

$$R_2(m) = \frac{g_2(m)}{\sum_{k\geq 1} g_k(m)}.$$

---

[7]In order to more efficiently compute the numbers, we also used further simplified formulas for specific k where possible. Root unifurcating Greg trees (here g_1) are especially easily computed. The root can only be labeled, since an unlabeled node as root must have degree at least two. Then, the number of possible root unifurcating Greg trees corresponds to $m\cdot g(m-1)$. Root-$(m-1)$-furcating rooted Greg trees for all $m\neq 2$ coincide with the pentagonal numbers (sequence $A000326$ in the OEIS), whose number is given by $\frac{3m^2-m}{2}$. This is so because there are only three principle architectures of root-$(m-1)$-furcating rooted Greg trees, the individual formulas for the enumeration of which sum to the same as the pentagonal numbers: $m + m(m-1) + \binom{m}{2}$. Finally, for a root m-furcation, there is always only one Greg tree.

15

m \ k	1	2	3	4	5	6	7	8	9	10	11	12	13	14	15	Σ
1	1	-	-	-	-	-	-	-	-	-	-	-	-	-	-	1
2	2	1	-	-	-	-	-	-	-	-	-	-	-	-	-	3
3	9	12	1	-	-	-	-	-	-	-	-	-	-	-	-	22
4	88	151	22	1	-	-	-	-	-	-	-	-	-	-	-	262
5	1 310	2 545	445	35	1	-	-	-	-	-	-	-	-	-	-	4 336
6	26 016	54 466	10 425	1 025	51	1	-	-	-	-	-	-	-	-	-	91 984
7	643 888	1 417 318	286 321	31 780	2 030	70	1	-	-	-	-	-	-	-	-	2 381 408
8	19 051 264	43 472 780	9 102 604	1 090 201	80 360	3 626	92	1	-	-	-	-	-	-	-	72 800 928
9	655 208 352	1 536 228 588	329 980 456	41 636 973	3 368 001	178 290	6 006	117	1	-	-	-	-	-	-	2 566 606 784
10	25 666 067 840	61 466 251 616	13 457 494 060	1 763 775 280	152 280 345	8 964 417	358 890	9 390	145	1	-	-	-	-	-	102 515 201 984
11	$1.13 * 10^{12}$	$2.75 * 10^{12}$	$6.1 * 10^{11}$	$8.23 * 10^{10}$	$7.46 * 10^{9}$	$4.74 * 10^{8}$	$2.13 * 10^{7}$	$6.61 * 10^{5}$	$1.4 * 10^{4}$	176	1	-	-	-	-	$4.58 * 10^{12}$
12	$5.49 * 10^{13}$	$1.36 * 10^{14}$	$3.05 * 10^{13}$	$4.21 * 10^{12}$	$3.96 * 10^{11}$	$2.66 * 10^{10}$	$1.3 * 10^{9}$	$4.64 * 10^{7}$	$1.18 * 10^{6}$	$2.02 * 10^{4}$	210	1	-	-	-	$2.26 * 10^{14}$
13	$2.93 * 10^{15}$	$7.33 * 10^{15}$	$1.66 * 10^{15}$	$2.34 * 10^{14}$	$2.27 * 10^{13}$	$1.59 * 10^{12}$	$8.31 * 10^{10}$	$3.25 * 10^{9}$	$9.41 * 10^{7}$	$1.97 * 10^{6}$	$2.82 * 10^{4}$	247	1	-	-	$1.22 * 10^{16}$
14	$1.71 * 10^{17}$	$4.31 * 10^{17}$	$9.86 * 10^{16}$	$1.41 * 10^{16}$	$1.39 * 10^{15}$	$1.02 * 10^{14}$	$5.58 * 10^{12}$	$2.34 * 10^{11}$	$7.47 * 10^{9}$	$1.71 * 10^{8}$	$3.16 * 10^{6}$	$3.83 * 10^{4}$	287	1	-	$7.15 * 10^{17}$
15	$1.07 * 10^{19}$	$2.73 * 10^{19}$	$6.29 * 10^{18}$	$9.01 * 10^{17}$	$9.2 * 10^{16}$	$6.89 * 10^{15}$	$3.94 * 10^{14}$	$1.75 * 10^{13}$	$6.03 * 10^{11}$	$1.61 * 10^{10}$	$3.27 * 10^{8}$	$4.89 * 10^{6}$	$5.01 * 10^{4}$	330	1	$4.53 * 10^{19}$

Table 2: Numbers of root k-furcations for all possible k for rooted Greg trees with m survivors up to $m = 15$, $g_k(m)$. Note that the first numbers until $m = 6$ occur in Hering (1967). Exact numbers are provided until $m = 10$ and for $(m - 1)$-trees and otherwise numbers are in scientific notation. The last column contains the sum $\sum_{k=1}^{m} g_k(m)$ which equals the number of all rooted Greg trees for the current m, compare Flight (1990). Code for computation is available from https://github.com/ArminHoenen/KFurcatingRootedGregTrees. The sequence of numbers for $k = 2$ is now integer sequence A286432 in the OEIS.

At $m = 2$, the proportion is one third, at $m = 3$, Maas' famous example, we witness a proportion of $R_2 = 0.54545$. For $m = 10$, R_2 is already 0.59958 with the increase slowing down. For $m = 20$, we have $R_2 = 0.60351$ and at $m = 100$ survivors the proportion is $R_2 = 0.60599$. Growth is further slowing down and at $m = 200$ the proportion is $R_2 = 0.60626$. While we are not able to prove it, we think it is a very safe conjecture that $R_2(m)$ converges to below 0.607, as m tends toward infinity. Figure 3 plots the proportions of trees with root degree $k = 1$, $k = 2$ and $k > 2$, as m becomes larger. Figure 4 plots the root degree distribution for fixed m.

Root bifurcations thus outweigh all other root degree patterns by far. Maas' argument was therefore generally true as what regards a large expectability of root bifurcations in reconstructible stemmata. Nevertheless, the observed proportions are considerably lower than Bédier's ones. However, a better fit occurs when we exclude all trees with root degree one from consideration. A root degree of one requires root to be labeled and thus surviving, a case which is empirically probably quite rare, although not impossible. In Bédier's collection presumably, there simply had not been any root unifurcating stemma with a surviving root and he does not comprehensively discuss this general possibility. In Castellani's (1957) and Haugen's (2010) collections there have been no counts of root unifurcations. At $m = 200$, the fraction of unifurcating trees is about 21.467%, which means that the fraction of trees with root degree two is

$$\tilde{R}_2(m) = \frac{0.60626}{1 - 0.21467} = 0.7719$$

at $m = 200$, when trees with root degree one are discarded. Comparing this number to those in Table 1, we observe that the empirically reported numbers for actual collections of stemmata are just slightly above this reference point. This would indicate that there seems to be a bias for root bifurcations, but that this bias is rather low.

While Bédier had looked at $R_2(m)$ or $\tilde{R}_2(m)$ (coinciding in his collection), Maas explicitly looked at

$$R_{k>2}(m) = \frac{\sum_{k>2} g_k(m)}{\sum_{k \geq 1} g_k(m)}$$

for $m = 3$, and based his counter argument to Bédier's conclusions on that. This has been criticized variously because $R_2(3)$ corresponds to 12 in 22, the complement of which is not Maas' 1 but 10 in 22, a ratio probably too small to base a counter argument on it. Neither Bédier nor Maas discuss root unifurcating cases extensively, but they could make a crucial difference in the ratios of interest since including root unifurcating trees, *non-root-bifurcating* would no longer be equivalent to *root multifurcating* in meaning. Thus, Maas' shift of focus from root bifurcating to root multifurcating introduces ambiguity. Responding to such ambiguity, we demonstrated a mathematically sound way of looking both at proportions of root degree patterns with $(R_2(m))$ and without root unifurcations $(\tilde{R}_2(m))$.

Hering (1967), probably aware that root degrees of $k = 1$ appear to be somewhat unrealistic in actually observed stemmas, stated that instead of following Maas' focus, one should rather look at

$$R_{\mathrm{HE}}(m) = \frac{\sum_{k>2} g_k(m)}{g_2(m)}$$

which Hering (1967) investigated until $m = 6$ and for which he speculated that it would probably never surpass 0.33 or lie even lower. Looking at the plot of the proportions, see Figure 3, we can see that Hering was right, the asymptote is however rather 0.3. The extraordinary role of root unifurcations is immediately visible, since they are the only k witnessing a decline. This naturally follows from their restrictions—for instance their root can only be labeled, meaning that only the first term in (1) will be relevant, while for all other root degree patterns both add up.

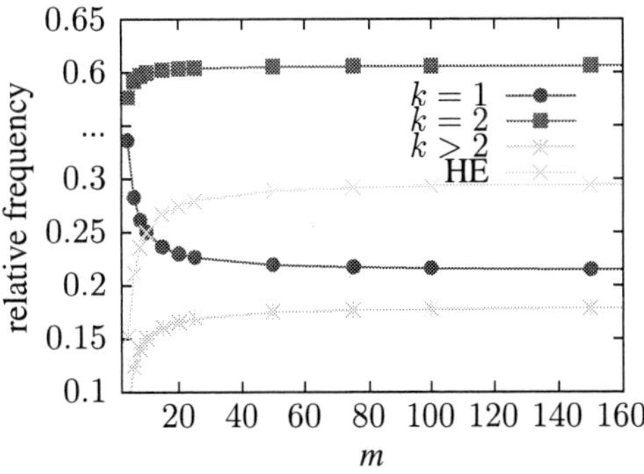

Figure 3: Proportions of root unifurcating and root bifurcating rooted Greg trees among all possible rooted Greg trees for a fixed m as well as $R_{k>2}(m)$ and $R_{\mathrm{HE}}(m)$. Note that the first three proportions add to 1.

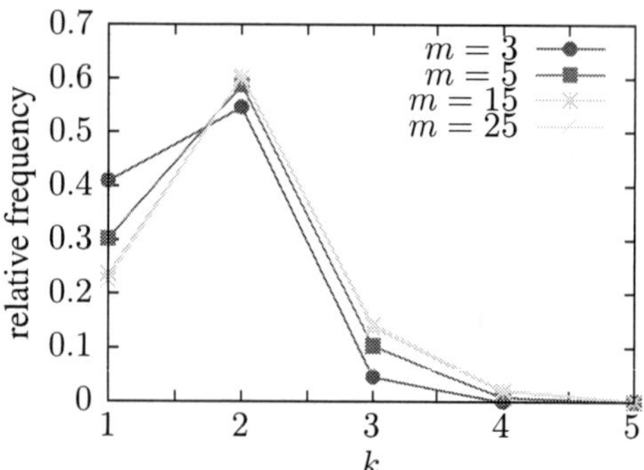

Figure 4: Root degree distribution for trees counted by $g_k(m)$ for fixed $m = 3, 5, 15, 25$.

In order to gain a deeper insight, we are now looking at another type of tree which plays an important role in stemmatology.

4 The Second Type of Manuscript Tree

While Maas had looked at possible trees a philologist can reconstruct, other studies looked at true historical trees and their proportionalities. The underlying process reflected in stemmatological trees is the generation of manuscripts and their copying. There is (in many cases) one original—which we can understand as a root node to a rooted tree—which gets copied a certain number of times (children in first generation). Each manuscript (including root) can be copied a certain number of times again (always including 0 times) and so forth. We assume each node to represent a unique text symbolized through a distinct label. In this way, the copy history can be understood/displayed as a rooted labeled tree. Since copying is a process from a vorlage[8] to a copy, the edges can be understood as directed.

Such a tree depicts the complete copy history of a text—and not as a stemma does, the reconstructible portion of it. It ignores loss of manuscripts (does not assume or know any unlabeled node) and extends to the entire copying history of a text. In order to avoid terminological confusion, the class of trees depicting this complete copy history of a tradition has been called an *arbre réel* in philology, a term coined by Fourquet (1946)—for convenience referred to as *arbre* in the rest of the paper.[9] Arbres were usually

used as hypothetical units of argumentation for outlining general scenarios of copying and proliferation in philological discourse, see for instance Castellani (1957). However, recently, they have gained actuality through artificial traditions, that is, complete copied sets with known ground truth (Spencer et al., 2004; Baret et al., 2006; Roos and Heikkilä, 2009; Hoenen, 2015), where arbres are used for evaluation, comparing them to computationally reconstructed stemmata.

In the following, we are looking at arbres themselves and provide an answer to the question how prevalent root bifurcation is in arbres. This may be useful for future research on the general effects of loss induced tree transformations (turning an arbre into a stemma), as has been exemplarily done for a restricted set of topologies by Trovato and Guidi (2004). Greg (1927) had already hypothesized that deformations arbres undergo through historical manuscript loss may be a reason for expectable root bifurcations in stemmata.[10]

We note that the following is a special case of our already derived results. In other words, we now evaluate $g_k(m, 0)$, in our above notation. However, this special case admits simpler closed-form formulas as well as a derivation of the asymptotic degree distribution.

5 Rooted Labeled Trees

By Cayley's formula (Cayley, 1889), the number T'_m of labeled trees on m nodes is given by m^{m-2}. The number T_m of rooted labeled trees is then given by m^{m-1} since each of the m nodes can be the root. Now, let's assume that the root has degree $k = 1, \ldots, m - 1$. How many such trees are there, $T_{m,k}$?

To answer this, we invoke our meta formula, Formula (1), with the following specification of $F(\mathbf{s}, \mathbf{p})$:

$$F(\mathbf{s}, \mathbf{p}) = g(s_1, 0) \cdots g(s_k, 0)$$

since $\mathbf{p} = (0, \ldots, 0)$, as we have no unlabeled nodes in this case. We have $g(s, 0) = T_s$ since $g(s, 0)$ retrieves the number of rooted labeled trees with s nodes.

---

for so-called R-trees, there is no conceptual overlap whatsoever.

[10]The kind of stemma we are talking about here is not a reconstructed stemma for any number of surviving manuscripts but rather the one single "true" stemma or *stemma reale* as termed by Timpanaro (2005).

---

[8]Vorlage is a loaned term for *original of a copy, not of a tradition* deriving from German used in philology.

[9]Although in French terminology the same term is used

Thus, combining this insight with the formula of Cayley, we find that there are exactly

$$\frac{m}{k!} \sum_{\mathbf{s}\in\mathcal{C}(m-1,k)} \binom{m-1}{\mathbf{s}} s_1^{s_1-1} \cdots s_k^{s_k-1} \qquad (2)$$

rooted labeled trees on m nodes with root degree k, where we let $\mathcal{C}(m-1,k)$ stand for $\mathcal{C}((m-1,0),k)$. An alternative, simpler formula for $T_{m,k}$ is given by:

$$T_{m,k} = m \cdot \binom{m-2}{k-1} \cdot (m-1)^{m-1-k}. \qquad (3)$$

For $k=1$ this formula is not difficult to show. For $k=2$ it has the following combinatorial interpretation. A rooted labeled tree has a root, for which we may choose any of the m nodes. Then there are $(m-1)$ vertices left. There are $(m-1)^{m-3}$ possible labeled trees on them. Since the $(m-1)$ vertices form a tree, there are $(m-2)$ edges connecting them. We may take any of these, and replace it by connections of their endpoints to the root. This yields all the rooted labeled trees in which the root node has degree 2. For $k>2$ a similar, but more involved argument applies (Moon, 1970, Theorem 3.2).

Next, we ask for the probability $P_m[k]$ that a randomly chosen rooted labeled tree from $\mathcal{T}_m$ has root degree $k=1,2,\ldots$. We find

$$P_m[k] = \frac{T_{m,k}}{T_m} = \frac{\binom{m-2}{k-1}}{(m-1)^{k-1}} \cdot \left(\frac{m-1}{m}\right)^{m-2}. \qquad (4)$$

The second factor in this product equals $(1-\frac{1}{m})^{m-2}$ and thus converges to $\exp(-1)$ as $m \to \infty$. For the first factor $A = \frac{\binom{m-2}{k-1}}{(m-1)^{k-1}}$, we find

- for $k=1$: $A = 1 \longrightarrow 1$,

- for $k=2$: $A = \frac{(m-2)}{(m-1)} \longrightarrow 1$,

- for $k=3$: $A = \frac{(m-2)(m-3)}{2} \frac{1}{(m-1)(m-1)} \longrightarrow \frac{1}{2}$

as $m \to \infty$. In general, we have for A:

$$A = \frac{(m-2)(m-3)\cdots(m-k)}{(m-1)(m-1)\cdots(m-1)} \frac{1}{(k-1)!}$$

When k is fixed and $m \to \infty$, then this converges to $\frac{1}{(k-1)!}$. Hence, the asymptotic distribution $P[k]$ of $P_m[k]$ is

$$P[k] = \frac{\exp(-1)}{(k-1)!}$$

which is a Poisson distribution with parameter $\lambda = 1$, denoted as Poisson(λ).

Figure 5 compares the asymptotic Poisson $P[k]$ distribution to the actual finite distributions $P_m[k]$. We see that convergence is rapid. For $m = 40$, $P_m[k]$ is visually already extremely close to Poisson($\lambda = 1$).

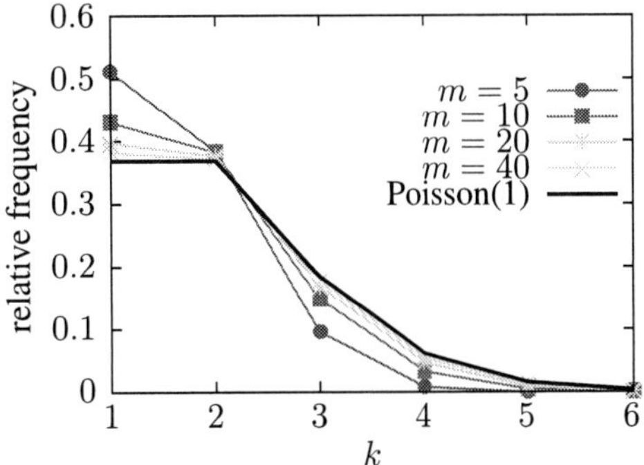

Figure 5: Asymptotic distribution Poisson($\lambda = 1$) and finite distributions $P_m[k]$ for $m = 5, 10, 20, 40$.

From $P[k]$, we infer that root bifurcations are asymptotically twice as likely as trifurcations but exactly as likely as unifurcations, and have a probability of roughly 0.37. Moreover, the larger k gets, the smaller the probability of root k-furcating trees—and this probability is rapidly decaying in k. As a side note, we emphasize that the asymptotic probability for bifurcations has a particularly beautiful mathematical form, namely, the inverse of Leonhard Euler's constant e.

These mathematical derivations, if they are based on a plausible description of reality, suggest that in history many original manuscripts may have been copied only once, the same number has been copied twice, half as many three times and a third of that number four times, a fourth of that number (for four) five times and so on. That is, if indeed a random process that selects each arbre for a fixed number of trees on m nodes with equal likelihood is a good model of true copy history. On this, any more sophisticated model can operate.

If root bifurcations are already very numerous, then an immediately related question would be what consequences this could have for a stemma when thinking about the transformations an arbre undergoes through historical loss. To this end, Weitzman (1982; 1987) has shown, and Trovato and Guidi (2004) come to a similar conclusion, that historically realistic scenarios of loss would imply a large quantity of bifurcations and root bifurcations in stemmata based on transformed arbres. Those do exceed $\frac{1}{e}$ and thus a possible effect of historical loss is to increase the percentage of root bifurcations, in which case $\frac{1}{e}$ would rather operate as a lower bound.

6 Conclusion

We have counted root k-furcating rooted labeled trees and root k-furcating rooted Greg trees. For the former, the asymptotic root degree distribution has been derived mathematically. For the latter, we have provided exact formulas that allow to approximate the asymptotic root degree distribution. From this, we (very strongly) conjecture that root bifurcating Greg trees have an asymptotic probability of above (and close to) 0.606.

In both cases, relating to a model of representation of arbres (true and complete historical manuscript genealogies) and stemmata (reconstructed genealogies from surviving nodes), the proportions of root bifurcating trees for historically relevant tradition sizes is the largest in respect to the other root degrees. Therefore, while previously other studies have shown that root bifurcations are expectable for true stemmata, we enhance this finding to reconstructible stemmata and arbres so that this statement now covers the three philologically relevant general types of trees discussed until today. Concerning stemmata, we have argued that the proportions of root bifurcating stemmata observed in real collections of genealogies is close to what is mathematically predicted, with a seemingly small bias for root bifurcations.

In the philological debate, where numerical arguments have been pursued since the very beginning, the formulas presented here contribute to clarify the basic combinatorial nature of the entities involved in the modeling of manuscript evolution. We believe that in an ever more computational stemmatological endeavour cultivating the mathematical foundations can only have positive effects.

While our findings with respect to root degrees of rooted labeled trees are certainly far from novel to the mathematics community, our formulas for Greg trees, which generalize rooted labeled trees, are, to our best knowledge, original.

References

P. Baret, C. Macé, and P. Robinson (eds.). 2006. Testing methods on an artificially created textual tradition. In *Linguistica Computationale XXIV-XXV*. Instituti Editoriali e Poligrafici Internationali, Pisa-Roma, volume XXIV-XXV, pages 255–281.

J. Bédier. 1890. *Jean Renart: Le lai de l'Ombre*. Saint-Paul. Reprint, New York: Johnson, 1968.

J. Bédier. 1913. *Jean Renart: Le lai de l'Ombre*. Firmin-Didot.

J. Bédier. 1928. La tradition manuscrite du 'Lai de l'Ombre': Réflexions sur l'Art d'Éditer les Anciens Textes. *Romania* 394:161–196, 321–356. (Rpt. Paris: Champion, 1970).

H. D. Cameron. 1987. The upside-down cladogram: problems in manuscript affiliation. In *Biological Metaphor and Cladistic Classification: an Interdiscplinary Approach*, University of Pennsylvania, pages 227–242.

A. E. Castellani. 1957. *Bédier avait-il raison?: La méthode de Lachmann dans les éditions de textes du moyen age: leçon inaugurale donnée à l'Université de Fribourg le 2 juin 1954*. Number 20 in Discours universitaires. Éditions universitaires.

A. Cayley. 1889. A theorem on trees. *Quarterly Journal of Mathematics* 23:376–378.

M. A. Covington. 2004. The number of distinct alignments of two strings. *Journal of Quantitative Linguistics* 11(3):173–182. https://doi.org/10.1080/0929617042000314921.

S. Eger. 2015. On the number of many-to-many alignments of multiple sequences. *Journal of Automata, Languages and Combinatorics* 20(1):53–65.

S. Eger. 2017. The combinatorics of weighted vector compositions. *ArXiv preprint* https://arxiv.org/abs/1704.04964.

J. Felsenstein. 1978. The number of evolutionary trees. *Systematic Zoology* 27(1):27–33.

J. Felsenstein. 2004. *Inferring phylogenies*. Sinauer Associates Sunderland.

C. Flight. 1990. How many stemmata? *Manuscripta* 34(2):122–128.

J. Fourquet. 1946. Le paradoxe de Bédier. *Mélanges* 1945(II):1–46.

W. W. Greg. 1927. *The calculus of variants: an essay on textual criticism*. Clarendon Press.

W. W. Greg. 1931. Recent theories of textual criticism. *Modern Philology* 28(4):401–404.

J. Grier. 1989. Lachmann, Bédier and the bipartite stemma: towards a responsible application of the common-error method. *Revue d'histoire des textes* 18(1988):263–278.

J. R. Griggs, P. Hanlon, A. M. Odlyzko, and M. S. Waterman. 1990. On the number of alignments of k sequences. *Graph. Comb.* 6(2):133–146. https://doi.org/10.1007/BF01787724.

O. E. Haugen. 2002. The Spirit of Lachmann, the Spirit of Bédier: Old Norse Textual Editing in the Electronic Age. In *Annual Meeting of The Viking Society, University College London*. volume 8.

O. E. Haugen. 2010. Is stemmatology inherently dichotomous? On the silva portentosa of Old Norse stemmata. *Studia Stemmatologica* .

O. E. Haugen. 2015. The silva portentosa of stemmatology Bifurcation in the recension of Old Norse manuscripts. *Digital Scholarship in the Humanities* 30(2).

W. Hering. 1967. Zweispaltige Stemmata. *Philologus-Zeitschrift für antike Literatur und ihre Rezeption* 111(1-2):170–185.

A. Hoenen. 2015. Das artifizielle Manuskriptkorpus TASCFE. In *DHd 2015 - Von Daten zu Erkenntnissen - Book of abstracts*, DHd. http://gams.uni-graz.at/o:dhd2015.abstracts-gesamt.

A. Hoenen. 2016. Silva Portentosissima Computer-Assisted Reflections on Bifurcativity in Stemmas. In *Digital Humanities 2016: Conference Abstracts. Jagiellonian University & Pedagogical University*. pages 557–560. http://dh2016.adho.org/abstracts/311.

D. H. Huson. 1998. Splitstree: analyzing and visualizing evolutionary data. *Bioinformatics* 14(1):68–73.

M. Josuat-Vergès. 2015. Derivatives of the tree function. *The Ramanujan Journal* 38(1):1–15.

A. Kleinlogel. 1968. Das Stemmaproblem. *Philologus-Zeitschrift für antike Literatur und ihre Rezeption* 112(1-2):63–82.

D. E. Knuth. 2005. The art of computer programming, volume 4: Generating all combinations and partitions, fascicle 3.

P. Maas. 1937. Leitfehler und Stemmatische Typen. *Byzantinische Zeitschrift* 37(2):289–294.

P. Maas. 1958. *Textual Criticism*. Clarendon Press.

J. W. Moon. 1970. *Counting labelled trees*. Canadian Mathematical Congress.

R. J. O'Hara. 1996. Trees of history in systematics and philology. *Memorie della Società Italiana di Scienze Naturali e del Museo Civico di Storia Naturale di Milano* 27(1):81–88.

G. Pasquali and D. Pieraccioni. 1952. *Storia della tradizione e critica del testo*. Le Monnier.

N. I. Platnick and H. D. Cameron. 1977. Cladistic methods in textual, linguistic, and phylogenetic analysis. *Systematic Zoology* 26(4):380–385.

T. Roos and T. Heikkilä. 2009. Evaluating methods for computer-assisted stemmatology using artificial benchmark data sets. *Literary and Linguistic Computing* 24:417–433.

M. Spencer, E. A. Davidson, A. C. Barbrook, and C. J. Howe. 2004. Phylogenetics of artificial manuscripts. *Journal of Theoretical Biology* 227:503–511.

D. L. Swofford. 1990. *PAUP: Phylogenetic Analysis Using Parsimony Version 3.0, May 1990*. Illinois Natural History Survey.

S. Timpanaro. 2005. *The Genesis of Lachmann's Method*. University of Chicago, Chicago.

P. Trovato and V. Guidi. 2004. Sugli stemmi bipartiti - decimazione, asimmetria e calcolo delle probabilità. *Filologia Italiana* 1:9–48.

M. P. Weitzman. 1982. Computer simulation of the development of manuscript traditions. *ALLC Bulletin. Association for Library and Linguistic Computing Bangor* 10(2):55–59.

M. P. Weitzman. 1987. The evolution of manuscript traditions. *Journal of the Royal Statistical Society. Series A (General)* pages 287–308.

M. L. West. 1973. *Textual Criticism and Editorial Technique: Applicable to Greek and Latin texts*. Teubner, Stuttgart.

F. Whitehead and C. E. Pickford. 1951. The two-branch stemma. *Bulletin Bibliographique de la Société Internationale Arthurienne* 3:83–90.

On the Logical Complexity of Autosegmental Representations

Adam Jardine
Rutgers University
`adam.jardine@rutgers.edu`

Abstract

Autosegmental mapping from disjoint strings of tones and tone-bearing units, a commonly used mechanism in phonological analyses of tone patterns, is shown to not be definable in monadic second-order logic. This is abnormally complex in comparison to other phonological mappings, which have been shown to be monadic second-order definable. In contrast, generation of autosegmental structures from strings is demonstrated to be first-order definable.

1 Introduction

This paper applies *logical transductions* as introduced by Courcelle (1994) to study the cognitive complexity of non-string representations and transformations in phonology. Generative phonology studies phonological patterns both in terms of *transformations*, or relations between input *underlying representations* (URs) and output *surface representations* (SRs), and *phonotactics*, or generalizations about the well-formedness of SRs. Studies of the computational complexity of these patterns have established clear bounds on the expressivity needed to describe them. For example, Johnson (1972) and Kaplan and Kay (1994) showed that the ordered rewrite-rule grammars of Chomsky and Halle (1968) describe exactly Regular string relations, and more recent cross-linguistic studies have shown that phonological transformations fall into more restrictive subclasses of the Regular class (Chandlee and Heinz, 2012; Chandlee, 2014; Heinz and Lai, 2013; Payne, 2014; Jardine, 2016a). Similarly, phonotactic patterns have been shown to fall into sub-Regular classes of formal languages (Heinz, 2009, 2010; Rogers et al., 2013). This has led to a

hypothesis that there is a sub-Regular bound on phonology (see, e.g., Heinz and Idsardi, 2013), which has clear connections to cognitive complexity (Rogers and Pullum, 2011; Rogers et al., 2013) and how humans learn sound patterns (Heinz, 2009, 2010; Lai, 2015).

However, these complexity classes are defined in terms of strings, and since the advent of autosegmental phonology (Goldsmith, 1976), generative phonology has commonly employed the use of non-string structures. Perhaps the most commonly used of these has been *autosegmental representations* (ARs), which represent words with graph structures in which disjoint strings are associated to one another in some fashion. For example, Fig. 1 shows an autosegmental derivation for the Mende word [félàmà] 'junction', which is comprised of a high-toned syllable followed by two low-toned syllables (following convention syllables are represented with σ).

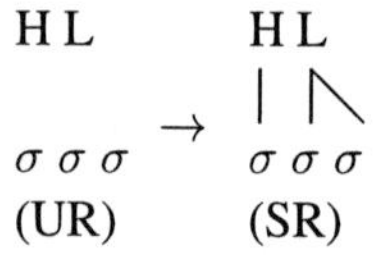

$$
\begin{array}{ccc}
\text{H L} & & \text{H L} \\
& \rightarrow & \;\,|\;\,\backslash\!\backslash \\
\sigma\ \sigma\ \sigma & & \sigma\ \sigma\ \sigma \\
(\text{UR}) & & (\text{SR})
\end{array}
$$

Figure 1: AR derivation for [félàmà] 'junction'

In the AR, the tone pattern of [félàmà] is represented as a HL (high-low) string associated to three syllables as depicted on the right-hand side of Fig. 1 (with association depicted by straight lines). While finite-state models of ARs and AR transformations exist, much of this work has found the need to use enriched automata with additional tapes (Kay, 1987; Wiebe, 1992; Kornai, 1995) or synchronized states (Bird and Ellison, 1994).

Instead, this paper takes a *logical* approach to studying the complexity of autosegmental representations (thus following the work of Bird and Klein, 1990; Jardine, 2014), as it allows

22

Proceedings of the 15th Meeting on the Mathematics of Language, pages 22–35,
London, UK, July 13–14, 2017. ©2017 Association for Computational Linguistics

for more flexibility with respect to the structures we can describe. This builds on a few key results. First, the Regular sets of strings are exactly those definable by *monadic second-order* (MSO) logic (Büchi, 1960; Elgot, 1961; Trakhtenbrot, 1961). Second, Courcelle (1994) introduced MSO *transductions* for graph structures, in which the output structure is determined as an MSO interpretation of the input structure. As MSO-definable string transductions subsume Regular functions (Filiot and Reynier, 2016), we can then recast the Regular hypothesis for phonology in logical terms: phonology is at most MSO-definable.

This leads to the two results of this paper, one negative and one positive. First, *tone mapping* transformations, in which an unassociated AR is mapped to a fully associated one—exemplified in Fig. 1—are not MSO-definable. Second, (at least some) ARs are *first-order* (FO) definable from strings; i.e., we can write a FO transduction from a string representing a sequence of toned syllables to its corresponding AR. Because this transduction is defined in the terms of the FO language of the input, this means that any FO formula we write over these ARs can be translated into the FO logic of their corresponding strings. This means that any FO constraint written over these ARs still describes a Regular set of strings—i.e., that ARs are not significantly more expressive than strings.

This paper is structured as follows. §2 introduces string models and logic, and §3 details how this relates to the study of phonology. §4 discusses the non-definability of tone mapping in MSO, and §5 discusses the FO-definability of ARs from strings. §6 concludes.

2 Preliminaries

2.1 String models and logics

Let an *alphabet* Σ be a finite set of symbols and a *string* w be a sequence of symbols in Σ; let $|w|$ denote the length of w. Let Σ^* represent all possible strings over Σ, including the *empty string* λ ($|\lambda| = 0$). A *stringset* (or *formal language*) is some subset of Σ^*. For some $\sigma \in \Sigma$, σ^n denotes the string consisting of n repetitions of σ.

A *relational model* $\langle U, R_1, R_2, ..., R_n \rangle$ is a representation of some structure with a universe U of elements and n relations $R_i \subseteq U^k$ for some finite k. We can represent a string $w \in \Sigma^*$ with a *finite* relational model $\mathcal{M}_w = \langle U, \prec, (P_\sigma)_{\sigma \in \Sigma} \rangle$ where $U = \{1, 2, ..., |w|\}$ is an initial segment of

the natural numbers representing the positions in the string, $\prec$ is a binary relation representing the natural order over the positions in the string, and each P_σ is a unary relation representing the set of positions containing the symbol σ. For example, for $\Sigma = \{a, b\}$ the model for the string aba is

$$\mathcal{M}_{aba} = \langle \{1, 2, 3\}_U, \{(1, 2), (1, 3), (2, 3)\}_\prec, \{1, 3\}_{P_a}, \{2\}_{P_b} \rangle.$$

We can use these models of this form to define a *first order* (FO) predicate logic over strings in Σ^*. Let $x, y, ...$ denote *variables* that range over positions in a string. For variables x and y, we can then use $x \prec y$ and $\sigma(x)$ for each $\sigma \in \Sigma$ as *atomic predicates* which are true when x and y are interpreted as positions related by $\prec$ in a string model and when x is interpreted as a position in the unary relation P_σ of a model, respectively. We also assume an additional atomic predicate $x = y$ which is true when x and y are interpreted as the same position. A FO logic is then the set of *formulas* built recursively out of these atomic predicates and the logical connectives $\neg, \wedge, \vee, \rightarrow$ and the quantifiers $\exists, \forall$ in the usual way. A *free variable* is a variable not bound by a quantifier; we write $\varphi(x_1, x_2, ..., x_n)$ to indicate that $x_1, x_2, ..., x_n$ is the exhaustive set of free variables in a FO formula φ. For example, we can define the following useful formulas with one free variable:

$$\mathtt{first}(x) \overset{\text{def}}{=} (\forall y)[\neg y \prec x]$$

$$\mathtt{last}(x) \overset{\text{def}}{=} (\forall y)[\neg x \prec y]$$

We also define a two-variable formula for the successor relation (using infix notation).

$$x \lhd y \overset{\text{def}}{=} x \prec y \wedge (\forall z)[\neg(x \prec z \wedge z \prec y)]$$

A formula φ with no free variables is called a *sentence*. Let satisfaction of a model $\mathcal{M}$ of φ, written $\mathcal{M} \models \varphi$, be defined in the usual way. The set of strings $L(\varphi)$ described by φ is the set of strings $\{w \in \Sigma^* | \mathcal{M}_w \models \varphi\}$. For example, if

$$\mathtt{last}_a \overset{\text{def}}{=} (\forall x)[\mathtt{last}(x) \rightarrow a(x)]$$

then $L(\mathtt{last}_a)$ is the set of strings that end in a.

A *monadic second order* (MSO) logic is a FO logic extended with the ability to quantify over arbitrary sets in the string. Let *set variables* $X, Y, ...$ which range over sets of positions in a string. A MSO logic is thus FO logic to which we add the atomic formulas $X(x)$, $Y(x)$, etc., which are true when x is interpreted as a position in the set assigned to X, Y, etc., and in which $\exists$ and $\forall$ can also bind set variables.

It is well-known that FO sentences over string models with $\prec$ describe exactly the Star-Free stringsets (McNaughton and Papert, 1971) whereas MSO sentences describe exactly the Regular stringsets (Büchi, 1960; Elgot, 1961; Trakhtenbrot, 1961).

2.2 Logically definable transductions

We can also use logic to define a *transduction* from an input structure to an output structure, as first introduced by Courcelle (1994) for graphs and later related to string transductions and their automata-theoretic characterizations (Engelfriet and Hoogeboom, 2001; Filiot, 2015). (For an overview of related work see Filiot and Reynier 2016.)

In such a logical transduction, the output structure is defined by an *interpretation* over a finite number of copies of the input structure (where 'interpretation' is used in the sense of a translation from the logical language of one structure into that of another; see, e.g. Hodges 1997). MSO and FO transductions are defined as follows.

Definition 1 (MSO/FO transduction) *Given some natural number k, an input alphabet Σ and an output alphabet Γ, an* MSO *(resp.* FO*) transduction is defined by*

- φ_{dom}, *a* domain formula, *or sentence in the MSO (FO) logic of the input that defines the domain of the transduction,*

- *For each $1 \leq n \leq k$ and $\gamma \in \Gamma$, a formula $\varphi_\gamma^n(x)$ in the MSO (FO) logic of the input with exactly one free variable, and*

- *For each $1 \leq n, m \leq k$, a formula $\varphi_{\prec_\Gamma}^{n,m}(x, y)$ in the MSO (FO) logic of the input with exactly two free variables*

To restrict our domain to strings in Σ^*, we include in our domain formula the sentence $\mathtt{string}_\Sigma$ as defined by

$$\mathtt{string}_\Sigma \overset{\mathrm{def}}{=} (\forall x)[\bigvee_{\sigma \in \Sigma} \sigma(x)] \wedge$$
$$(\forall x)[\bigwedge_{\sigma \neq \sigma' \in \Sigma} \sigma(x) \to \neg \sigma'(x)] \wedge$$
$$(\forall x, y, z)[x \lhd y \wedge x \lhd z \to y = z] \wedge$$
$$(\forall x, y)[\mathtt{first}(x) \wedge \mathtt{first}(y) \to$$
$$x = y]$$

The output of such a transduction is defined as follows. For each position x in the input and for each n for which exactly one $\varphi_\gamma^n(x)$ is true, a copy of x labeled γ appears in the output. For each pair of positions x, y and for each pair n, m for which $\varphi_{\prec_\Gamma}^{n,m}$ is true, the nth copy of x precedes the mth copy of y with respect to the output ordering $\prec_\Gamma$ on positions in the output. For example, to rewrite strings over the alphabet $\Sigma = \Gamma = \{a, b\}$ such that each b immediately following another b in the input is written out as an a then we set $k = 1$ and

$$\varphi_{\mathrm{dom}} \overset{\mathrm{def}}{=} \mathtt{string}_\Sigma$$
$$\varphi_a^1(x) \overset{\mathrm{def}}{=} a(x) \vee (b(x) \wedge (\exists y)[y \lhd x \wedge b(y)])$$
$$\varphi_b^1(x) \overset{\mathrm{def}}{=} b(x) \wedge \neg(\exists y)[y \lhd x \wedge b(y)]$$
$$\varphi_{\prec_\Gamma}^{1,1}(x, y) \overset{\mathrm{def}}{=} x \prec y$$

This transduction is illustrated in Fig. 2 for an input string $abba$.

First, $\varphi_a^1(x)$ is defined to be true in the output for all positions x in the input that are either labeled a or are labeled b but also succeed some input b. Thus, the copies corresponding to each a in the input are labeled as in the output, as well as the copy for the second input b in the input. Likewise, $\varphi_b^1(x)$ is defined to be true in the output for all positions x in the input that are labeled b and do not succeed another b. In Fig. 2, this is true for the first b in the input, so its copy is also labeled b. As the output order $\varphi_{\prec_\Gamma}^{1,1}(x, y)$ is defined to be true when the input order $x \prec y$ is true, the order is preserved exactly in the output.

Input: (a) → (b) → (b) → (a)

Output: (a) → (b) → (a) → (a)

Figure 2: Replacing b with a following an input b

Note as there is only one interpretation per input, these transductions are functional. (For non-functional MSO transductions, see Engelfriet and Hoogeboom 2001.)

To give one more example, we can define a transduction that 'doubles' a string, i.e. given an input $w \in \Sigma^*$ outputs ww. We set $k = 2$, $\varphi_{\mathrm{dom}} \overset{\mathrm{def}}{=} \mathtt{string}_\Sigma$, and

$$\varphi_a^1(x) \overset{\mathrm{def}}{=} \varphi_a^2(x) \overset{\mathrm{def}}{=} a(x) \quad \varphi_{\prec_\Gamma}^{1,2}(x, y) \overset{\mathrm{def}}{=} x < y \,\mathtt{True}$$
$$\varphi_b^1(x) \overset{\mathrm{def}}{=} \varphi_b^2(x) \overset{\mathrm{def}}{=} b(x) \quad \varphi_{\prec_\Gamma}^{2,1}(x, y) \overset{\mathrm{def}}{=} \mathtt{False}$$
$$\varphi_{\prec_\Gamma}^{1,1}(x, y) \overset{\mathrm{def}}{=} \varphi_{\prec_\Gamma}^{2,2}(x, y) \overset{\mathrm{def}}{=} x \prec y$$

(`True` and `False` indicate a formula is evaluated to true or false for any input positions.)

This is interpreted as follows. Each a and b in the input is given two identical copies in the output. As both $\varphi^{1,1}_{\prec_\Gamma}(x,y)$ and $\varphi^{2,2}_{\prec_\Gamma}(x,y)$ are set equal to $x \prec y$, the first set of copies has the same order as the input, as does the second. That $\varphi^{1,2}_{\prec_\Gamma}(x,y)$ is set to `True` states that all copies in the first set precede all copies in the second set; this establishes an order between the two sets of copies. Finally, setting $\varphi^{2,1}_{\prec_\Gamma}(x,y)$ to `False` ensures that second copies never precede first copies. An example with *abba* in the input is depicted in Fig. 3.

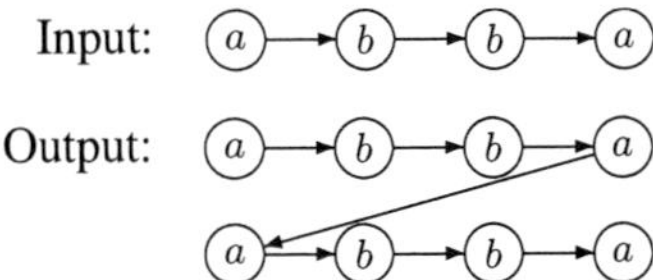

Figure 3: Doubling *abba*

This second example shows that we can freely manipulate the order of elements in the output; indeed, the output need not be a string. In fact, we can define the output structure to have a new binary relation R not present in the input structure by definining a predicate $\varphi^{m,n}_R(x,y)$ in terms of the MSO logic of the input structure. We make use of this in §4.2. Importantly, as they are defined in terms of interpretations, both MSO and FO transductions are closed under composition for graph structures in general (Courcelle, 1994).

3 Logic and phonology

Because of its well-known connections to computational complexity, we can apply logic to the study of the complexity of phonological patterns. This section reviews relevant results from the study of phonotactic (phonological surface well-formedness) patterns as stringsets and phonological transformations (mappings from URs to SRs) as transductions. Both show that MSO-definability is a clear, if loose, bound on the complexity of phonology.

3.1 Stringsets

Phonotactics are language-specific well-formedness constraints on how sounds can be combined to create words. An example from Kagoshima Japanese is given in Table 1: all words have a high tone either on the final or penultimate mora (Kubozono, 2012). (The high tone is marked with an acute accent [á] on the vowel.)

hána	'nose'	HL
sakúra	'cherry blossom'	LHL
kagaríbi	'watch fire'	LLHL
...		...
haná	'flower'	LH
usagí	'rabbit'	LLH
kakimonó	'document'	LLLH
...		...

Table 1: Kagoshima Japanese tone well-formedness

Such constraints can be modeled as stringsets. For example, given the alphabet $\{\mathrm{H}, \mathrm{L}\}$, the Kagoshima pattern can be modeled as the set of strings $\{\mathrm{HL}, \mathrm{LHL}, \mathrm{LLHL}, ..., \mathrm{LH}, \mathrm{LLH}, \mathrm{LLLH}, ...\}$, where Ls and Hs represent low- and high-toned moras, respectively.

All previous work on natural language phonotactics as stringsets has found these patterns to be at most Regular stringsets, with all but a few exceptions being sub-Star Free (Heinz, 2007, 2009, 2010; Heinz et al., 2011; Rogers et al., 2013). In logical terms, this means that definability in MSO is a clear bound on the complexity of phonotactics, with most patterns being FO-definable. To illustrate, the stringset representing the Kagoshima tone pattern can be defined with the FO sentence

$$(\mathtt{last}_\mathrm{H} \vee \mathtt{penult}_\mathrm{H}) \wedge$$
$$(\forall x, y)[(\mathrm{H}(x) \wedge \mathrm{H}(y)) \to x = y],$$

where $\mathtt{last}_\mathrm{H}$ is defined as $\mathtt{last}_a$ above and $\mathtt{penult}_\mathrm{H}$ is defined as $(\forall x, \exists y)[(x \triangleleft y \wedge \mathtt{last}(y)) \to \mathrm{H}(x)]$. This sentence describes the set of strings that has exactly one H either in final or penultimate position.

These results are important for a theory of phonology because they allow for the hypothesis that phonotactics are at most MSO-definable, a hypothesis which can be interpreted in terms of cognitive complexity (Rogers and Pullum, 2011; Rogers et al., 2013) and how humans learn phonotactics (Heinz, 2010; Lai, 2015; McMullin and Hansson, 2015). More restrictive characterizations exist, based on subclasses of the Star-Free stringsets (see, e.g., Heinz, 2010), but for the present purposes it is enough to consider FO- and MSO-definability.

3.2 Transductions

We can also fruitfully apply logical transductions to phonological theory (Heinz, forthcoming), as mainstream theories of generative phonology aim to explain linguistic sound patterns through a transformation from an input UR to an output SR (Chomsky and Halle, 1968; Prince and Smolensky, 2004). Indeed, these transformations have been studied from an automata-theoretic perspective, leading to restrictive characterizations of phonology. Johnson (1972) and Kaplan and Kay (1994) show that the phonological rewrite rules of Chomsky and Halle (1968) are describable with finite state machines; that is, that they describe Regular relations. Subsequent work on phonological transformations has demonstrated for a wide variety of processes—including local assimilation, deletion, and epenthesis (Chandlee, 2014), dissimilation (Payne, 2014), metathesis (Chandlee and Heinz, 2012), and vowel harmony (Heinz and Lai, 2013)—to lie in even more restrictive subclasses of Regular string transductions. The single known possible exception to this is full reduplication—i.e. the copying over of an entire input form, as in the Indonesian *buku-buku* 'books', lit. 'book-book' (Sneddon et al., 2010). This is not a Regular relation, although it can be argued that this process is morphological and not phonology proper (for discussion see Chandlee and Heinz, 2012).

From the logical perspective, all of these results place phonological transformations squarely within the class of MSO-definable transductions. Any functional Regular relation is MSO-definable (Filiot and Reynier, 2016), so any phonological transformation describable with a (functional) rewrite rule is MSO-definable. Even full reduplication is FO-definable, as witnessed by the string doubling transduction defined in §2.2. Thus, MSO-definability appears to be a loose, yet clear, bound on the computational complexity of phonological transformations.

3.3 Interim summary

The above has reviewed the evidence for MSO-definability as a complexity bound on phonology. The advantage of viewing such a complexity bound in logical terms is that we are able to view the complexity of both phonotactics and transformations in unified terms.

A further advantage of the logical perspective is that it allows us to study the complexity of non-string representations in the same terms. The remainder of the paper studies *autosegmental representations* (ARs) in the same terms.

4 The complexity of tone mapping

This section motivates *tone mappings* and ARs using a well-known empirical case, then it is shown that tone mapping is not MSO-definable.

4.1 Tone mapping in Mende

Mende (Leben, 1973, 1978) is a classic example of a tone pattern which has been argued to be best analyzed in terms of autosegmental mapping of tones to syllables. Mende nouns fall in to one of five categories: 1) words for which all syllables are pronounced with a high tone (e.g. [háwámá] 'waist'), 2) words for which all syllables are pronounced with a low tone (e.g. [kpàkàlì] 'three-legged chair'), 3) words which begin high but end low (e.g. [mbû] 'owl' and [félàmà] 'junction'), 4) words which begin low but end high (e.g. [mbǎ] 'rice' and [ndàvúlá] 'sling'), and 5) words which show a rising-falling pattern (e.g. [mbã̌] 'companion' and [nìkílì] 'peanut'). These five categories are exemplified by the forms in Table 2, where tones are indicated as diacritics on the vowels as follows: [á] = high tone, [à] = low tone, [â] = falling tone, [ǎ] = rising tone, and [ã̌] = rising-falling tone. The columns are of 1-, 2-, and 3-syllable words.

1. H	kɔ́	pélé	háwámá
	'war'	'house'	'waist'
2. L	kpà	bèlè	kpàkàlì
	'debt'	'pants'	'chair'
3. HL	mbû	ngílà	félàmà
	'owl'	'dog'	'junction'
4. LH	mbǎ	fàndé	ndàvúlá
	'rice'	'cotton'	'sling'
5. LHL	mbã̌	nyàhâ	nìkílì
	'companion'	'woman'	'peanut'

Table 2: Mende noun tone (Leben, 1978)

Of interest is the fact that *contour tones*—that is, the rising and falling toned-syllables—and *plateaus*, or sequences of like-toned syllables, only occur on the right edge of the word. For example, [nyáhâ] 'woman' is attested, but a word like *[nyâhá], with a rising toned-syllable on the left edge, is not attested. Likewise, words like [félàmà] 'junction', with a sequence of two low-toned syllables on the right edge, are commonly attested, whereas words like *[fèlàmá], are rare.[1] Furthermore, Mende words conform to one of the five tonal shapes exemplified in Table 2; words showing a falling-rising pattern, for example, are unattested.

Furthermore, these tonal shapes are maintained when toneless suffixes are affixed to the noun, resulting in the tone of the suffix varying depending on the tone pattern of its root. The following data illustrate this with the toneless suffix /-ma/ 'on'.

	Isolation	Suffixed	
H	pélé	pélé-má	'war'
HL	mbû	mbú-mà	'owl'
	ngílà	ngílà-mà	'dog'
LHL	nyàhâ	nyàhá-mà	'companion'

Table 3: Mende /-ma/ suffix tone (Leben, 1978)

Note that the suffixed forms also preserve the generalizations noted above restricting contours and plateaus to the right edge, and so the tone pattern 'stretches' to accomodate the new syllable. Thus [mbû] 'owl', which has a contour falling tone in isolation, is realized with a sequence of pure high and low-toned syllables as [mbú-mà] 'on owl' when suffixed.

Following a proposal by Leben (1973), and subsequent work in autosegmental phonology (e.g., Goldsmith, 1976; Pulleyblank, 1986; Yip, 2002) tone patterns like Mende's have been explained by a left-to-right mapping of a *melody*, or string of tonal units, to a string of syllables. These disjoint strings are referred to as *tiers*, and the representation as a whole is an AR. For example, the words in Table 2, Row 3 share a HL (high-low) melody, which is then mapped to the syllables in the words as depicted in Fig. 4. Following convention, syllables are denoted with σ and the *association* relation depicted as lines drawn between units on distinct tiers.

Thus, the contour falling tone of [mbû] 'owl is the result of an HL sequence associating to a single syllable; likewise the plateau of low-toned syllables in [félàmà] 'junction' is the result of an L tone associating to multiple syllables. The question then is how to restrict association such that this multiple association occurs only on the right edge

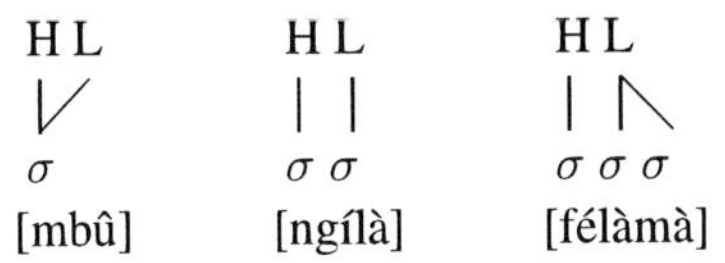

<hr>

[1]See Dwyer (1978) and rebuttal by Leben (1978) for discussion about exceptions to the generalizations stated here.

H L   H L   H L
σ   σ σ   σ σ σ
[mbû]   [ngílà] [félàmà]

Figure 4: ARs illustrating mapping of HL melody to words of various syllable length

of the word. Because this association pattern holds for all lexical items, including the suffixed forms in Table 3, it is thus entirely predictable and taken not to be present in the UR of a word. Thus, it must be created by some phonological transformation that associates tones to syllables. This transformation has been analyzed as proceeding according to laws often referred to as the *well-formedness conditions* (WFCs). The following definition is due to Yip (2002).

Definition 2 *The well-formedness conditions*

a. *Every syllable must have a tone.*

b. *Every tone must be associated to some syllable.*

c. *Association proceeds one-to-one, left-to-right.*

d. *Association lines must not cross.*

Intuitively, the WFCs ensure that in the SR, every tone is associated to some syllable, and vice-versa, by a step-by-step process in which first tone and first syllable are associated, then the second tone and second syllable, and so on. (It bears mentioning that 'one-to-one' here is used not as it is to describe mathematical functions, but in terms of how pairs of tones and syllables are associated one after another.) If there are remaining tones or syllables on the right edge of a tier that have not been paired off, WFCs (a), (b), and (d) associate them to the rightmost unit on the opposite tier: (a) and (b) require all units to be associated, but (d) forbids the crossing of any existing associations to do so. Fig. 5 shows how this process works for [nyàhâ] 'woman', [nyàhá-mà] 'on woman', both of which have an underlying LHL melody, and [félàmà] 'junction', which has an underlying HL melody. This figure demonstrates that the WFCs explain the generalization in Mende that contours and plateaus only occur on the right edge of the word through a transformation from a UR with no association to a fully associated SR.

The WFCs have been shown not to be strictly universal; whether or not tones are as-

UR   SR

L H L   L H L   L H L   L H L
      → | → | | → | ⋁
σ σ   σ σ   σ σ   σ σ
              [nyàhâ] 'woman'

L H L   L H L   L H L   L H L
      → | → | | → | | |
σ σ σ   σ σ σ   σ σ σ   σ σ σ
            [nyàhá-mà] 'on woman'

H L   H L   H L   H L
    → | → | | → | ⟍
σ σ σ   σ σ σ   σ σ σ   σ σ σ
            [félàmà] 'junction'

Figure 5: Step-by-step breakdown of the association transformation

sociated from left-to-right or right-to-left, or whether or not contours are built out of left-over tones or are simply left unpronounced, have been shown to vary from language to language (Goldsmith, 1976; Newman, 1986; Pulleyblank, 1986; Hewitt and Prince, 1989; Archangeli and Pulleyblank, 1994; Yip, 2002). Also, there have since been non-derivational approaches to tone mapping (Zoll, 2003). However, all generative explanations of tone mapping patterns that use ARs rely on the idea of an unassociated UR being transformed into a SR with associations, with one-to-one association forming the basis of the transformation. The following shows that such a transformation is not a MSO-definable transduction.

4.2 Tone mapping as a logical transduction

We can characterize this transformation, as depicted in Fig. 6, as a transduction that takes two strings of length n and m, respectively ($n, m > 0$), as input and adds an association relation between the positions in the strings that follows the WFCs outlined in Def. 2. As long as we can assume there is some property that distinguishes between units on each tier, we can abstract away from distinctions between units on a particular tier and instead focus on predicates $a(x)$ and $b(x)$ which are true if and only if x is on the 'upper' and 'lower' tier, respectively. (For example, $a(x)$ can mean 'x is a tone' and $b(x)$ can mean 'x is a syllable'.)

In terms of relational models, the transduction takes models of the form

$$\langle U, \prec, P_a, P_b \rangle$$

Input: Output:

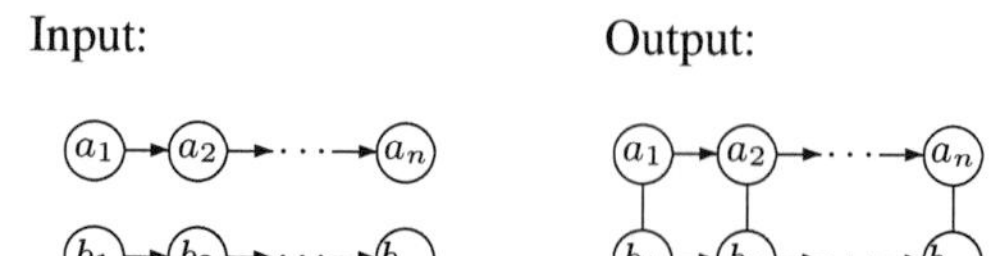

Figure 6: The association transduction

and describes a model of the form

$$\langle U', \prec', \circ, R_a, R_b \rangle,$$

where $\circ$ denotes a new relation representing association. This relation must conform to the WFCs in Def. 2; this is formalized in Def. 3 of the association relation.

Definition 3 *For an AR whose tiers are a pair of disjoint strings $a_1 a_2 ... a_n$ and $b_1 b_2 ... b_m$, and for ℓ being the lesser of n and m, the association relation $\circ$ is the unique relation comprised of the symmetric closure of all pairs (a_i, b_i) for $1 \leq i \leq \ell$ unioned with $(a_n, b_{n+1}), ..., (a_n, b_m)$ if $n = \ell$ or $(a_{m+1}, b_m), ..., (a_n, b_m)$ if $m = \ell$.*

To see why this definition matches that in Def. 2, note that WFC (c) in Def. 2, which stipulates one-to-one, left-to-right association, requires that $(a_1, b_1), (a_2, b_2), ..., (a_\ell, b_\ell) \in \circ$. If $n \neq m$, this leaves either a final segment $a_{\ell+1} a_{\ell+2} ... a_n$ (if $m = \ell$) or $b_{\ell+1} b_{\ell+2} ... b_m$ (if $n = \ell$) that must also be associated, per WFCs (a) and (b). WFC (d), which bans line crossing, stipulates that these final elements cannot 'reach back' and associate to anything except for the final element on the opposite tier. So if $m = \ell$ then $(a_{m+1}, b_m), (a_{m+2}, b_m)..., (a_n, b_m) \in \circ$ and if $n = \ell$ then $(a_n, b_{n+1}), (a_n, b_{n+2})..., (a_n, b_m) \in \circ$. We take the symmetric closure as association is usually regarded as symmetric (Kornai, 1995).

No MSO definition of this relation is possible. To show how, we can rely on two facts established above: 1) MSO transductions are essentially interpretations where relations in the output structure are defined in terms of the logical language of the input structure; and 2) MSO transductions are closed under composition.

First, we can instead consider as input strings of the form $a^n b^m$ (again, with $n, m > 0$). This set is MSO definable, as witnessed by the formula $\text{string}_{a^n b^m}$ defined below:

$$\text{string}_{a^n b^m} \overset{\text{def}}{=}$$
$$(\forall x, y)[\neg(b(x) \land a(y) \land x \prec y)] \land$$
$$(\exists x, y)[a(x) \land b(y)]$$

These strings are equivalent, in terms of MSO, to pairs of disjoint strings of shape a^n and b^m. The reason is that we can write a MSO transduction from one to another. If $\prec$ is the ordering in the input $a^n b^m$ string, we simply define the order $\prec_\Gamma$ for the output structure such that it omits all precedence between a positions and b positions.

$$
\begin{aligned}
\varphi_{\text{dom}} &\stackrel{\text{def}}{=} \texttt{string}_\Sigma \wedge \texttt{string}_{a^n b^m} \\
\varphi_a^1(x) &\stackrel{\text{def}}{=} a(x) \\
\varphi_b^1(x) &\stackrel{\text{def}}{=} b(x) \\
\varphi_{\prec_\Gamma}^{1,1}(x,y) &\stackrel{\text{def}}{=} x \prec y \wedge \neg(a(x) \wedge b(y))
\end{aligned}
$$

An example of this is given in Fig. 7.

Input: $\textcircled{a}\!\rightarrow\!\textcircled{a}\!\rightarrow\!\textcircled{a}\!\rightarrow\!\textcircled{b}\!\rightarrow\!\textcircled{b}\!\rightarrow\!\textcircled{b}\!\rightarrow\!\textcircled{b}$

Output: $\textcircled{a}\!\rightarrow\!\textcircled{a}\!\rightarrow\!\textcircled{a}$

$\textcircled{b}\!\rightarrow\!\textcircled{b}\!\rightarrow\!\textcircled{b}\!\rightarrow\!\textcircled{b}$

Figure 7: Relating $a^n b^m$ strings to disjoint tiers a^n and b^m.

Thus, we know that MSO statements over pairs of disjoint strings a^n and b^m are equivalent to statements over strings in $a^n b^m$. We can then use this to prove Theorem 1.

Theorem 1 *Association between two disjoint strings according to the WFCs in Definition 2 is not MSO definable.*

Proof: The proof is by contradiction. Assume the converse, that we can define using MSO a transduction that takes disjoint pairs of strings a^n and b^m and outputs them as associated autosegmental representations with some association relation $\circ$ that obeys the WFCs as defined in Def. 3.

With this relation we can write a sentence $\varphi_{\text{eq}} \stackrel{\text{def}}{=} (\forall x, \exists y, \forall z)[x \circ y \wedge (x \circ z \rightarrow z = y)]$, which holds that every position is associated to exactly one position. If $\circ$ obeys the WFCs, then per the discussion of the structure of $\circ$ in Def. 3, φ_{eq} is only true for structures whose association relation is the set of pairs of the form (a_i, b_i) (and their converse). That is, the only structures for which φ_{eq} are true are pairs of disjoint strings a^n and b^m for which $n = m$.

Now consider strings of the form $a^n b^m$. As shown above, there is a MSO transduction from these strings to disjoint pairs of a^n and b^m. By assumption, there is then a MSO transduction

that adds an association relation $\circ$ to these pairs. Because MSO transductions are closed under composition, then φ_{eq} can be written as the MSO language of strings $\{a, b\}^*$. Thus the sentence $\texttt{string}_{a^n b^m} \wedge \varphi_{\text{eq}}$ restricts us exactly the set of strings $a^n b^m$ for which $n = m$. It is well-known that this set is not regular (see, e.g., Hopcroft et al., 2006), and thus not MSO definable. Thus we have a contradiction, and so the assumption must be false. $\square$

Importantly, because MSO transductions are closed under composition, there is no breakdown of this process into a finite set of composite steps (such as those illustrated in Fig. 5) that are themselves MSO-definable. Note also that the proof highlights that in particular it is the *one-to-one* requirement on association that makes it not MSO-definable: it is this property that introduces the ability to check the parity of the a and b tiers.

4.3 Interpreting the result

We have thus shown that, as a transduction from an unassociated pair of tiers to an associated one following the WFCs in Def. 2, tone mapping is not MSO-definable. This puts it in sharp contrast to all other phonological UR-SR transductions for whom complexity results exist: as discussed in Sec. 3.2, these processes have been shown to be well within MSO-definability. This makes tone mapping aberrant in terms of its computational complexity. How do we interpret this result?

One answer is to take this as evidence that tone has access to more computational power than other parts of phonology. In fact, this has been argued by Jardine (2016a) on the basis of comparisons between the complexity of tonal phenomena and segmental phenomena when viewed as string transductions. However, an issue with this interpretation is that the tonal phenomena that Jardine cites are still Regular relations and thus MSO-definable. Thus tone mapping is still highly complex, even compared to other tonal phenomena.

Another possible interpretation is that tone mapping is simply an incorrect characterization of the data available. For example, both Dwyer (1978) and Shih and Inkelas (2015) take issue with Leben (1973)'s tone mapping characterization of Mende, and offer alternative explanations using representational assumptions that do not require tone mapping (or an analogue thereof). However, other tone patterns that have been successfully accounted for using tone mapping include those of Hausa (Newman, 1986, 2000), Kukuya (Hyman, 1987), and the wide variety of languages analyzed in Goldsmith (1976), Pulleyblank (1986), and Zoll (2003). The alternative explanations mentioned above have yet to be shown to enjoy the same broad empirical coverage (though future work may indeed show this).

Another explanation is that it is wrong to assume that the mapping generalization holds for tiers of unbounded length. The proof above relies on the fact that the input strings are of the form $a^n b^m$ for any n, m—if either n or m had some bound, the proof would no longer hold. Indeed, Yli-Jyrä (2013) gives a finite-state (and thus MSO-definable) implementation of tone mapping given the assumption that the tonal tier is bounded. However, it is not clear that this can be assumed for all cases. For example, in Kikuyu (Clements and Ford, 1979), morphological concatenation can extend the tonal tier before mapping occurs. Regardless, "tonal tiers must be bounded" is a hypothesis worth further testing, as the result here shows it has consequences for the complexity of phonology.

A final interpretation of the result is to posit that the one-to-one property of tone mapping as a phonological *universal* and thus is not relevant to the study of the complexity of *language-specific* phonological phenomena. As noted in §4, there are languages whose patterns have been shown to violate the WFCs in Def. 2 with respect to directionality and whether or not all tones or syllables are associated. However, the one-to-one property appears to be shared by all such patterns. Jardine (2016b, to appear) demonstrates for many of these patterns that, if one-to-one association is assumed in the representation, these language-specific constraints on association can be described with a restricted propositional logic, well within the complexity of MSO. Thus, if we isolate the one-to-one property of association, shown in the proof of Thm. 1 to be responsible for its non-definability in MSO, from the other aspects of tone mapping, then we can maintain MSO-definability as a cohesive bound on the complexity of language-specific phonological phenomena. How this separation might be implemented in a theory of tone will be left for future work.

5 Deriving autosegmental representations from single strings

The result in the previous section raises an important question: How powerful are ARs? Specifically, does invoking ARs allow for grammars that are too expressive to provide a reasonable theory of phonological patterns? To put it in more concrete terms, we can represent the tone pattern of a word either as a string of toned syllables or as an AR.[2] Table 4 gives some examples, two from Mende and one hypothetical, where strings are over an alphabet $\{H, L, F, R\}$ whose symbols represent high-, low-, falling-, and rising-toned syllables, respectively.

Form	String	AR
[félàmà] 'junction'	HLL	H L connected to σ σ σ (H to first σ; L to second and third σ)
[nyàhâ] 'woman'	LF	L H L connected to σ σ (L to first σ; H and L to second σ)
(hypothetical) LLRH	LLRH	L H connected to σ σ σ σ (L to first, second, third σ; H to third and fourth σ)

Table 4: Strings and ARs

Thus, for any string over $\{H, L, F, R\}$, there is a corresponding autosegmental representation. Note that these ARs obey WFCs (a), (b), and (d) from Def. 2, in that each tone is associated to a syllable and vice versa, and these association lines do not cross. However, the AR for LLRH violates WFC (c), as the tones have not associated in a left-to-right manner. We can thus talk about ARs that obey (a), (b), and (d), but will ignore (c), as the latter would restrict us to a subset $\{H, L, F, R\}^*$.

We can then talk about strings and their corresponding ARs. Jardine and Heinz (2015) define such a relationship in terms of concatenation, but they do not address how this relationship connects to complexity. As discussed in §3.1, natural language phonotactics are largely describable with FO-definable stringsets. The question then is, given, for example, an FO logic over ARs, can we describe sets of strings that are not FO-definable? This is a valid question as, for example, even restrictions on FO over trees can generate Context

<hr>

[2] We abstract away from the issue of whether tones are a property of moras, syllables, or some other unit. See, e.g., Yip (2002) for more on this issue.

Free stringsets (Rogers, 1997).

The following demonstrates otherwise: ARs are FO-definable from strings, and thus any FO formula over ARs is translatable into a FO formula in strings (given $\prec$). It should be noted that autosegmental phonology is not a monolithic theory, and in practice various definitions of ARs have been proposed (one formal overview can be found in Coleman and Local, 1991). A full survey of these and how they might be defined is beyond the scope of this paper; instead, this section focuses on demonstrating that basic ARs obeying the WFCs (a), (b), and (d) in Def. 2 are FO-definable from strings over $\{H, L, F, R\}$. In other words, we formalize the relationship between strings and ARs exemplified in Table 4. This illustrates that the fundamental ideas of autosegmental structure—distinct tiers associated to one another according to some well-formedness conditions—is FO-definable from strings.

5.1 Definition of transduction

We define a transduction from string models of the form

$$\langle U, \prec, P_{\mathrm{H}}, P_{\mathrm{L}}, P_{\mathrm{F}}, P_{\mathrm{R}} \rangle$$

to autosegmental models of the form

$$\langle U', \lhd, \circ, R_{\mathrm{H}}, R_{\mathrm{L}}, R_{\sigma} \rangle .$$

Essentially, we define the transformation from the second column of Table 4 to the third, for all strings in $\{H, L, F, R\}^*$.

We do this by defining the notion of a *span*, or a series of consecutive positions in U that share the same tone and thus will be associated to the same tone on the melody tier in the output AR. We then create extra copies of each element in U that represents a change in spans. These extra copies become the tones in the melody tier. Note that the order in the output is a successor relation $\lhd$; this is not essential, but was chosen for two reasons. One, it is more straightforward to depict in the examples below. Two, its definition gives formal weight to an idea long noted by phonologists: local relationships (i.e. those over $\lhd$) in ARs correspond to long-distance relationships (i.e. those over $\prec$) in strings (see, e.g., Odden, 1994).

First, we define some useful formula and notational shortcuts. The first denotes when y lies between some x and z.

$$x \prec y \prec z \overset{\text{def}}{=} x \prec y \wedge y \prec z$$

We then define formulas in the logic of the input string that represent the tonal relationships between units in the string. The following formula $\mathtt{sametone}(x, y)$ is true when x ends with the same tone that y begins with (thus it is true for an H and an F pair as F starts high).

$$\mathtt{sametone}(x, y) \overset{\text{def}}{=}$$
$$(\mathrm{H}(x) \wedge \mathrm{H}(y)) \vee (\mathrm{L}(x) \wedge \mathrm{L}(y)) \vee$$
$$(\mathrm{H}(x) \wedge \mathrm{F}(y)) \vee (\mathrm{L}(x) \wedge \mathrm{R}(y)) \vee$$
$$(\mathrm{R}(x) \wedge \mathrm{H}(y)) \vee (\mathrm{F}(x) \wedge \mathrm{L}(y))$$

We do this because we will create tones on the melody tier exactly at syllables where there is a change in tone. This marks the beginning of a *span* of one or more like-toned syllables.

$$\mathtt{spanfirst}(x) \overset{\text{def}}{=}$$
$$(\forall y)[y \lhd x \rightarrow \neg \mathtt{sametone}(x, y)]$$

$$\mathtt{span}(x, y) \overset{\text{def}}{=} x \prec y \wedge \mathtt{sametone}(x, y) \wedge$$
$$(\forall z)[x \prec z \prec y \rightarrow \mathtt{sametone}(x, z)]$$

For example, the positions in the following strings that satisfy $\mathtt{spanfirst}(x)$ are underlined: <u>H</u>LL, <u>L</u>F, and <u>L</u>L<u>R</u>H. Note that neither the R nor the H in LLRH satisfy this formula because $\mathtt{sametone}(x, y)$ returns true when x is L and y is R and likewise when x is R and y is H.

We can then define the transduction from strings to autosegmental representations by setting $k = 3$. One set of copies transfers over the syllables, the next initial tones. The third set of copies is for creating the additional tones in the F and R contours. (In general, for strings whose symbols represent contours of at most length n, $k = n + 1$.)

We define the unary labeling relations in the autosegmental representation as follows.

$$\begin{aligned}
\varphi_{\sigma}^{1}(x) &\overset{\text{def}}{=} \text{True} \\
\varphi_{\mathrm{H}}^{2}(x) &\overset{\text{def}}{=} (\mathrm{H}(x) \vee \mathrm{F}(x)) \wedge \mathtt{spanfirst}(x) \\
\varphi_{\mathrm{L}}^{2}(x) &\overset{\text{def}}{=} (\mathrm{L}(x) \vee \mathrm{R}(x)) \wedge \mathtt{spanfirst}(x) \\
\varphi_{\mathrm{H}}^{3}(x) &\overset{\text{def}}{=} \mathrm{R}(x) \qquad \varphi_{\mathrm{L}}^{3}(x) \overset{\text{def}}{=} \mathrm{F}(x)
\end{aligned}$$

All other unary formulas are set to false; that is, $\varphi_{\mathrm{H}}^{1}(x) \overset{\text{def}}{=} \varphi_{\mathrm{L}}^{1}(x) \overset{\text{def}}{=} \varphi_{\sigma}^{2}(x) \overset{\text{def}}{=} \varphi_{\sigma}^{3}(x) \overset{\text{def}}{=} \text{False}$. This works as depicted in Fig. 8 for the strings HLL, LF, and LLRH. Each set of copies in the output is organized into a labeled row.

As $\varphi_{\sigma}^{1}(x)$ is set to True, every element in the input has a copy in set 1 labeled σ. This corresponds to the intuition that each position in the input string

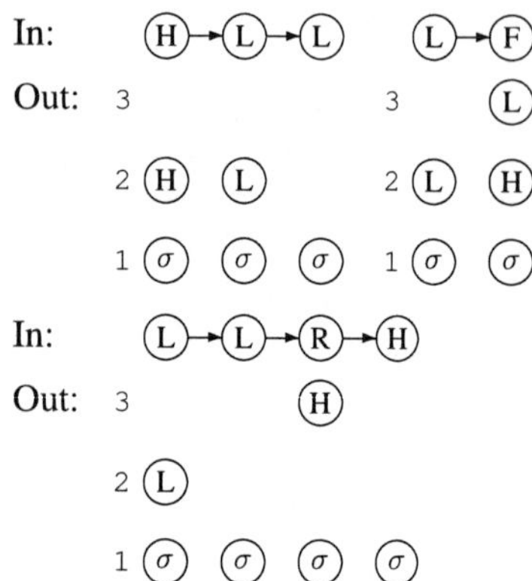

Figure 8: Creating elements in autosegmental representations.

represents a syllable. In copy set 2, Hs and Ls are copied over only at the positions in a string representing a change in tone, as $\varphi_H^2(x)$ and $\varphi_L^2(x)$ are defined to be true only when $\texttt{spanfirst}(x)$ is true. (Note again that for the string LLRH, this is false for the positions labeled R and H.) Finally, $\varphi_H^3(x)$ and $\varphi_L^3(x)$ ensure that the contour-toned syllables F and R in the input are given a third copy L and H, respectively, in the output.

The next step is to define an order on the elements of the output. Again, we define the successor relation $\vartriangleleft'$. Like the definition of the unary relations, this definition will make use of the $\texttt{spanfirst}(x)$ and $\texttt{span}(x,y)$ formulas.

$$\varphi_{\vartriangleleft'}^{1,1}(x,y) \overset{\text{def}}{=} x \vartriangleleft y$$
$$\varphi_{\vartriangleleft'}^{2,2}(x,y) \overset{\text{def}}{=} x \prec y \wedge$$
$$\texttt{spanfirst}(x) \wedge \texttt{spanfirst}(y) \wedge$$
$$(\forall z)[x \prec z \prec y \rightarrow \texttt{span}(x,z)]$$
$$\varphi_{\vartriangleleft'}^{2,3}(x,y) \overset{\text{def}}{=}$$
$$\left(\texttt{spanfirst}(x) \wedge (F(y) \vee R(y)) \wedge x = y\right) \vee$$
$$\left(\texttt{spanfirst}(x) \wedge (F(x) \vee R(x)) \wedge \texttt{span}(x,y)\right)$$
$$\varphi_{\vartriangleleft'}^{3,2}(x,y) \overset{\text{def}}{=} (F(x) \vee R(x)) \wedge \texttt{spanfirst}(y) \wedge$$
$$\texttt{span}(x,y)$$
$$\varphi_{\vartriangleleft'}^{3,3}(x,y) \overset{\text{def}}{=} (F(x) \vee R(x)) \wedge (F(y) \vee R(y)) \wedge$$
$$\texttt{span}(x,y)$$

We set $\varphi_{\vartriangleleft'}^{m,n}(x,y)$ for all other values of m and n to false. These definitions work, as illustrated in Fig. 9, as follows. The formula $\varphi_{\vartriangleleft'}^{1,1}(x,y)$ copies the successor relation from the input faithfully for the initial copies (i.e. those labeled σ). Next, $\varphi_{\vartriangleleft'}^{2,2}(x,y)$ draws a successor relation between the second copies of the initial positions for adjacent spans. Finally, $\varphi_{\vartriangleleft'}^{2,3}(x,y)$ and $\varphi_{\vartriangleleft'}^{3,2}(x,y)$

deal with the extra elements in a contour. The formula $\varphi_{\vartriangleleft'}^{2,3}(x,y)$ draws a successor relation from the second copy of a contour-toned syllable to its third (i.e., between the two tones in the contour) when the contour is first in a span (e.g. F in LF). In case the first part of a contour is part of a previous span (e.g. in the case of R in LLRH), it draws a successor relation from the first position in the previous span to the second part of the contour. The formulas $\varphi_{\vartriangleleft'}^{3,2}(x,y)$ and $\varphi_{\vartriangleleft'}^{3,3}(x,y)$ then similarly draw a successor relation from the second part of a contour that is the start of a span to the initial position of the next successive span (these latter two formulas are not used in the examples).

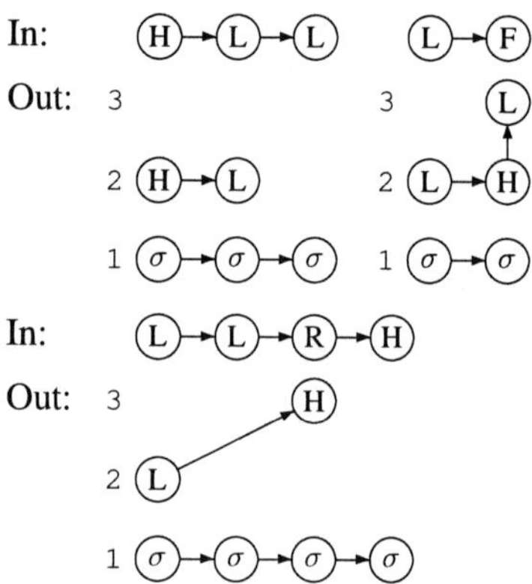

Figure 9: Creating the successor relation in autosegmental representations.

Finally, we draw the associations between the tiers. This is relatively simple: we define formulas that relate a second or third copy of a node with its own first copy as well as any first copies in its span (and vice versa, to obtain a symmetric relation).

$$\varphi_{\circ}^{2,1}(x,y) \overset{\text{def}}{=} \texttt{spanfirst}(x) \wedge$$
$$(x = y \vee \texttt{span}(x,y))$$
$$\varphi_{\circ}^{1,2}(x,y) \overset{\text{def}}{=} \texttt{spanfirst}(y) \wedge$$
$$(x = y \vee \texttt{span}(y,x))$$
$$\varphi_{\circ}^{3,1}(x,y) \overset{\text{def}}{=} (F(x) \vee R(x)) \wedge$$
$$(x = y \vee \texttt{span}(x,y))$$
$$\varphi_{\circ}^{1,3}(x,y) \overset{\text{def}}{=} (F(y) \vee R(y)) \wedge$$
$$(x = y \vee \texttt{span}(y,x))$$

We set $\varphi_{\circ}^{m,n}(x,y)$ for all other values of m and n to false. This obtains the association relations as depicted in Fig. 10.

As the reader can confirm, we have thus obtained the relationship between strings and ARs as originally examplified in Table 4.

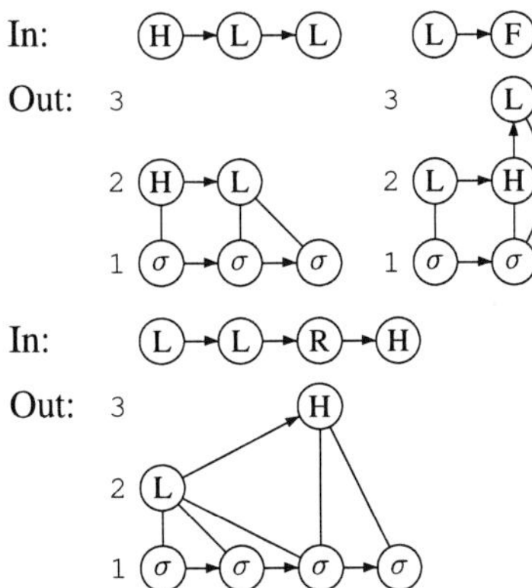

Figure 10: Creating the association relation in autosegmental representations.

Theorem 2 *The above FO transduction maps every string over* $\{H, L, R, F\}^*$ *to an AR following WFCs (a), (b), and (d).*

Proof: (Sketch.) WFCs (a) and (b) in Def. 2 stipulate that every syllable is associated to a tone and vice versa. Note that for any position x in the input string, either $\texttt{spanfirst}(x)$ will be true of this position or $\texttt{span}(y, x)$ is true of this position in some y. The formulas defining the $\circ$ relation ensures all such pairs are associated and each $\texttt{spanfirst}(x)$ is associated with its copy on the melody tier, and that this relation is symmetric. For WFC (d), that the definitions depend on both $\texttt{spanfirst}(x)$ and $\texttt{span}(x, y)$ means that no association line will 'cross' into a new span. $\square$

5.2 Discussion

We have thus demonstrated a set of ARs that are FO-definable from strings representing sequences of toned syllables. Because this transduction is defined as an *interpretation* of the input structure, the relations in the AR model are equivalent to FO-statements in the string model. For example, the atomic formula $x \triangleleft' y$ in the AR model is true when either of the formulas $\varphi_{\triangleleft'}^{n,m}$ is true. In other words,

$$x \triangleleft' y \equiv \varphi_{\triangleleft'}^{1,1}(x, y) \vee \varphi_{\triangleleft'}^{2,2}(x, y) \vee \varphi_{\triangleleft'}^{2,3}(x, y) \vee \\ \varphi_{\triangleleft'}^{3,2}(x, y) \vee \varphi_{\triangleleft'}^{3,3}(x, y).$$

The same is true for the other atomic formulas $x \circ y$, $H(x)$, $L(x)$, and $\sigma(x)$ in the FO logic of the AR model. This means that any FO formula in the logic of the AR model can be translated into the FO logic of the string model. Thus, FO over these ARs is equivalent to FO in the string model.

One caveat is that in the definition for the AR successor order, $\varphi_{\triangleleft'}^{2,2}(x, y)$, $\varphi_{\triangleleft'}^{2,3}(x, y)$, $\varphi_{\triangleleft'}^{3,2}(x, y)$, and $\varphi_{\triangleleft'}^{3,3}(x, y)$ all used the string precedence predicate $x \prec y$, either directly in the definition or through the use of the predicate $\texttt{span}(x, y)$. While concerns for space preclude a full proof, it is easy to see that these same predicates could not be defined using the string successor $x \triangleleft y$ and still account for spans of arbitrary length. This means that including the precedence relation $\prec$ in the string model is crucial for the definition of the AR successor $\triangleleft'$ (note that $x \triangleleft y$ is FO-definable from $x \prec y$ but the reverse is not true). As mentioned above, this means that successor in the AR, specifically successor on the melody tier, corresponds to precedence in the string model.

To summarize, this section has introduced a method for defining ARs in FO from strings representing sequences of toned syllables. Thus, FO statements over ARs are no more powerful than FO statements over strings (with $\prec$). Note again that this definition is categorically different from the tone-mapping transformation discussed in the previous section, which was shown to not be MSO-definable.

6 Conclusion

This paper has presented two new results, one negative and one positive, regarding complexity and autosegmental representations in phonology. The first result is that tone mapping transformations assigning units on one tier to units on another tier in a one-to-one fashion are not MSO-definable. This is in sharp contrast to other phonological patterns, which have been shown to be at least MSO definable and, in most cases, FO-definable. The second, positive, result is that ARs *are* FO-definable from strings, showing that they do not significantly increase the expressive power of phonotactic grammars. It is thus also likely that they do not significantly increase the expressive power of string mappings, although the logical study of phonological transformations is still ongoing (see, e.g., Heinz, forthcoming). This work thus represents one of many steps towards an understanding of phonological computation and representation.

Acknowledgments

The author thanks Jeffrey Heinz, Thomas Graf, Jane Chandlee, and three anonymous reviewers for their thoughtful comments and discussion.

References

Diana Archangeli and Douglas Pulleyblank. 1994. *Grounded Phonology.* Cambridge: MIT Press.

Steven Bird and T. Mark Ellison. 1994. One-level phonology: Autosegmental representations and rules as finite automata. *Computational Linguistics* 20.

Steven Bird and E. Klein. 1990. Phonological events. *Journal of Linguistics* 26:33–56.

J. Richard Büchi. 1960. Weak second-order arithmetic and finite automata. *Zeitschrift für Mathematische Logik und Grundlagen der Mathmatik* 6:66—92.

Jane Chandlee. 2014. *Strictly Local Phonological Processes.* Ph.D. thesis, University of Delaware.

Jane Chandlee and Jeffrey Heinz. 2012. Bounded copying is subsequential: Implications for metathesis and reduplication. In *Proceedings of the 12th Meeting of the ACL Special Interest Group on Computational Morphology and Phonology.* Association for Computational Linguistics, Montreal, Canada, pages 42–51.

Noam Chomsky and Morris Halle. 1968. *The Sound Pattern of English.* Harper & Row.

George N. Clements and Kevin C. Ford. 1979. Kikuyu tone shift and its synchronic consequences. *Linguistic Inquiry* 10:179–210.

John Coleman and John Local. 1991. The "No Crossing Constraint" in autosegmental phonology. *Linguistics and Philosophy* 14:295–338.

Bruno Courcelle. 1994. Monadic second-order definable graph transductions: a survey. *Theoretical Computer Science* 126:53–75.

David Dwyer. 1978. What sort of tone language is Mende? *Studies in African Linguistics* 9:167–208.

C. C. Elgot. 1961. Decision problems of finite automata design and related arithmetics. *Transactions of the American Mathematical Society* 98(1):21–51.

Joost Engelfriet and Hendrik Jan Hoogeboom. 2001. MSO definable string transductions and two-way finite-state transducers. *ACM Transations on Computational Logic* 2:216–254.

Emmanual Filiot. 2015. Logic-automata connections for transformations. In *Logic and Its Applications (ICLA)*, Springer, pages 30–57.

Emmanuel Filiot and Pierre-Alain Reynier. 2016. Transducers, logic, and algebra for functions of finite words. *ACM SIGLOG News* 3(3):4–19.

John Goldsmith. 1976. *Autosegmental Phonology.* Ph.D. thesis, Massachussets Institute of Technology.

Jeffrey Heinz. 2007. *The Inductive Learning of Phonotactic Patterns.* Ph.D. thesis, UCLA.

Jeffrey Heinz. 2009. On the role of locality in learning stress patterns. *Phonology* 26:303–351.

Jeffrey Heinz. 2010. Learning long-distance phonotactics. *LI* 41:623–661.

Jeffrey Heinz, editor. forthcoming. *Doing Computational Phonology.* Oxford: Oxford University Press.

Jeffrey Heinz and William Idsardi. 2013. What complexity differences reveal about domains in language. *Topics in Cognitive Science* 5(1):111–131.

Jeffrey Heinz and Regine Lai. 2013. Vowel harmony and subsequentiality. In Andras Kornai and Marco Kuhlmann, editors, *Proceedings of the 13th Meeting on Mathematics of Language.* Sofia, Bulgaria.

Jeffrey Heinz, Chetan Rawal, and Herbert G. Tanner. 2011. Tier-based strictly local constraints for phonology. In *Proceedings of the 49th Annual Meeting of the Association for Computational Linguistics.* Association for Computational Linguistics, Portland, Oregon, USA, pages 58–64.

Mark Hewitt and Alan Prince. 1989. OCP, locality, and linking: the N. Karanga verb. In E. Jane Fee and Katherine Hunt, editors, *WCCFL 8.* Stanford: CSLI Publications, pages 176–191.

Wilfred Hodges. 1997. *A Shorter Model Theory.* Cambridge: Cambridge University Press.

John Hopcroft, Rajeev Motwani, and Jeffrey Ullman. 2006. *Introduction to Automata Theory, Languages, and Computation.* Addison-Wesley, third edition.

Larry Hyman. 1987. Prosodic domains in Kukuya. *NLLT* 5:311–333.

Adam Jardine. 2014. Logic and the generative power of Autosegmental Phonology. In John Kingston, Claire Moore-Cantwell, Joe Pater, and Robert Staubs, editors, *Supplemental proceedings of the 2013 Meeting on Phonology (UMass Amherst).* LSA, Proceedings of the Annual Meetings on Phonology.

Adam Jardine. 2016a. Computationally, tone is different. *Phonology* 33:247–283.

Adam Jardine. 2016b. *Locality and non-linear representations in tonal phonology.* Ph.D. thesis, University of Delaware.

Adam Jardine. to appear. The local nature of tone association patterns. *Phonology*.

Adam Jardine and Jeffrey Heinz. 2015. A concatenation operation to derive autosegmental graphs. In *Proceedings of the 14th Meeting on the Mathematics of Language (MoL 2015).* Association for Computational Linguistics, Chicago, USA, pages 139–151.

C. Douglas Johnson. 1972. *Formal aspects of phonological description.* Mouton.

Ronald Kaplan and Martin Kay. 1994. Regular models of phonological rule systems. *Computational Linguistics* 20:331–78.

Martin Kay. 1987. Nonconcatenative finite-state morphology. In *Proceedings, Third Meeting of the European Chapter of the Association for Computational Linguistics*. pages 2–10.

András Kornai. 1995. *Formal Phonology*. Garland Publication.

Haruo Kubozono. 2012. Varieties of pitch accent systems in Japanese. *Lingua* 122:1395–1414.

Regine Lai. 2015. Learnable versus unlearnable harmony patterns. *LI* 46:425–451.

W. R. Leben. 1973. *Suprasegmental phonology*. Ph.D. thesis, Massachussets Institute of Technology.

W. R. Leben. 1978. The representation of tone. In Victoria Fromkin, editor, *Tone—A Linguistic Survey*, Academic Press, pages 177–219.

Kevin McMullin and Gunnar Ólafur Hansson. 2015. Locality in long-distance phonotactics: evidence for modular learning. In *NELS 44, Amherst, MA*. In press.

Robert McNaughton and Seymour Papert. 1971. *Counter-Free Automata*. MIT Press.

Paul Newman. 1986. Tone and affixation in Hausa. *Studies in African Linguistics* 17(3).

Paul Newman. 2000. *The Hausa Language: An encyclopedic reference grammar*. New Haven: Yale University Press.

David Odden. 1994. Adjacency parameters in phonology. *Language* 70(2):289–330.

Amanda Payne. 2014. Dissimilation as a subsequential process. In *Proceedings of the 44th Meeting of the North East Linguistic Society (NELS)*.

Alan Prince and Paul Smolensky. 2004. *Optimality Theory: Constraint Interaction in Generative Grammar*. Blackwell Publishing.

Douglas Pulleyblank. 1986. *Tone in Lexical Phonology*. Dordrecht: D. Reidel.

James Rogers. 1997. Strict lt2 : Regular :: Local : Recognizable. In Christian Retoré, editor, *Logical Aspects of Computational Linguistics: First International Conference, LACL '96 Nancy, France, September 23–25, 1996 Selected Papers*, Springer Berlin Heidelberg, Berlin, Heidelberg, pages 366–385.

James Rogers, Jeffrey Heinz, Margaret Fero, Jeremy Hurst, Dakotah Lambert, and Sean Wibel. 2013. Cognitive and sub-regular complexity. In *Formal Grammar*, Springer, volume 8036 of *Lecture Notes in Computer Science*, pages 90–108.

James Rogers and Geoffrey Pullum. 2011. Aural pattern recognition experiments and the subregular hierarchy. *Journal of Logic, Language and Information* 20:329–342.

Stephanie Shih and Sharon Inkelas. 2015. Autosegmental aims in surface optimizing phonology. Ms., available at lingbuzz/002520.

James Neil Sneddon, K. Alexnader Adelaar, Dwi N. Djenar, and Michael Ewing. 2010. *Indonesian: A comprehensive grammar*. New York: Routledge.

Boris Avraamovich Trakhtenbrot. 1961. Finite automata and logic of monadic predicates. *Doklady Akademii Nauk SSSR* 140:326–329.

Bruce Wiebe. 1992. *Modelling Autosegmental Phonology with Multi-Tape Finite State Transducers*. Master's thesis, Simon Fraser University.

Moira Yip. 2002. *Tone*. Cambridge University Press.

Anssi Yli-Jyrä. 2013. On finite-state tonology with autosegmental representations. In *Proceedings of the 11th International Conference on Finite State Methods and Natural Language Processing*. Association for Computational Linguistics, pages 90–98.

Cheryl Zoll. 2003. Optimal tone mapping. *LI* 34(2):225–268.

Extracting Forbidden Factors from Regular Stringsets

James Rogers
Dept. of Computer Science
Earlham College
Richmond, IN, USA
jrogers@cs.earlham.edu

Dakotah Lambert
Depts. of Mathematics and Computer Science
Earlham College
Richmond, IN, USA
djlambe11@earlham.edu

Abstract

The work presented here continues a program of completely characterizing the constraints on the distribution of stress in human languages that are documented in the StressTyp2 database with respect to the Local and Piecewise sub-regular hierarchies.

We introduce algorithms that, given a Finite-State Automaton, compute a set of forbidden words, units, initial factors, free factors and final factors that define a Strictly Local (SL) approximation of the stringset recognized by the FSA, along with a minimal DFA that recognizes the residue set: the set of strings in the approximation that are not in the stringset recognized by the FSA. If the FSA recognizes an SL stringset, then the approximation is exact (otherwise it overgenerates).

We have applied these tools to the 106 lects that have associated DFAs in the StressTyp2 database, a wide-coverage corpus of stress patterns that are attested in human languages. The results include a large number of strictly local constraints that have not been included in prior work categorizing these patterns with respect to the Local and Piecewise Sub-Regular hierarchies of Rogers et al. (2012), although, of course, they do not contradict the central result of that work, which establishes an upper bound on their complexity that includes strictly local constraints.

1   Introduction

A stringset L is Strictly k-Local if and only if (iff) it is completely determined by its k-factors: the substrings of length at most k that occur in strings $\rtimes \cdot w \cdot \ltimes$ for $w \in L$. (The '$\rtimes$' and '$\ltimes$' are endmarkers.) That is to say, L contains all and only the strings that are generated by the substring relation from that set of k-factors. The class of stringsets that are Strictly k-local for some k is known as SL. This is at the bottom of the local side of a collection of classes of stringsets, all strict subclasses of the class of Regular stringsets, which are hierarchically related and are characterized by finite sets of either substrings (the Local Hierarchy) or subsequences (the Piecewise Hierarchy) or by combinations of the two. In Rogers et al. (2012) we argue that these hierarchies provide a robust notion of cognitive complexity for constraints on strings.

The long-term project of our group is to characterize all of the stress patterns collected in Goedemans et al. (2015)—a wide-coverage database of stress patterns occurring in human languages—with respect to this hierarchy. In Edlefsen et al. (2008), we established that roughly 75% of these patterns are SL_k for $k \leq 6$ and that half are SL_k for $k \leq 3$. Subsequently, we derived a set of "primitive" constraints sufficient to define all of the patterns by co-occurrence and classified them into abstract categories (Fero et al., 2014). Most of these constraints were, in fact, SL, and it turned out that all of the patterns could be defined by co-occurrence of constraints at the bottom two levels of the hierarchies. This is significant, since at these levels it is possible to determine whether a string satisfies a constraint solely on the basis of the information that is explicitly contained in the string, without inferring any additional structure. Recent work by Heinz and his co-workers (Heinz, forthcoming; Heinz, 2010; Chandlee, 2014; Jardine, 2016) suggests that much of phonology may be characterizable by correspondingly simple sets of structures or functions.

The work on primitive constraints, however,

Proceedings of the 15th Meeting on the Mathematics of Language, pages 36–46,
London, UK, July 13–14, 2017. ©2017 Association for Computational Linguistics

did not include any of the factors from the SL stringsets because the algorithm for determining if a given Finite State Automaton (FSA) recognizes an SL stringset, and determining k if it does, does not yield the set of k-factors that define the stringset. We resolve that problem in this work.

In Section 2 we introduce our notation and basic formal definitions. In Section 3 we formally define Strictly Local stringsets and discuss their formal properties. In Section 4 we distinguish five types of forbidden factors—factors in the complement of the set of factors that generate the stringset. In Section 5 we develop our algorithms for extracting those factors given a Finite State Automaton. In Section 7 we extend these algorithms in a way that allows them to be used to partition non-SL stringsets in a way that provides a set of SL constraints that approximates it (to varying degrees of closeness) and an automaton that captures the non-SL aspects of the stringset. We close with thoughts about where these results lead.

2 Formal Preliminaries

A finite state automaton (FSA) is an edge-labeled directed graph with distinguished vertices that we will represent by a five-tuple $\langle \Sigma, Q, \delta, I, F \rangle$ where Σ is the alphabet of the language of the automaton, Q is the set of states, $\delta \subseteq (\Sigma \times Q \times Q)$ is a transition relation where $\langle \sigma, q_1, q_2 \rangle \in \delta$ iff there is an edge labeled σ from q_1 to q_2, I is the set of initial states, and F is the set of accepting states. Let $\mathcal{A} = \langle \Sigma, Q, \delta, I, F \rangle$.

Let $w = \sigma_1 \sigma_2 \ldots \sigma_n \in \Sigma^*$ be a string and let $q_1, q_n \in Q$. Then there is a path $q_1 \overset{w}{\leadsto} q_n$ iff there exists some sequence of edges

$$\langle \langle \sigma_i, q_i, q_{i+1} \rangle \in \delta \mid \quad 0 < i < n,$$
$$w = \sigma_1 \sigma_2 \ldots \sigma_{n-1} \rangle .$$

This is an accepting path on w if q_n is in F, else it is a non-accepting path.

The automaton $\mathcal{A}$ is *total* iff for every symbol $\sigma \in \Sigma$ and for every state $q \in Q$, there exists some q' such that $\langle \sigma, q, q' \rangle \in \delta$. It is (partial) *functional* iff δ is functional in its first two places. That is, given a state $q \in Q$ and a symbol $\sigma \in \Sigma$, there is at most one $q' \in Q$ such that $\langle \sigma, q, q' \rangle \in \delta$.

An FSA is (fully) *deterministic* (a proper DFA) iff it has exactly one initial state and it is both total and functional. We also consider trim functional automata to be deterministic, where $\mathcal{A}$ is *trim* iff for all states $q \in Q$ there is some accepting path from q.

An automaton is *minimal* iff it is deterministic and no two states are Nerode-equivalent[1]. Further, it is *normalized* iff it is both minimal and trim.

Given a string w, the factors of w are those v that are substrings of w (notation: $v \preccurlyeq w$). If k is the length of v (notation: $|v| = k$) then v is a k-factor of w.

The powerset graph of the automaton $\mathcal{A}$, $\mathrm{PSG}(\mathcal{A}) = \langle V, E \rangle$, is another edge-labeled directed graph where:

$$V = \mathcal{P}(Q) \quad \text{and}$$
$$E = \{\langle \sigma, S_1, S_2 \rangle \mid \sigma \in \Sigma,$$
$$S_2 = \{q' \in Q \mid (\exists q \in S_1)[\langle \sigma, q, q' \rangle \in \delta]\}\}$$

Often we are interested only in the subgraph of this generated from a given set of initial states.

Lemma 1 *If $\mathcal{A}$ is deterministic, then the sizes of the sets along any path in $PSG(\mathcal{A})$ are monotonically non-increasing.*

This is because if $\mathcal{A}$ is deterministic δ maps each state in S_1 to at most one state in S_2.

Corollary 1 *All sets in any cycle are equal in size.*

Corollary 2 *All in-edges to Q and all out-edges from $\emptyset$ are self-edges.*

3 Strictly Local Stringsets

L is *Strictly k-Local* ($L \in \mathrm{SL}_k$) iff it is completely characterized by its k-factors. Let Σ be the alphabet of L and define $\mathrm{F}_k(\Sigma) = \{v \in \Sigma^* \mid |v| = k\}$ and $\mathrm{F}_{\leq k}(\Sigma) = \bigcup_{1 \leq i \leq k}[\mathrm{F}_i(\Sigma)]$. For any string $w \in \Sigma^*$, the k-factors of w are

$$\mathrm{F}_k(w) = \begin{cases} \{w\} & \text{if } |w| <= k, \\ \{v \in \mathrm{F}_k(\Sigma) \mid \\ \qquad w = w_1 v w_2, \ w_1, w_2 \in \Sigma^*\} \\ \text{otherwise.} \end{cases}$$

Similarly for $\mathrm{F}_{\leq k}(w)$. This lifts to sets of strings in the obvious way.

Let $G \subseteq \mathrm{F}_{\leq k}(\{\rtimes\} \cdot \Sigma^* \cdot \{\ltimes\})$ be the set of permitted factors in L. Then the stringset generated by G is

$$L(G) = \{w \in \Sigma^* \mid \mathrm{F}_{\leq k}(\rtimes \cdot w \cdot \ltimes) \subseteq G\} .$$

Since Σ is assumed to be finite, $\mathrm{F}_{\leq k}(\Sigma)$ is also finite, and an SL_k language can equivalently be defined in terms of its forbidden factors: $\overline{G} =$

---

[1] q_1 and q_2 are Nerode-equivalent iff for all strings w, there is an accepting path on w from q_2 iff there is an accepting path on w from q_1

$F_{\leq k}(\Sigma) - G$. This is more natural in many applications, including many linguistic ones (as in "no pair of unstressed syllables occur adjacently").

A stringset is said to be SL if it is SL_k for any finite k.

The following proposition characterizes SL_k.

Proposition 1 (Suffix Substitution Closure) (SSC)

$L \in SL_k$ iff
$$(\forall x \in F_{k-1}(\Sigma))[\quad \begin{aligned} &\textit{if } w_1 = u_1 \cdot x \cdot v_1 \in L \\ &\textit{and } w_2 = u_2 \cdot x \cdot v_2 \in L \\ &\textit{then } u_1 \cdot x \cdot v_2 \in L \quad \end{aligned}].$$

This is because if a symbol σ can follow x in some string of $L(\mathcal{A})$ then $x \cdot \sigma$ is a permitted factor and σ can follow x in any string of $L(\mathcal{A})$.

One consequence of this is that if $L(\mathcal{A}) \in SL_k$ and $\mathcal{A}$ is deterministic, then for each length $k-1$ string x, all states in the set

$$\{q' \in Q \mid (\exists q \in Q)[q \overset{x}{\leadsto} q']\}$$

are Nerode Equivalent. If $\mathcal{A}$ is minimal as well, then all paths that end with the same $(k-1)$-factor lead to the same state. The computations of the automaton synchronize after at most $k-1$ steps.

This is the basis of the algorithm used by Edlefsen et al. (2008)[2] to determine if a given $\mathcal{A}$ recognizes an SL stringset and, if it does, to find the parameter k.

Proposition 2 *Suppose $\mathcal{A}$ is a normalized DFA. Then $L(\mathcal{A}) \in SL_k$ iff every path from Q in $PSG(\mathcal{A})$ that is of length $k-1$ leads to a singleton vertex. If that is the case, then k is one plus the length of the longest path from Q to a singleton (that does not include other singletons). If there is no such longest path (i.e., there is an infinite path) then there is some cycle of non-singleton vertices, $L(\mathcal{A})$ does not satisfy SSC for any k and it is not SL.*

In practice, it is not necessary to build even just the subgraph of $PSG(\mathcal{A})$ generated by Q. All that one needs for a counter-example to SSC is a single pair of strings in which SSC fails. So it suffices to just explore the subgraph of $PSG(\mathcal{A})$ that is generated by doubleton subsets of Q. The size of this subgraph is only $\Theta(\mathbf{card}(Q)^2)$, in contrast to the subgraph generated by Q, which is $\Theta(2^{\mathbf{card}(Q)})$.

The following is an immediate consequence of this proposition.

---

[2]The pair-graph algorithm was first published in Caron (2000).

Lemma 2 *If if $\mathcal{A}$ is a normalized DFA and $L(\mathcal{A}) \in SL_k$ then all cycles in $PSG(\mathcal{A})$ are cycles of singletons.*

4 Classes of Forbidden Factors

Factors may or may not include either a left-end marker at the beginning or a right-end marker at the end. In the case that a factor contains neither, it can occur anywhere in a string (including, possibly, at the beginning or end) and we say that it is a *free factor* or, if forbidden, *free forbidden factor*. If the length of a free forbidden factor is one, then it has somewhat different status than free forbidden factors of greater length; it is, in essence, a restriction to the alphabet. We will refer to these as *forbidden units*. If the first symbol of a forbidden factor is '⋊', then it can only occur at the left end of the word; this is an *initial forbidden factor*. If the last symbol is '⋉', then it can only occur at the right end of the word; it is a *final forbidden factor*. Note that the length of the string that these anchored factors match is $k-1$. An SL_k definition can restrict length $k-1$ prefixes and suffixes, but not, in general length k prefixes and suffixes.[3] Finally, if a factor contains both end-markers it is a *forbidden word*, where the word it forbids is actually of length $k-2$.

5 Forbidden Factors of SL Stringsets

5.1 Free Forbidden Factors

Suppose $\mathcal{A}$ is a DFA. A factor w is a free forbidden factor of $L(\mathcal{A})$ iff there is no path in the transition graph of $\mathcal{A}$ from q_0 to an accepting state that includes w as a substring. If $\mathcal{A}$ is normalized, this will be the case iff there is no path at all that is labeled w from any state of $\mathcal{A}$, as all such paths would necessarily lead to the sink state which has been trimmed. Thus, in $PSG(\mathcal{A})$ the path from Q that is labeled w leads to $\emptyset$. Again, the converse holds.

So the set of all labels of paths Q to $\emptyset$ in $PSG(\mathcal{A})$ are free forbidden factors of $L(\mathcal{A})$, moreover, that set includes all free forbidden factors of $L(\mathcal{A})$. Since in general $PSG(\mathcal{A})$ may include cycles and even in the case that $L(\mathcal{A})$ is SL it may include cycles of singleton vertices, in general this

---

[3]In the original definition of SL_k (McNaughton and Papert, 1971) prefix and suffix factors and forbidden words could be of length k. But the definition we use is equivalent in all significant aspects and accounts for the information contained in an anchored factor; it has become the prevailing definition in most of the literature.

set of paths will be infinite. (In fact, since $\mathrm{PSG}(\mathcal{A})$ invariably includes a trivial cycle on $\emptyset$ for each $\sigma \in \Sigma$, it will *always* be infinite.) The paths including trivial cycles on $\emptyset$ are labeled with strings in $w \cdot \Sigma^*$, where w is a free forbidden factor. We are interested in the set of paths that are minimal in the sense that the label of the path does not include the label of any other such path as a substring.

Note that, by Corollary 2, any such path that includes an in-edge to Q or an out-edge from $\emptyset$ includes another path from Q to $\emptyset$ that is strictly shorter. Thus none of those paths are minimal free forbidden factors. Note, also, that if $L(\mathcal{A}) \in \mathrm{SL}$, then there are no cycles on Q, although there will always be trivial cycles on $\emptyset$ for each $\sigma \in \Sigma$.

The next two lemmas establish that if $L(\mathcal{A})$ is SL then there is some bound such that all cyclic paths from Q to $\emptyset$ in $\mathrm{PSG}(\mathcal{A})$ with length greater than that bound will be labeled with a string that includes, as a suffix, the label of an acyclic path from Q to $\emptyset$. Thus the set of minimal free forbidden factors of $L(\mathcal{A})$ is just the set of labels from paths from Q to $\emptyset$ in $\mathrm{PSG}(\mathcal{A})$ that do not include the label of any other such path as a suffix and that do not include self-edges on $\emptyset$. This allows us to collect forbidden factors with a breadth-first bottom-up traversal of $\mathrm{PSG}(\mathcal{A})$.

Lemma 3 *If v and w label* acyclic *paths from Q to $\emptyset$ in $PSG(\mathcal{A})$ and $v \preccurlyeq w$, then $w = uv$ for some $u \in \Sigma^*$.*

Proof: $v \preccurlyeq w$ implies that $w = uvx$ for some $u, x \in \Sigma^*$. Since $Q \overset{v}{\rightsquigarrow} \emptyset$ and all vertices $S \subseteq Q$, for all vertices S, $S \overset{v}{\rightsquigarrow} \emptyset$ as well, and, in particular, $Q \overset{u}{\rightsquigarrow} S \overset{v}{\rightsquigarrow} \emptyset$. Hence x is either ε or the path it labels is a self-loop on $\emptyset$, contradicting the assumption of acyclicity. $\dashv$

Lemma 4 *If a path from Q to $\emptyset$ in $PSG(\mathcal{A})$, with $L(\mathcal{A}) \in SL$ includes a cycle other than a trivial cycle on Q or $\emptyset$, then there is a finite bound on the number of times the cycle can be taken before the label of the path includes the label of an acyclic path from Q to $\emptyset$ as a suffix.*

Proof: Since $L(\mathcal{A})$ is SL, any cycle must be a cycle of singletons. Suppose, then that there is a path:

$$Q \overset{u}{\rightsquigarrow} \{q_0\} \overset{v}{\rightsquigarrow} \{q_1\} \overset{w}{\rightsquigarrow} \{q_0\} \overset{x}{\rightsquigarrow} \emptyset$$

where, possibly, v may be a prefix of x. Since $q_0, q_1 \in Q$ there must be a path:

$$Q = S_0 \overset{v}{\rightsquigarrow} S_1 \overset{w}{\rightsquigarrow} S_2 \overset{v}{\rightsquigarrow} S_3 \cdots$$

where $q_0 \in S_{2i}$ and $q_1 \in S_{2i+1}$ for $i \geq 0$. Since there are no cycles of non-singletons, by Lemma 1 the sequence of S_is must ultimately be decreasing in size. Thus, for some n it resolves to:

$$Q \overset{v}{\rightsquigarrow} S_1 \overset{w}{\rightsquigarrow} S_2 \overset{v}{\rightsquigarrow} S_3 \cdots \overset{w}{\rightsquigarrow} S_{2n} = \{q_0\} \overset{x}{\rightsquigarrow} Q$$

So $(vw)^n x$ labels a path from Q to $\emptyset$ and will be a suffix of all paths Q to $\emptyset$ that take the $\{q_0\} \rightsquigarrow \{q_1\}$ cycle at least $2n$ times. $\dashv$

Theorem 1 *If $L(\mathcal{A}) \in SL$ then a string w is a free forbidden factor of $L(\mathcal{A}) \in SL$ iff it labels a path in $PSG(\mathcal{A})$ from Q to $\emptyset$. It is minimal if that path does not include any cycles other than cycles of singletons and w does not include the label of any other such path as a suffix.*

Note that if $L(\mathcal{A}) \in \mathrm{SL}$ then the only cycles of non-singletons will be trivial cycles on $\emptyset$. Labels of paths including these will include some free forbidden factor as a prefix and are, thus, not minimal.

Paths including cycles of singletons are necessary since none of the paths labeled $u(vw)^i x$ as in the proof of Lemma 4 is labeled with a factor of any of the others; they are minimal with respect to each other. It is only the label of the acyclic path that subsumes the labels of further iterations.

5.2 Final Forbidden Factors

Suppose $\mathcal{A}$ is a DFA. A factor w is a final forbidden factor of $L(\mathcal{A})$ iff there is no path from q_0 to an accepting state in the transition graph of $\mathcal{A}$ that includes w as a suffix but there is some path from q_0 to an accepting state that includes w as a proper substring. (If no there is no such accepting path, then w is a free forbidden factor.) If $\mathcal{A}$ is normalized then w is a final forbidden factor iff all paths labeled w from any state in Q end at a non-accepting state and there is some such path. This will be the case iff the path from Q in $\mathrm{PSG}(\mathcal{A})$ labeled w ends at a non-empty vertex that is disjoint with F.

Lemma 5 *Suppose $\mathcal{A}$ is a DFA. No final forbidden factor of $L(\mathcal{A})$ includes a free forbidden factor of $L(\mathcal{A})$ as a substring.*

This is because if v is a free forbidden factor of $L(\mathcal{A})$ then the path from Q in $\mathrm{PSG}(\mathcal{A})$ leads to $\emptyset$ and, hence, the path labeled v from any vertex of $\mathrm{PSG}(\mathcal{A})$ leads to $\emptyset$ as well.

Note that a final forbidden factor may include another as a suffix. (It is irrelevant whether it includes an final forbidden factor as a non-suffix, since final forbidden factors are, by definition, only relevant as suffixes.)

Theorem 2 *If a path from Q to a non-empty vertex disjoint from F in $\mathrm{PSG}(\mathcal{A})$, with $L(\mathcal{A}) \in SL$, includes a cycle other than a trivial cycle on Q, then there is a finite bound on the number of times the cycle can be taken before the label of the path includes the label of an acyclic path from Q to a non-empty vertex disjoint from F as a suffix.*

The proof is essentially the same as the proof of Lemma 4.

5.3 Initial Forbidden Factors

Suppose $\mathcal{A}$ is a DFA. A string w is an initial forbidden factor of $L(\mathcal{A})$ iff it is w^{R} (w reversed) for some w, a final forbidden factor of $L(\mathcal{A}^{\mathrm{R}})$, where $\mathcal{A}^{\mathrm{R}}$ is the DFA that recognizes the reversal of $L(\mathcal{A})$.

5.4 Forbidden Words

Suppose $\mathcal{A}$ is a DFA and $L(\mathcal{A}) \in \mathrm{SL}_k$. Then w is a forbidden word of $L(\mathcal{A})$ iff it labels a path of length less than or equal to k that leads from q_0 to a state in $Q - F$.

6 Algorithms

Theorem 1 guarantees that if we do a breadth-first bottom-up traversal of $\mathrm{PSG}(\mathcal{A})$ then we will discover each minimal forbidden factor before we discover any of its proper suffixes. Expanding the frontier of the search in discrete stages, every (reverse) path from $\emptyset$ to Q found in the kth stage will be a minimal forbidden k-factor.

There may be more than one such path so we do need to avoid gathering more than one instance of the factor. In general, there will be open paths (not reaching Q) that are labeled with the same factor. Extended to Q, they would include the factor as a proper suffix. So we exclude these from the frontier for the next stage.

We structure the bottom-up traversal of $\mathrm{PSG}(\mathcal{A})$ as a top-down traversal of $\mathrm{PSG}^{\mathrm{R}}(\mathcal{A})$, in which each of the edges of $\mathrm{PSG}(\mathcal{A})$ is reversed. For convenience (and convergence) we trim self-edges on

$\emptyset$ and Q while reversing the graph. Since we are traversing bottom-up, we actually find w^{R} of each factor w, but we gather these in a list structure, inserting at the head, which reverses the factor again as we construct it.

For the purposes of the algorithm, a *Path* in an edge-labeled graph $\langle V, E \rangle$ as a computational structure, is a 3-Tuple: $\langle v, S, w \rangle$, where $v \in V$ is the final vertex of the path, $S \subseteq V$ is the (unordered) set of vertices along the path and $w \in \Sigma^*$ is the sequence of labels of the edges in the path, in reverse order. A *Frontier* is a set of paths. Forbidden factors are gathered in stages, with Stage_i expanding $\mathrm{Frontier}_{i-1}$ to $\mathrm{Frontier}_i$, gathering the set FF_i of all minimal forbidden i-Factors in the process.

The initial frontier $\mathrm{Frontier}_0$ for finding free forbidden factors includes just the trivial (0-length) path from $\emptyset$. For finding final forbidden factors $\mathrm{Frontier}_0$ includes the trivial path from each vertex that is a subset of $Q - F$.

Theorem 1 guarantees that, if we eliminate paths labeled with a forbidden i-Factor from $\mathrm{Frontier}_i$ the search will converge after finitely many iterations, k, with $\mathrm{Frontier}_k$ empty. (Note it is an empty set of Paths, not a set including a path ending at $\emptyset$.) The set of minimal free forbidden factors will be the union of the sets of factors gathered at stages 2 through k, where $L(\mathcal{A}) \in \mathrm{SL}_k$. (Forbidden 1-factors are not included, since they are forbidden units.) The search for final forbidden factors will terminate after $k - 1$ iterations, with the minimal k-final forbidden factors including the right-end marker.

Pseudo-code for the algorithms is given in Figures 1 and 2.

6.1 Forbidden Words for SL Stringsets

If $L(\mathcal{A}) \in \mathrm{SL}_k$ and $\mathcal{A}$ is deterministic, then the words it forbids are just the labels of paths of length $k - 2$ (to allow for the endmarkers) from the (single) initial state to a state in $Q - F$. These can be gathered by doing a bounded traversal of $\mathcal{A}$.

6.2 Forbidden Units

If $\mathcal{A}$ is normalized (minimal and trim), the forbidden units of $L(\mathcal{A})$ are just the symbols of Σ that do not label any edge in δ. In $\mathrm{PSG}(\mathcal{A})$ these will label edges Q to $\emptyset$ and will be gathered in Stage_1 while gathering free forbidden factors. But these may not be the only forbidden units of interest. In many applications there will be an alphabet that includes all symbols that occur in any of a collection of stringsets and the subset of that alphabet that is not included in the alphabet of the FSA will also be significant. This is the case in most linguistic applications, for example (as in "this lect forbids unstressed heavy syllables").

In those applications we need to include the difference between some default alphabet and the set of symbols that label edges in $\mathcal{A}$. Since we are building $\text{PSG}(\mathcal{A})$ anyway, the simplest way of doing this is to just take the difference between the default alphabet and the labels of the out-edges from Q. If we union that with the labels of the subset of those edges that lead to $\emptyset$ we get the free forbidden 1-factors as well. We can avoid gathering the latter in both the set of free forbidden factors and the set of forbidden units by not including the forbidden factors gathered in Stage_1. (Or, in order to simplify the code, by removing them from the set of free forbidden factors.)

7 Forbidden Factors of non-SL Stringsets

Every non-SL stringset can be fully defined by the conjunction of a set of SL constraints (possibly trivial: Σ^*, $\emptyset$ and Σ^+ are SL_1 and SL_2, respectively) along with a set of properly non-SL constraints. In applications that are exploring constraints across a collection of stringsets, most linguistic applications for instance, these SL constraints are significant. We would like to be able to factor the constraints so that the non-SL constraints capture, to the extent possible, just the non-strictly-local aspects of the patterns.

The problem isn't finding factors that characterize the stringset, the problem is that there are too many of them. $\Sigma^* - L(\mathcal{A})$, augmented with left and right endmarkers, is a set of forbidden factors that characterizes $L(\mathcal{A})$ exactly. It is, of course, in general infinite and necessarily so if $L(\mathcal{A})$ is not SL.

The algorithms for SL stringsets are still partially correct for non-SL stringsets. The problem is that if $L(\mathcal{A})$ is non-SL then there will be non-singleton cycles (in addition to those on $\emptyset$) and the traversal will not terminate.

These non-singleton cycles actually localize the reason that the stringset is not SL. They capture circumstances under which the automaton fails to synchronize ever; they identify places in which

SSC (Proposition 1) fails for $L(\mathcal{A})$.

As with the set of forbidden words, the set of labels of the paths in $\text{PSG}(\mathcal{A})$ that include non-singleton cycles are all legitimate forbidden factors of $L(\mathcal{A})$, but again there are infinitely many of them. The stringset they define is what we would like to isolate as the non-SL fragment of $L(\mathcal{A})$.

It is tempting to try modifying the traversal so it follows only singleton cycles. But, unfortunately, if there are non-singleton cycles the chain of the proof of Lemma 4 may be infinite, so there is no guarantee of termination even when following only singleton cycles.

Another approach would be to modify $\mathcal{A}$, working backward from $\text{PSG}(\mathcal{A})$, in a way that would eliminate the non-singleton cycles. We have not really pursued this idea, but our sense is that it is likely to fail for the same reason as simply not following non-singleton cycles fails.

In any case, we are looking for a set of forbidden factors that approximates $L(\mathcal{A})$. Since none of our algorithms introduces constraints that are not manifest in the automaton, the approximation will overgenerate. The issue is how close do we need it to be.

7.1 SL Approximations

First of all, as we noted above, Σ^* is an SL approximation of every stringset over Σ. But it's a particularly licentious one. Another possibility is to only gather the forbidden factors that label non-cyclic paths in $\text{PSG}(\mathcal{A})$. This will miss many forbidden factors that may well be significant—all those factors labeling paths with singleton cycles that would have eventually been subsumed if there were no non-singleton cycles. On the other hand, it gives the smallest set of forbidden factors that comprise a reasonable approximation of $L(\mathcal{A})$.

Another way of bounding the traversal is to note that no acyclic path from Q to $\emptyset$ in $\text{PSG}(\mathcal{A})$ can be $2^{\mathbf{card}(Q)} - 2$ or longer. But the set of factors gathered by a traversal with this bound, although arguably the largest justifiable set of forbidden factors, is almost certainly unreasonably large.

SL approximations that are too large are misleading both in terms of the apparent complexity of the SL aspects of the constraints and in terms of the their non-SL aspects, which will appear to need to include many exceptions in order to account for the strings excluded by the SL approximation. When the SL approximation overesti-

41

mates, the non-SL residue undergeneralizes.

In some applications, there may be a theoretically justified bound on how long the relevant factors are, that is, on how many times a cycle should be followed in the traversal. As we noted in the introduction, all of the SL stress patterns in StressTyp2 are SL_k for $k \leq 6$. Thus one may well be justified in limiting the SL fragment to factors of length no more than six. Even assuming the bound is well-justified, this is still likely to generate too close an approximation. Forbidden factors that should properly be captured by the non-SL constraints, that involve non-singleton cycles that are not needed to terminate the traversal of singleton cycles, will be included. If the goal is to explore the nature of the constraints across a collection of stringsets these will likely be misleading, particularly since half of the patterns in StressTyp2 are SL_3 (or less, SL is an inclusive hierarchy in k).

It is straightforward to modify the algorithms given above for either of these approaches. Cycles can be completely excluded by modifying the definition of Extensions. Limits on the size of the factor are just depth limits on the traversal. It is also straightforward to combine these, only following singleton cycles and only doing it up to a depth limit. To bound the search for forbidden words we first compute the sets of forbidden initial, free and final factors and then bound the depth to $\max(|\text{frFF}| - 2, |\text{inFF}| - 1, |\text{fiFF}| - 1)$, where $|\text{frFF}|$, $|\text{inFF}|$, $|\text{fiFF}|$ are the maximum width of the free, initial and final factors, respectively.

As our goal in developing these algorithms is to provide tools that phonologists can use productively in exploring systems of phonotactic constraints the third approach to bounding the traversal seems most useful, although we have currently only implemented the acyclic path approach.

7.2 Residue Automata

When the algorithms are run on automata that recognize non-SL stringsets the result is a set of forbidden factors for the approximated stringset. We are just as interested in the characteristics of the stringset that these forbidden factors miss. Most work on approximating stringsets with stringsets in a weaker complexity class has focused on approximating CFLs with regular stringsets (Nederhof (2000) includes a good survey) or Tree-Adjoining Stringsets (TALs) with CFLs (Schabes and Waters, 1993; Rogers, 1994). Whenever the class of stringsets that is being approximated includes CFLs the (symmetric) difference between the approximation and the target will not be a decidable set. Consequently, there is little that can be determined about that difference.

We have the advantage that all of our stringsets are regular and so the difference is not only decidable but an automaton recognizing it is effectively constructible. Moreover, in this case, we know that every string excluded by the approximation is necessarily excluded by the target. The approximation never undergenerates. To isolate the non-SL characteristics of the target we construct an automaton that recognizes exactly the set of strings that are overgenerated by the SL approximation.

Using well-known algorithms for combining automata, it is straightforward to construct an automaton $\mathcal{A}_{\text{FF}}$ that recognizes the set of strings licensed by the set of forbidden factors. One starts with deterministic automata that recognize each of the given factors, complements them and then builds the automaton that recognizes the intersection of those complements. It is then straightforward to construct $\mathcal{A}_{\text{res}}$, the residue automaton[4] which recognizes exactly $L(\mathcal{A}_{\text{FF}}) - L(\mathcal{A})$. This residue automaton captures exactly the non-SL aspects of $L(\mathcal{A})$, up to the degree to which the forbidden factors approximate the strictly SL aspects of $L(\mathcal{A})$.

8 Results and Prospectus

We have designed and implemented algorithms that, given a Finite-State Automaton, compute a set of forbidden words, units, initial factors, free factors and final factors that define an SL approximation of the stringset recognized by the FSA, along with a minimal DFA that recognizes the residue set: the set of strings in the approximation that are not in the stringset recognized by the FSA. If the FSA recognizes a stringset that is SL, then the approximation is exact.

As we explain in Section 7.1, the closeness of the approximation is a parameter that may be varied depending on the application. As we have implemented it, we obtain the smallest set of factors that is arguably a reasonable approximation.

---

[4]The term "residue" is motivated from the perspective of factoring constraints. These automata should not be confused with the *residual automata* of Denis et al. (2002), NFAs in which every state corresponds to the residual stringset wrt some prefix. "Residual" in that context is justified from the perspective of factoring strings.

We have also implemented an algorithm that collects the union of the forbidden factors of each type from a collection of these results, although we don't present it here, the algorithm being obvious.

We have applied these tools to the 106 lects that have associated DFAs in the StressTyp2 database. For the individual lects the maximum number of forbidden words is 20. Since the size of our default alphabet is 15 (five degrees of weight and three degrees of stress) and some lects have only one weight and two levels of stress, the maximum number of forbidden units is 13. The maximum number of forbidden initial factors is 15. The maximum number of forbidden free and final factors is 386 and 117, respectively, but these are all due to Pirahã, an outlier. Without Pirahã they are 185 and 32, respectively.

For the union factor types, there are 14 distinct forbidden units (only unstressed light syllables occur in every lect), 44 distinct forbidden words, 35 distinct forbidden initial factors, 904 distinct forbidden free factors and 230 distinct forbidden final factors. The maximum width of forbidden words, initial factors and free factors is 5. The maximum width of forbidden final factor is 6, due to a single lect (Içuã Tupi) which is also the only example of a properly SL_6 stringset, the other SL patterns all being SL_4 or less.

That is still a lot of factors, too many to draw much insight from. But these are all in ground form, with each syllable type represented by a distinct alphabet symbol. In future work we plan to adapt the alphabet type to be tuples of features or perhaps non-re-entrant feature structures (adding full feature structures we will leave for others), which will provide opportunities to generalize across those features. We know, just from the phonology, that this will reduce the total number of exemplars significantly.

The algorithms we have presented here, are asymptotically exponential-time in the size of the automaton, but that is actually optimal for algorithms that construct sets of ground factors: the worst case size of the set of factors of the stringset of an automaton with $\mathbf{card}(Q)$ states is $\Omega(\mathbf{card}(\Sigma)^{\mathbf{card}(Q)})$. Nevertheless these algorithms are actually quite effective in practice. We have incorporated them into a Haskell workbench for manipulating automata with a particular focus on logical descriptions of sub-regular constraints. With only minimal optimization the algorithm computes the forbidden factors and the residue automaton for all 106 lects in our corpus in less than an hour, which is practical as it stands, but can be improved significantly. The asymptotic bound is due to the potential size of the powerset graph as well as the potential size of the set of factors. These are not, however, the dominant factor in the practical performance. Rather it is the time it takes to generate a minimal DFA from the forbidden factors. This is an easy target for optimization; the intersection step, a critical path in the construction, can be done in time logarithmic in the number of factors, for example. There are many other easy opportunities for optimization and Haskell provides a particularly powerful platform form implementing them.

Acknowledgments

The work reported here builds on the work of at least a dozen undergraduate students and alumni of Earlham College over the course of about ten years. We are greatly indebted to their willingness to think carefully about hard problems and to collaborate effectively both within the group and over time. We are also indebted to the collaboration of Jeff Heinz and his students in Linguistics and Cognitive Science at the University of Delaware and to the helpful suggestions of the anonymous reviewers.

References

Pascal Caron. 2000. Families of locally testable languages. *Theoretical Computer Science*, 242:361–376.

Jane Chandlee. 2014. *Strictly Local Phonological Processes*. Ph.D. thesis, The University of Delaware.

François Denis, Aurélien Lemay, and Alain Terlutte. 2002. Residual finite state automata. *Fundamenta Informaticae*, 51(4):339–368.

Matt Edlefsen, Dylan Leeman, Nathan Myers, Nathaniel Smith, Molly Visscher, and David Wellcome. 2008. Deciding strictly local (SL) languages. In Jon Breitenbucher, editor, *Proceedings of the Midstates Conference for Undergraduate Research in Computer Science and Mathematics*, pages 66–73.

Margaret Fero, Dakotah Lambert, Sean Wibel, and James Rogers. 2014. Abstract categories of phonotactic constraints. `https://ncurdb.cur.org/ncur2017/archive/`

`Display_NCUR.aspx?id=83628`, Retrieved 19 May 2017, April. Presented at the National Conference on Undergraduate Research (NCUR'14).

R. W. Goedemans, Jeffrey Heinz, and Harry van der Hulst. 2015. `http://st2. ullet.net\slashfiles\slashfiles\ slashst2-v1-archive-0415.tar.gz`, April. Retrieved 24 Jun 2015.

Jeffrey Heinz. 2010. Learning long-distance phonotactics. *Linguistic Inquiry*, 41(4):623–661.

Jeffrey Heinz. forthcoming. The computational nature of phonological generalizations. In Larry Hyman and Frans Plank, editors, *Phonological Typology*, Phonetics and Phonology. Mouton. Final version submitted November 2015. Expected publication in 2017.

Adam Jardine. 2016. *Locality and non-linear representations in tonal phonology*. Ph.D. thesis, University of Delaware.

Robert McNaughton and Seymour Papert. 1971. *Counter-Free Automata*. MIT Press.

Mark-Jan Nederhof. 2000. Practical experiments with regular approximation of context-free languages. *Computational Linguistics, Volume 26, Number 1, March 2000*.

James Rogers, Jeff Heinz, Margaret Fero, Jeremy Hurst, Dakotah Lambert, and Sean Wibel. 2012. Cognitive and sub-regular complexity. In Glyn Morrill and Mark-Jan Nederhof, editors, *Formal Grammar 2012*, volume 8036 of *Lecture Notes in Computer Science*, pages 90–108. Springer.

James Rogers. 1994. Capturing CFLs with tree adjoining grammars. In *Proceedings of the 32th Annual Meeting of the Association for Computational Linguistics*, pages 155—162, Las Cruces, NM. Association for Computational Linguistics.

Yves Schabes and Richard C. Waters. 1993. Lexicalized context-free grammars. In *31st Annual Meeting of the Association for Computational Linguistics (ACL'93)*, pages 121–129, Columbus, OH. Association for Computational Linguistics.

Figure 1: Main procedures for Free and Final forbidden factors.

FreeFFs

Given: $PSG^R = \langle V, E \rangle$
Where: $V \subseteq \mathcal{P}(Q),\ E \subseteq \Sigma \times \mathcal{P}(Q) \times \mathcal{P}(Q)$
    $\triangleright$ The reversed powerset graph of a DFA $\mathcal{A} = \langle \Sigma, \delta, I, F \rangle$

1   **Let:**
2      $Front_0 = \{\langle \emptyset, \{\emptyset\}, \varepsilon \rangle\}$
    $\triangleright$ The initial frontier,
3      $Goals = \{Q\}$
4      $Extensions(\langle v, S, w \rangle) = \{\langle v', S \cup \{v\}, \sigma \cdot w \rangle \mid \langle \sigma, v, v' \rangle \in E$ and either $v' \notin S$ or v is singleton$\}$
    $\triangleright$ Outedges that are acyclic if not from a singleton
5   **while** $Front_i \neq \emptyset$
     Construct:
6      $FrFF = \bigcup[FF_1, FF_2, \ldots],$
    $\triangleright$ the set of free forbidden factors of $L(\mathcal{A})$
7   in stages: $\textsc{Stage}_1, \textsc{Stage}_2, \ldots$
    $\triangleright$ as given in Figure 2

FinalFFs

Given: $PSG^R = \langle V, E \rangle$
    $\triangleright$ as in **FreeFFs**

1   **Let:**
2      $Front_0 = \{\langle S, \{\emptyset\}, \varepsilon \rangle \mid S \subseteq Q - F\}$
    $\triangleright$ The set of trivial paths from vertices disjoint with F
3      $Goals = \{Q\}$
4      $Extensions(\langle v, S, w \rangle)$
    $\triangleright$ as in **FreeFFs**
5   **while** $Front_i \neq \emptyset$
     Construct:
6      $FiFF = \bigcup[FF_1, FF_2, \ldots],$
    $\triangleright$ the set of final forbidden factors of $L(\mathcal{A})$
7   in stages: $\textsc{Stage}_1, \textsc{Stage}_2, \ldots$
    $\triangleright$ as given in Figure 2

Figure 1: Main procedures for Free and Final forbidden factors.

$\textsc{Stage}_i$

Given:

$Front_{i-1} \subseteq \{\langle v, S, w\rangle \mid v \in V, S \subseteq V, w \in \Sigma^*\}$
$\triangleright$ The frontier of the search, a set of $Path$
Where: $v \in V$ is the final vertex,
$S \subseteq V$ is the (unordered) set of vertices in the path,
$w \in \Sigma^*$ is the sequence of labels of the edges in the path, in reverse order
$Goals \subseteq V$ is the set of goal vertices
$Extensions$ is a function taking a $Path$ to its qualified extensions
Construct: $Front_i$, FF_i

```
1   ForEach Path ∈ Front_{i-1}
2       ForEach ⟨v', S ∪ {v}, σ · w⟩ ∈ Extensions(Path)
3           if σ · w ∉ FF_i        ▷ σ · w has not already been found to be an i-FF
4           then if v' ∈ Goals      ▷ σ · w is an i-FF
                then
5                       Front_i ← Front_i − {⟨−, −, σ · w⟩ ∈ Front_i}
                            ▷ Remove any paths labeled with this factor from Front_i
6                       FF_i ← FF_i ∪ {σ · w}
                            ▷ Add σ · w to FF_i
                else
7                       Front_i ← Front_i ∪ {⟨v', S ∪ {v}, σ · w⟩}
                            ▷ Add extension to Front_i
    End of STAGE_i.
```

Figure 2: Gathering Forbidden Factors

Latent-Variable PCFGs: Background and Applications

Shay B. Cohen
School of Informatics
University of Edinburgh
Edinburgh, EH8 9AB, UK
scohen@inf.ed.ac.uk

Abstract

Latent-variable probabilistic context-free grammars are latent-variable models that are based on context-free grammars. Nonterminals are associated with latent states that provide contextual information during the top-down rewriting process of the grammar. We survey a few of the techniques used to estimate such grammars and to parse text with them. We also give an overview of what the latent states represent for English Penn treebank parsing, and provide an overview of extensions and related models to these grammars.

1 Introduction

Probabilistic grammars have been one of the most important modeling tools available in the natural language processing toolkit. They are often humanly interpretable because of their symbolic backbone, while their probabilistic component helps with reasoning under uncertainty. Probabilistic grammars have mostly been used for syntactic analysis in NLP (Charniak, 1997; Collins, 2003; Hockenmaier and Steedman, 2002), but they are also useful for other problems both in and outside of NLP (Sakakibara et al., 1994; Guerra and Aloimonos, 2005; Lin et al., 2009).

Latent-variable models, on the other hand, are also a modeling tool of great importance in natural language processing. They have been used for many applications, including machine translation, natural language generation, question answering and semantics. Latent-variable models are centered around learning from incomplete data. This means that the underlying statistical model is defined over latent random variables that are not observed in the data used for learning. The latent variables explain correlations between observed random variables.

As such, it is not surprising that latent-variable models were combined with probabilistic grammars to train strong models that detect unobserved patterns in data, while retaining the interpretability and symbolic backbone contained within grammars. Latent-variable grammars have been mostly used for syntactic parsing, most prominently through the use of *latent-variable probabilistic context-free grammars* (L-PCFGs) – PCFGs that are augmented with latent states.

In this paper, we survey the use of L-PCFGs for syntactic parsing and other applications. We also survey the two main families of algorithms used for learning L-PCFGs: expectation-maximization algorithms and spectral algorithms. We analyze the latent state representations that one learns with L-PCFGs, and also describe extensions of L-PCFGs and related models (such as those that appear in deep learning).

2 Latent-Variable PCFGs

Latent-variable PCFGs (L-PCFGs) are PCFGs with additional latent states that decorate each nonterminal in each rule. While the backbone of an L-PCFG is simply a context-free grammar (because the decoration of the nonterminal with a latent state together with the nonterminal itself can be thought of as a new composite nonterminal), the use of L-PCFGs also implies a specific process of learning them from data: the decoration of the nonterminals with latent states is assumed to be absent from the sampled data from which we learn the model.

More formally, a latent-variable probabilistic context-free grammar (L-PCFG; in Chomsky normal form) is 5-tuple $(\mathcal{N}, \mathcal{R}, m, n, p)$ where:

- $\mathcal{N}$ is the set of nonterminal symbols in the

Proceedings of the 15th Meeting on the Mathematics of Language, pages 47–58,
London, UK, July 13–14, 2017. ©2017 Association for Computational Linguistics

grammar.

- $[m]$ is the set of possible hidden states where $[m]$ is defined as $\{1, \ldots, m\}$.

- $[n]$ is the set of possible words.

- $\mathcal{R}$ is a set of context-free rules in the form of $a \rightarrow b\,c$ or $a \rightarrow x$, where $a, b, c \in \mathcal{N}$ are nonterminals and $x \in [n]$ is a word.

- For all $a \rightarrow b \quad c \in \mathcal{R}$, $h_1, h_2, h_3 \in [m]$, we have a context-free rule $a(h_1) \rightarrow b(h_2) \quad c(h_3)$ and a parameter $p(a(h_1) \rightarrow b(h_2) \quad c(h_3) \mid a(h_1))$.

- For all $a \rightarrow x \in \mathcal{R}$, $h \in [m]$, we have a context-free rule $a(h) \rightarrow x$ and a parameter $p(a(h) \rightarrow x \mid a(h))$.

- For all $a \in \mathcal{N}$ and $h \in [m]$, we have a parameter $p(a(h))$ which is the probability of nonterminal a paired with hidden variable h being at the root of the tree.

The parameters satisfy the following normalization constraints:

$$\sum_{a,h} p(a(h)) = 1,$$

and for all $a \in \mathcal{N}$ and $h \in [m]$:

$$\sum_{a(h) \rightarrow b(h_2)\ c(h_3)} p(a(h) \rightarrow b(h_2)\ c(h_3))$$
$$+ \sum_{a(h) \rightarrow x} p(a(h) \rightarrow x) = 1$$

Note that for simplicity, we consider the case where every nonterminal symbol has the same number m of hidden state values. It is simple to generalize the method to allow different numbers of hidden states for each nonterminal. In both cases, though, the space of latent states for each nonterminal is separate. This means that latent state 2 for example for an NP has no relationship to latent state 2 for VP.

The generative story that such an L-PCFG model induces is similar to one of PCFGs. We begin with the top node of the derivation tree with its latent state by drawing a nonterminal and a latent state from $p(a(h))$, and then recursively draw rules in the form of $a(h) \rightarrow x$ or $a(h_1) \rightarrow$

$b(h_2)\ c(h_3)$ conditioned on the parent node, until all nodes at the bottom of the tree are terminal nodes from $[n]$. Figure 1 provides an example of such a derivation.

We refer to a rule of the form $a \rightarrow b \quad c$ as a "skeletal" rule from the "skeletal grammar." We note that while we provided the formulation of L-PCFGs in Chomsky normal form, there is a natural extension to arbitrary PCFGs, where rules of the form $a \rightarrow \alpha$ for $\alpha \in \mathcal{N}^*$ would be decorated with $|\alpha| + 1$ latent states, a state per nonterminal that appears in the rule.

The usual independence assumption made by a PCFG is that "inside" and "outside" trees (shown in Figure 2) are conditionally independent from each other if we know the node that connects them. More formally, if we denote a tree τ and a node β in that tree as a decomposition $\tau = (\tau_0, \tau_1, \beta)$ where τ_0 is the outside tree at node β and τ_1 is the inside tree at node β, then it holds that

$$p(\tau_1 \mid \tau_0, \beta) = p(\tau_1 \mid \beta).$$

This independence assumption of PCFGs can be quite restrictive in modeling syntax of language or in general. Essentially, no local context is modeled for syntactic categories, context which should be carried from different parts of the tree.

Latent-variable PCFGs weaken these independence assumption by introducing a latent state at every node in the tree. Now an inside and an outside tree are conditionally independent of each other given both the nonterminal node in the tree that connects these two subtrees *and* the latent state that is associated with that node.

In the rest of the paper, we denote a skeletal tree by τ and a full derivation with latent state assignments $h = (h_1, \ldots, h_N)$ by $\tau(h)$ where N is the number of nodes in τ.

3 Evolution of Latent-Variable PCFGs

The idea of decorating nonterminals with additional information, and breaking the statistical independence assumptions that PCFGs typically make, has a long history in natural language processing. Johnson (1998) introduced a variety of tree transformations on the Penn treebank, with an aim to improve the parsing accuracy of a PCFG extracted from that treebank. One of the transformations introduced was that of "parent annotation" where each nonterminal is annotated with its parent symbol.

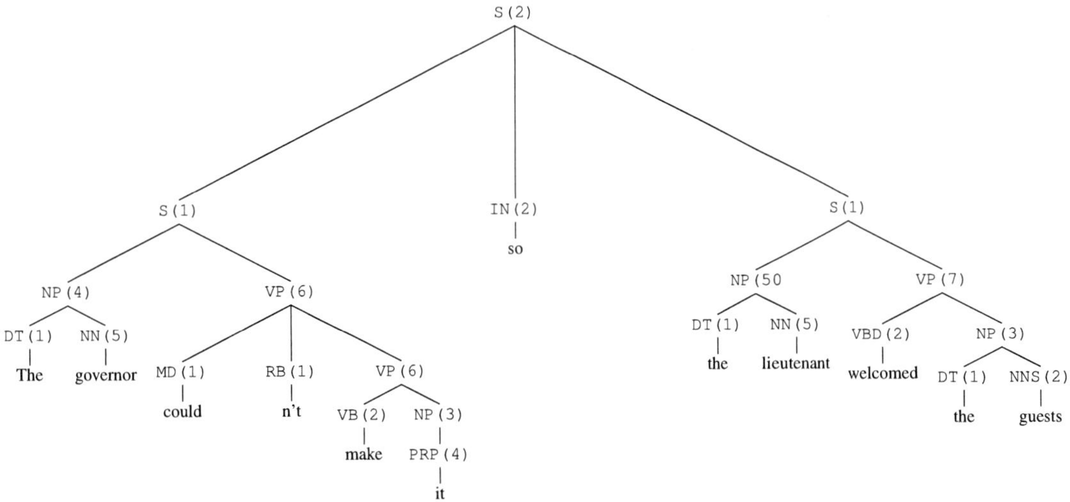

Figure 1: An example of a phrase-structure tree in English inspired by the Penn treebank (Marcus et al., 1993), potentially generated from an L-PCFG model. The indices next to each nonterminal in the tree denote the latent states associated with that node in the derivation. (Punctuation omitted.)

This idea is also strongly related to *lexicalized grammars*,[1] in which nonterminals are decorated with a head word propagated from the bottom of the tree. Most often, the head word is propagated using *head rules*, which decide which child of a given node is the head node based on linguistically motivated rules.

The context-free grammar formalism that corresponds to head lexicalization is bilexical grammar, which was introduced by Eisner and Satta (1999). Head lexicalization was used by Charniak (1997) and Collins (2003) to achieve state-of-the-art parsing results for English. Head lexicalization of grammars served as the basis for much of the subsequent parsing work.

Klein and Manning (2003) further built on the idea of tree transformations, and created linguistically motivated nonterminal refinements to parse the English treebank. Their work avoided the use of head lexicalization, but still produced a relatively high level of accuracy (though not state of the art) for parsing the Penn treebank. Some of the refinements they proposed generalize parent annotation (to higher order "vertical" Markovization) in a rather generic manner, but other refinements

rely heavily on linguistic knowledge of English, and as such they do not generalize to treebanks in other languages.

With all of this previous work, nonterminal refinement is central to the underlying parsing formalism. However, these decorations are extracted from the treebank by means of transformations on trees. It was not until the work by Matsuzaki et al. (2005) and Prescher (2005) that the decoration became a "latent annotation." At that point, L-PCFGs were performing close to state of the art in syntactic parsing. Dreyer and Eisner (2006) suggested a more complex training algorithm for L-PCFGs to improve their accuracy. Then, Petrov et al. (2006) further improved the parsing results of L-PCFGs to match state of the art and also suggested a coarse-to-fine approach that made parsing much more efficient (the asymptotic computational complexity of parsing with L-PCFGs, in their vanilla form, grows cubically with the number of latent states). It was at this time that many other researchers started to make use of L-PCFGs for a variety of syntax parsers in different languages, some of which are described in the rest of the paper.

4   Learning of L-PCFGs

Given that we assume with L-PCFGs that the latent states decorating the nonterminals are not ob-

---

[1]This is different than the lexicalization of grammars where all derivation rules are put into the lexicon, such as with combinatory categorial grammars (Steedman, 2000) or head-driven phrase structure grammars (Pollard and Sag, 1994).

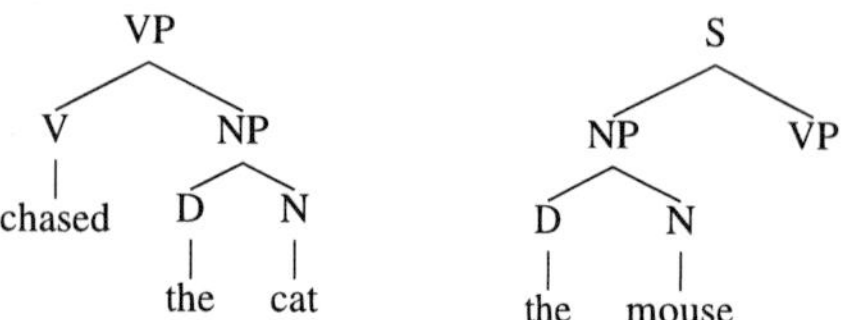

Figure 2: The inside tree (left) and outside tree (right) for the nonterminal VP in the parse tree (S (NP (D the) (N mouse)) (VP (V chased) (NP (D the) (N cat)))) for the sentence *the mouse chased the cat.*

served as part of the data, the learning problem posed by L-PCFGs is challenging and non-trivial. It requires learning the rule probabilities of an L-PCFG – such rules have latent states attached to them, as described in §2 – without knowing any of the latent states that are attached to the trees sampled from the underlying L-PCFG distribution.

Formally, we are given M training examples in the form of skeletal trees, $\tau^{(1)}, \ldots, \tau^{(M)}$ and our goal is to assign probabilities to the grammar rules with latent states. Implicitly, we assume that the skeletal grammar is given, and indeed it can be read off a treebank by extracting parents together with their immediate children (possibly after a "binarization" process).

4.1 Expectation-Maximization Learning

The first attempt at learning the rule probabilities of an L-PCFG from a treebank was carried out by Matsuzaki et al. (2005), who used the expectation-maximization (EM) algorithm. The EM algorithm is tailored to this specific problem of estimating L-PCFGs. It iterates between two steps, the E-step and the M-step. In the E-step it collects statistics about the latent state distributions for all nodes in all trees in the treebank (using a dynamic programming algorithm akin to CKY) and in the M-step it re-estimates the model based on these collected statistics. Before the first E-step is executed, the L-PCFG rule probabilities are initialized randomly.

While the EM algorithm is a highly influential algorithm that changed the way we reason about learning from incomplete data, it has some weaknesses, both practical and theoretical. The major practical weakness is that it requires running an "inference" algorithm multiple times over the data to collect the statistics in each E-step, which can be computationally expensive. This inference, as mentioned above, comes in the form of running a

dynamic programming algorithm that computes a marginal probability for each node in each tree of the form $\mu(a, h_k, i, j)$ where i and j are indices in the string the dynamic programming is run on, a is a nonterminal and h_k is a latent state for node k in a tree.

This marginal probability corresponds to:

$$\mu(a, h_k, i, j) = \sum_{\tau(h) \,:\, (a, h_k, i, j) \in \tau(h)} p(\tau(h))$$

i.e. the sum of all probabilities of full tree derivations (that include latent states) such that $a(h_k)$ spans the substring between index i and j. It is important to note that while the dynamic programming algorithm is akin to parsing algorithms such as CKY, its complexity is not cubic, but linear. This is true because during the E-step, the skeletal tree is fixed. We are not actually parsing a string, but instead marginalize the latent states in the fixed skeletal tree. When parsing a string with latent-variable PCFG (see §5), the complexity indeed becomes cubic because the skeletal tree needs to be inferred.

By estimating the parameters of an L-PCFG, the EM algorithm finds a maximum for the following objective function:

$$L\big(\tau^{(1)}, \ldots, \tau^{(M)}\big) = \sum_{i=1}^{M} \log \left(\sum_{h} p(\tau^{(i)}(h)) \right).$$
$$(1)$$

This is the *log-likelihood* of the observed data, marginalizing out the latent state from the underlying L-PCFG distribution. This log-likelihood function can be thought of as a measure of how well a specific set of parameters fit the data. The higher the log-likelihood is for a specific set of parameters, the more "likely" these parameters make the observed data.

Maximizing the log-likelihood function has deep roots in frequentist statistical theory, and in many cases, it can be shown that finding the global maximum of the log-likelihood will lead to "consistent" parameter estimates – meaning, as the amount of data increases, the parameters will get closer to the parameters from which the data was sampled.

The problem with the EM algorithm is that it only identifies a *local* maximum of this log-likelihood function, as the function has a "bumpy" surface that includes several maxima, some better than others. Finding the global one (i.e. the local maximum that gives the highest value to the log-likelihood) is a computationally difficult problem for L-PCFGs, and there are no known solutions that guarantee such identification in an efficient manner. As a result, much of the theory of maximum likelihood estimation that is mentioned above does not apply to the EM algorithm with L-PCFGs.

It is important to note that this issue with the EM algorithm and L-PCFGs is more of a theoretical concern than a practical concern. In practice, if the EM algorithm is initialized in the way specified by Matsuzaki et al. (2005), it converges to a local maximum that provides a relatively high parsing accuracy for syntactic parsing of English and other languages. In addition, the log-likelihood is not fully correlated with parsing performance (such as that measured by the PARSEVAL metric; Black et al., 1991) and therefore identifying the global maximum of the log-likelihood function does not guarantee optimal parsing performance.

Coarse-to-Fine Techniques Building on the expectation-maximization algorithm, Petrov et al. (2006) introduced a coarse-to-fine technique (Charniak and Johnson, 2005) for estimating L-PCFGs. This technique uses the EM algorithm as a subroutine. It first starts by running the EM algorithm with a small number of latent states for each nonterminal. It then works by successively "splitting" and "merging" nonterminals. In a split step, more latent states are added to the nonterminals (usually by multiplying the number of latent states associated with them by two). To avoid overfitting the model to the training data, this is accompanied by a merge step, in which latent states are merged together to decrease the number of latent states associated with each nonterminal.

After each split and merge step (which are done in "split-merge cycles"), EM is initialized using the last model and re-run again to obtain a new set of parameters for the split/merge grammar. As such, this coarse-to-fine technique aims to locally maximize the marginal log-likelihood as given in Eq. 1 while also controlling for model size (the number of latent variables associated with each nonterminal).

4.2 Spectral Learning

At their core, spectral algorithms exploit the conditional independence that L-PCFGs makes to extract the parameters with the latent states (Cohen et al., 2013, 2014). More specifically, L-PCFGs assume that an "inside" tree and an "outside" tree, shown in Figure 2 are conditionally independent of each other given the nonterminal and latent state that attaches them to each other. As such, the correlation between patterns in the inside tree and outside tree distributions dictate the identity of the latent states and their distribution. To identify such a correlation, one can extract the latent state parameters by building a co-occurrence matrix (or a cross-covariance matrix) of inside and outside trees (in skeletal form; these are represented by feature vectors over such trees; see below), and then apply singular value decomposition (SVD; Strang et al., 1993) on this matrix. This approach was originally introduced for hidden Markov models (Hsu et al., 2012) and has been used for other types of grammars and parsing formalisms as well (Bailly et al., 2010; Luque et al., 2012; Dhillon et al., 2012).

As mentioned above, the inside and outside trees are represented by *feature vectors* in the co-occurrence matrix. This means that the inside and outside trees are mapped to real vectors. This is a common way to reduce a structured object into a manageable mathematical object that can be statistically processed. In the case of spectral algorithms for parsing, the feature functions indicate local neighborhood surrounding the top node (for inside trees) or footer node (for outside trees). As such, these methods distill information that was previously used by approaches such as parent annotation, annotation with linguistic features or Markovization (see §3) into latent states. The EM algorithm, on the other hand, does not make any use of feature functions in the process of learning.

Another advantage of spectral algorithms over the expectation-maximization algorithm is that they provide a natural way to select the number of latent states for each nonterminal. The singular values of the inside-outside co-occurrence matrix offer a criterion to do that. Each singular value is associated with a latent state, and to retain a good approximation of this matrix with SVD, one needs to select only the largest singular values. The smaller ones can be removed from the SVD approximation.

That being said, Narayan and Cohen (2016) showed that the number of latent states can be further optimized with spectral algorithms by using coarse-to-fine techniques such as Petrov et al. (2006) used. This means that while the criterion above with top singular values is natural and easy to implement, further refinements can be made to improve it.

It is important to note that unlike spectral algorithms, the EM algorithm has an interpretation that is valid even when the data it is applied on is not generated from an L-PCFG in the family we are estimating from. It can be viewed as minimizing Kullback-Leibler (KL) divergence, a measure for distributional divergence, between the empirical distribution and the family of possible L-PCFGs from which a model is selected. To date, the theoretical guarantees of L-PCFGs with spectral algorithms require the assumption that the data is generated from an L-PCFG distribution. Still, the EM algorithm and spectral algorithms yield similar results on a variety of benchmarks for multilingual parsing, even for data that are clearly not sampled from an L-PCFG (as one might argue is true for most natural language data).

4.3 Other Learning Algorithms

Other scenarios and algorithms for learning L-PCFGs have been proposed. One such example is that of the work on self-training (McClosky et al., 2006) of L-PCFGs by Huang et al. (2010) and Huang and Harper (2009). With self-training, a parser is trained from seed annotated data (such as the Penn treebank), and then the parser learned is used to parse a large amount of unlabeled data. After that step, a new parser is learned from both the annotated data and the parsed data, as if the latter are gold-standard data.

Petrov (2010) exploited the fact that the EM algorithm is sensitive to its initialization point (and returns a different model in each execution) and estimated an L-PCFG multiple times from annotated data using a coarse-to-fine EM technique. Once all grammars were learned, they were combined in a product-of-experts style, and then they were used to parse unseen data. Similar experiments were done later with spectral algorithms, and showed that essentially using multiple grammars has the effect of regularization.

Petrov and Klein (2008) extended L-PCFGs to log-linear latent grammars – this means that the

rule weights learned are no longer constrained to be probabilities. Instead, the model has an additional normalization constant that ensures that it defines a probability of derivation trees even though the rule weights are no longer probabilities. They used *discriminative training* to estimate such L-PCFGs. Instead of finding a local maximum of the marginal log-likelihood as EM does, they find a local maximum for the *conditional* marginal log-likelihood.

One of the earlier training algorithms of L-PCFGs selectively refines nonterminals by using EM with annealing (Dreyer and Eisner, 2006). The authors slightly modify the L-PCFG model to pass features between nodes in the parse tree, motivated by prior linguistic work on feature passing. Finally, Stanojević and Sima'an (2015) used a refinement model for the induction of a Reordering Grammar for machine translation. They used the EM algorithm for estimating their model.

5 L-PCFGs for Multilingual Parsing

In this section, we discuss the use of L-PCFGs for syntactic parsing.

5.1 Parsing with L-PCFGs

Parsing with L-PCFGs usually entails finding a skeletal tree for a given string. While the latent variables assist in the modeling by adding contextual information to the derivation, they are not necessarily a target for prediction, and therefore we are interested in marginalizing them out during parsing.

Given a string w, this means that we are interested in finding the following tree:

$$\tau^* = \arg\max_\tau \sum_h p(\tau(h) \mid w).$$

The maximization of a sum of products in this form (the product originates in the probability $p(\tau(h) \mid w)$, which is proportional to a product of rule probabilities) is computationally intractable. As such, other approaches to parsing with L-PCFGs have been developed. The most common one used is based on *minimum Bayes risk decoding* (MBR; Goodman, 1996). With MBR, the maximization problem turns into maximizing the sum of the marginal probabilities of each node that appear in the tree. It is motivated by maximizing the recall of correct constituents in the predicted tree.

MBR can be quite expensive to run with a large number of latent states, as the dynamic programming algorithm that performs the parsing scales cubically as a function of the number of latent states. This is where coarse-to-fine techniques have an advantage – one can parse incrementally with coarser models until reaching the most refined model, at each point pruning the parsing chart with low probability items from the coarser model. Tensor decomposition can also be used to speed up L-PCFG parsing (Cohen and Collins, 2012).

The main application for L-PCFGs is multilingual syntactic parsing. In their early version (Matsuzaki et al., 2005), L-PCFGs were used to parse the English Penn treebank with some success – the results were not state of the art, but close to it. In subsequent work, Petrov et al. (2006) used coarse-to-fine techniques to further improve EM estimation of L-PCFGs. This led to the development of the Berkeley parser, which has given state-of-the-art results for English and other languages. Spectral algorithms also yield results which are close to state of the art in a multilingual setting.

Since their inception, L-PCFGs have been used for syntactic parsing in multiple studies for a variety of languages such as English, French, German, Chinese, Arabic and other morphologically rich languages (Candito et al., 2010; Attia et al., 2010; Green and Manning, 2010; Tounsi and Van Genabith, 2010; Goldberg and Elhadad, 2011; Dehdari et al., 2011; Björkelund et al., 2014; Zeng et al., 2014; Sun et al., 2014; Huang et al., 2014; Narayan and Cohen, 2016)

5.2 Interpretation for Latent States

One of the advantages of using latent variable models is that often the latent variables can be assigned an interpretation once they are inferred. This post-hoc interpretation can be revealing about linguistic patterns that are present in the data and are learned automatically.

To do this kind of analysis, we computed the marginals for each node in each tree in the Penn treebank after training a model with the spectral algorithm of Narayan and Cohen (2015). The results are available in the LPCFGVIEWER tool.[2]

<hr>

[2]Available in http://cohort.inf.ed.ac.uk/lpcfgviewer. The online tool includes analysis of other languages, including French, German, Hebrew, Hungarian, Korean, Polish, Swedish and Basque.

State	Frequent words
	IN (preposition)
0	of ×323
1	about ×248
2	than ×661, as ×648, because ×209
3	from ×313, at ×324
4	into ×178
5	over ×122
6	Under ×127
	DT (determiners)
0	These ×105
1	Some ×204
2	that ×190
3	both ×102
4	any ×613
5	the ×574
6	those ×247, all ×242
7	all ×105
8	another ×276, no ×211
	CD (numbers)
0	8 ×132
1	million ×451, billion ×248
	RB (adverb)
0	up ×175
1	as ×271
2	not ×490, n't ×2695
3	not ×236
4	only ×159
5	well ×129
	CC (conjunction)
0	But ×255
1	and ×101
2	and ×218
3	But ×196
4	or ×162
5	and ×478

Table 1: A list of part-of-speech tags and most frequent words associated with latent states that were learned for them using the algorithm of Narayan and Cohen (2015). The numbers next to each word indicate a strength (the number of times that word appeared with that POS tag and latent state in the Penn treebank).

There are a few observations that can be concluded when inspecting these results:

- **Lexicalization of closed-word tags:** For the closed-word part-of-speech (POS) tags, both the EM algorithm and the spectral algorithm associate a latent state with mostly a single word. For example, for determiners, there would be a latent state for "the" and for "a." Often words with different casing or contracted form joined in the same cluster. Table 1, for example, shows that latent state 2 for RB (adverbs) is associated with "not" and "n't."

- **Semantic clustering of phrases:** Consider the noun phrases in Table 2, which are the most frequently ones associated with each latent state in the learned L-PCFG model. We see that there is in certain cases semantic clustering of such noun phrases, in cases where such semantic clustering can be directed by syntactic information in the parse trees. For example, the first latent state of the noun phrases is mostly associated with dates in the form of month, day, year. The seventh latent state is mostly associated with dollar amount.
- **Dependence on domain:** It is clear from Table 2 and similar statistics for other nonterminals that the association of latent states with nonterminal phrases is highly dependent on the domain on which the L-PCFG was trained. The L-PCFG in Table 2 was trained on the Penn treebank, and as such, latent states are associated with financial terms that have some similarity (such as shares and currency).

When inspecting the phrases associated with specific latent states and nonterminals (in the LPCFGVIEWER tool), one might argue that there is a weak notion of substitutability that exists with L-PCFGs. By this, we are referring to the idea that phrases associated with an identical latent state with high probability are more likely to be substitutable in different contexts, not just syntactically, but also semantically. The latent state associated with dates (in noun phrases) in Table 1 is one example of that. This hypothesis remains to be proven empirically in a methodical way.

6 Extensions of L-PCFGs and Related Models

Extensions Natural generalizations of PCFGs, such as probabilistic linear-context free rewriting systems (LCFRS; Kallmeyer and Maier, 2010) and synchronous grammars can also be turned into probabilistic grammars with latent states. As long as the backbone structure of the skeletal grammar is of the form $a \to \alpha$ where α includes nonterminals in one form or the other, the nonterminals can be decorated with additional latent state information.

Work about using other grammar formalisms with latent states includes the work of Fowler and Penn (2010) who introduced latent states into a combinatory categorial grammar (CCG) for syntactic parsing, the work of Saluja et al. (2014), who generalized L-PCFGs to synchronous L-PCFGs and proposed to estimate them using both a spectral algorithm and EM for machine translation and the work of Louis and Cohen (2015) who modeled online forum topic structure by using LCFRS with latent states (the latent states corresponded to topics that need to be inferred from data). Models similar to L-PCFGs have been used for parsing with discontinuous elements (Nederhof and Yli-Jyrä, 2017). They have also been used to describe transition-based systems for dependency formalisms (Nederhof, 2016).

Related Models As mentioned above, the traditional L-PCFG parsing algorithm requires computing marginals for each node in the tree. They are computed using an inside-outside algorithm – an inside pass that works bottom up in the tree, similarly to the CKY algorithm, and an outside pass that works top down.

When computing marginals with the skeletal tree being fixed, the bottom up inside algorithm can be viewed as an algorithm that *propagates* vector representations of each node up in the tree by using tensor contraction. The general form of that algorithm is:

$$v_{\text{parent}} = F(v_{\text{lc}}, v_{\text{rc}}). \tag{2}$$

where $v_{\text{parent}}, v_{\text{lc}}, v_{\text{rc}} \in \mathbb{R}^m$ denoting vectors associated with a parent node, its left child and its right child, and $F \colon \mathbb{R}^m \times \mathbb{R}^m \to \mathbb{R}^m$ takes as input two children node vectors and returns the output node vector. In the case of L-PCFGs, F is a function that is associated with a rule $a \to b\ c$, $T^{a \to b\ c}$, where b is the nonterminal for the left child, c is the nonterminal for the right child and a is the parent nonterminal. $T^{a \to b\ c}$ is then defined as:

$$[T^{a \to b\ c}(v_{\text{lc}}, v_{\text{rc}})]_{h_1}$$
$$= \sum_{h_2, h_3} p(a(h_1) \to b(h_2)\ c(h_3) \mid a(h_1))[v_{\text{lc}}]_{h_2}[v_{\text{rc}}]_{h_3}.$$

0	”Aug. 30 , 1988”, ”Aug. 31 , 1987”, ”Dec. 31 , 1988”, ”Oct. 16 , 1996”, ”Oct. 1 , 1999”, ”Oct. 1 , 2019”, ”Nov. 8 , 1996”, ”Oct. 15 , 1999”, ”April 30 , 1988”, ”Nov. 8 , 1994”
1	”12,000 miles”, ”About 20,000 years”, “this year”, “A year”, “a year” ×7
2	“FROG-7B missiles , the bomber version of the An-12 , MiG-23BN high-altitude aircraft , MiG-29s , which can outfly Pakistan 's U.S.-built F16s ,”, “AMERICAN BUILDING MAINTENANCE INDUSTRIES Inc. , San Francisco , provider of maintenance services , annual revenue of $ 582 million , NYSE ,”, “DIASONICS INC. , South San Francisco , maker of magnetic resonance imaging equipment , annual sales of $ 281 million , Amex ,”, “EVEREX SYSTEMS INC. , Fremont , maker of personal computers and peripherals , annual sales of $ 377 million , OTC ,”, “ANTHEM ELECTRONICS INC. , San Jose , distributor of electronic parts , annual sales of about $ 300 million , NYSE ,”, “APPLIED MATERIALS INC. , Santa Clara , maker of computer-chip machine systems , annual sales of $ 490 million , OTC ,”
3	“James McCall , vice president , materials , at Battelle , a technology and management-research giant based in Columbus , Ohio”, “Frank Kline Jr. , partner in Lambda Funds , a Beverly Hills , Calif. , venture capital concern”, “Allen Hadhazy , senior analyst at the Institute for Econometric Research , Fort Lauderdale , Fla. , which publishes the New Issues newsletter on IPOs”, “a group of investment banks headed by First Boston Corp. and co-managed by Goldman , Sachs & Co. , Merrill Lynch Capital Markets , Morgan Stanley & Co. , and Salomon Brothers Inc”, “Charles J. O'Connell , deputy district director in Los Angeles of the California Department of Transportation , nicknamed Caltrans”, “Francis J. McNeil , who , as deputy assistant secretary of state for inter-American affairs , first ran across reports about Mr. Noriega in 1977”
4	“TREASURY BILLS : Results of the Monday , October 16 , 1989 , auction of short-term U.S. government bills , sold at a discount from face value in units of $ 10,000 to $ 1 million : 7.37 % 13 weeks ; 7.42 % 26 weeks .”, “California Health Facilities Financing Authority – $ 144.35 million of revenue bonds for Kaiser Permanente , due 19931999 , 2004 , 2008 , 2018 and 2019 , tentatively priced by a PaineWebber Inc. group to yield from 6.25 % in 1993 to 7.227 % in 2018 .”, “TREASURY BILLS : Results of the Monday , October 16 , 1989 , auction of short-term U.S. government bills , sold at a discount from face value in units of $ 10,000 to $ 1 million : 7.37 % 13 weeks ; 7.42 % 26 weeks .”, “Health Care Property Investors Inc. , offering of 2,250,000 shares of common stock , via Merrill Lynch Capital Markets , Alex . Brown & Sons Inc. and Dean Witter Reynolds Inc .”, ”SUN MICROSYSTEMS INC. , Mountain View , maker of desktop computers , annual sales $ 1.77 billion , OTC , no injured employees and very little damage to buildings .”
5	“$ 615,000 face amount”, “a share” × 7, “a revival”
6	“bonds due Nov. 2 , 1993 , with equity-purchase warrants”, “bonds due Nov. 8 , 1994 , with equity-purchase warrants”, “20,000 to 30,000 Soviet Central Asian KGB Border Guards”, “bonds due Nov. 1 , 1993 , with equity-purchase warrants”, “a van Gogh , a Monet , other paintings , furniture”, “3,111,000 common shares”, “offering of 2,250,000 shares of common stock”, “a Newark , N.J. , textile businessman”
7	“$ 1,150,000”, “166,900,000 shares”, “$ 124,732”, “$ 45,000”, “$ 1,500”, “$ 20,000”, “$ 342,122”, “$ 1,000”, “$ 3,000”, “S$ 500,000”
8	“$ 20,000 a year”, “$ 342,122 last year”, “as many as 60,000 additional tourists a day”, “$ 80,000 a year”, “1,000 flights a day”, “26,000 units next year”, “200,000 cars a year”, “$ 200 a share” ×3

Table 2: Examples of most likely phrases for the noun phrase category (NP) for a latent-variable PCFG extracted using the algorithm of Narayan and Cohen (2015) from the Penn treebank. Numbers next to the phrases indicate that the phrase appeared multiple times in the list.

This general formulation as in Eq. 2 gives rise to generalizations of other formulations of latent representations that are propagated in a (parse) tree structure. Perhaps the most related one to L-PCFGs is the recursive neural network of Socher et al. (2010). This recursive neural network propagates word vectors (Turian et al., 2010) from the bottom of an unlabeled tree all the way to its top using the function:

$$F(v_{\mathrm{lc}}, v_{\mathrm{rc}}) = \tanh(W[v_{\mathrm{lc}}, v_{\mathrm{rc}}] + b),$$

where $W \in \mathbb{R}^{m \times 2m}$ and $b \in \mathbb{R}^m$ are weight matrices and biases that are learned when training the neural network, $[u_1, u_2]$ denotes the concatenation of two vectors u_1 and u_2, and $\tanh \colon \mathbb{R}^m \to \mathbb{R}^m$ is a function that applies the hyperbolic tangent function coordinate-wise. Even more closely related to L-PCFGs is the further refinement of these recursive neural networks by Socher et al. (2013) where the weights in the neural network are parametrized by labels in the tree, corresponding to syntactic categories.

Similarly to a formulation of the inside tree as a vector propagation procedure (in Eq. 2), there is also a formulation for the outside algorithm (Cohen et al., 2014). Le and Zuidema (2014) also extended the recursive neural networks mentioned above to make use of the outside tree information.

Finally, it is also important to note that L-PCFGs are related to probabilistic regular tree grammars (PRTGs; Knight and Graehl, 2005) where the righthand side trees of the PRTG rules are of depth 1. With general PRTGs, the righthand side can be of arbitrary depth, where the leaf nodes of these trees correspond to latent states in the L-PCFG formulation above and the internal nodes of these trees correspond to interminal symbols in the L-PCFG formulation.

7 Conclusion

Latent-variable PCFGs are a flexible model for modeling syntax and other problems in NLP. Their backbone is a symbolic grammar, and as such they can be easily interpreted, while they are augmented with probabilities and latent states, allowing the modeler to reason under uncertainty.

We gave an overview of the parsing and learning algorithms used with L-PCFGs and described some natural extensions and related models.

Acknowledgments

I would like to thank Shashi Narayan, Nikos Papasarantopoulos and Giorgio Satta for useful feedback and comments. This work was supported by an EU H2020 grant (688139/H2020-ICT-2015; SUMMA) and a grant from Bloomberg.

References

M. Attia, J. Foster, D. Hogan, J. Le Roux, L. Tounsi, and J. Van Genabith. 2010. Handling unknown words in statistical latent-variable parsing models for Arabic, English and French. In *Proceedings of the NAACL-HLT 2010 First Workshop on Statistical Parsing of Morphologically-Rich Languages*.

R. Bailly, A. Habrard, and F. Denis. 2010. A spectral approach for probabilistic grammatical inference on trees. In *Proceedings of ALT*.

A. Björkelund, Ö. Çetinoğlu, A. Faleńska, R. Farkas, T. Müller, W. Seeker, and Z. Szántó. 2014. Introducing the IMS-Wrocław-Szeged-CIS entry at the SPMRL 2014 shared task: Reranking and morphosyntax meet unlabeled data. In *Proceedings of the First Joint Workshop on Statistical Parsing of Morphologically Rich Languages and Syntactic Analysis of Non-Canonical Languages*.

E. Black, S. Abney, D. Flickenger, C. Gdaniec, R. Grishman, P Harrison, D. Hindle, R. Ingria, F. Jelinek, J. Klavans, M. Liberman, M. Marcus, S. Roukos, B. Santorini, and T. Strzalkowski. 1991. A procedure for quantitatively comparing the syntactic coverage of English grammars. In *Proceedings of DARPA Workshop on Speech and Natural Language*.

M. Candito, J. Nivre, P. Denis, and E. H. Anguiano. 2010. Benchmarking of statistical dependency parsers for french. In *Proceedings of COLING*.

E. Charniak. 1997. Statistical parsing with a context-free grammar and word statistics. In *Proceedings of AAAI*.

E. Charniak and M. Johnson. 2005. Coarse-to-fine n-best parsing and maxent discriminative reranking. In *Proceedings of ACL*.

S. B. Cohen and M. Collins. 2012. Tensor decomposition for fast parsing with latent-variable PCFGs. In *Proceedings of NIPS*.

S. B. Cohen, K. Stratos, M. Collins, D. P. Foster, and L. Ungar. 2013. Experiments with spectral learning of latent-variable PCFGs. In *Proceedings of NAACL*.

S. B. Cohen, K. Stratos, M. Collins, D. P. Foster, and L. Ungar. 2014. Spectral learning of latent-variable PCFGs: Algorithms and sample complexity. *Journal of Machine Learning Research* .

M. Collins. 2003. Head-driven statistical models for natural language processing. *Computational Linguistics* 29:589–637.

J. Dehdari, L. Tounsi, and J. van Genabith. 2011. Morphological features for parsing morphologically-rich languages: A case of Arabic. In *Proceedings of the Second Workshop on Statistical Parsing of Morphologically Rich Languages*.

P. S. Dhillon, J. Rodu, M. Collins, D. P. Foster, and L. H. Ungar. 2012. Spectral dependency parsing with latent variables. In *Proceedings of CoNLL-EMNLP*.

M. Dreyer and J. Eisner. 2006. Better informed training of latent syntactic features. In *Proceedings of EMNLP*.

J. Eisner and G. Satta. 1999. Efficient parsing for bilexical context-free grammars and head automaton grammars. In *Proceedings of ACL*.

T. AD Fowler and G. Penn. 2010. Accurate context-free parsing with combinatory categorial grammar. In *Proceedings of ACL*.

Y. Goldberg and M. Elhadad. 2011. Joint Hebrew segmentation and parsing using a PCFG-LA lattice parser. In *Proceedings of ACL (short papers)*.

J. Goodman. 1996. Parsing algorithms and metrics. In *Proceedings of ACL*.

S. Green and C. D. Manning. 2010. Better Arabic parsing: Baselines, evaluations, and analysis. In *Proceedings of COLING*.

G. Guerra and Y. Aloimonos. 2005. Discovering a language for human activity. In *Proceedings of AAAI Workshop on Anticipation in Cognitive Systems*.

J. Hockenmaier and M. Steedman. 2002. Generative models for statistical parsing with combinatory categorial grammar. In *Proceedings of ACL*.

D. Hsu, S. M. Kakade, and T. Zhang. 2012. A spectral algorithm for learning hidden Markov models. *Journal of Computer and System Sciences* 78(5):1460–1480.

Q. Huang, L. He, D. F. Wong, and L. S. Chao. 2014. Chinese unknown word recognition for PCFG-LA parsing. *The Scientific World Journal* 2014.

Z. Huang and M. Harper. 2009. Self-training pcfg grammars with latent annotations across languages. In *Proceedings of EMNLP*.

Z. Huang, M. Harper, and S. Petrov. 2010. Self-training with products of latent variable grammars. In *Proceedings of EMNLP*.

M. Johnson. 1998. PCFG models of linguistic tree representations. *Computational Linguistics* 24(4):613–632.

L. Kallmeyer and W. Maier. 2010. Data-driven parsing with probabilistic linear context-free rewriting systems. In *Proceedings of COLING*.

D. Klein and C. D. Manning. 2003. Accurate unlexicalized parsing. In *Proceedings of ACL*.

K. Knight and J. Graehl. 2005. An overview of probabilistic tree transducers for natural language processing. In *Computational linguistics and intelligent text processing*, Springer, volume 3406 of *Lecture Notes in Computer Science*, pages 1–24.

P. Le and W. Zuidema. 2014. The inside-outside recursive neural network model for dependency parsing. In *Proceedings of EMNLP*.

L. Lin, T. Wu, J. Porway, and Z. Xu. 2009. A stochastic graph grammar for compositional object representation and recognition. *Pattern Recognition* 8.

A. Louis and S. B. Cohen. 2015. Conversation trees: A grammar model for topic structure in forums. In *Proceedings of EMNLP*.

F. M. Luque, A. Quattoni, B. Balle, and X. Carreras. 2012. Spectral learning for non-deterministic dependency parsing. In *Proceedings of EACL*.

M. P. Marcus, B. Santorini, and M. A. Marcinkiewicz. 1993. Building a large annotated corpus of English: The Penn treebank. *Computational Linguistics* 19:313–330.

T. Matsuzaki, Y. Miyao, and J. Tsujii. 2005. Probabilistic CFG with latent annotations. In *Proceedings of ACL*.

D. McClosky, E. Charniak, and M. Johnson. 2006. Effective self-training for parsing. In *Proceedings of HLT-NAACL*.

S. Narayan and S. B. Cohen. 2015. Diversity in spectral learning for natural language parsing. In *Proceedings of EMNLP*.

S. Narayan and S. B. Cohen. 2016. Optimizing spectral learning for parsing. In *Proceedings of ACL*.

M.-J. Nederhof. 2016. Transition-based dependency parsing as latent-variable constituent parsing. In *Proceedings of the SIGFSM Workshop on Statistical NLP and Weighted Automata*.

M.-J. Nederhof and A. Yli-Jyrä. 2017. A derivational model of discontinuous parsing. In *International Conference on Language and Automata Theory and Applications*. Springer, pages 299–310.

S. Petrov. 2010. Products of random latent variable grammars. In *Proceedings of NAACL*.

S. Petrov, L. Barrett, R. Thibaux, and D. Klein. 2006. Learning accurate, compact, and interpretable tree annotation. In *Proceedings of COLING-ACL*.

S. Petrov and D. Klein. 2008. Discriminative log-linear grammars with latent variables. In *Proceedings of NIPS*.

C. Pollard and I. A. Sag. 1994. *Head-driven phrase structure grammar*. University of Chicago Press.

D. Prescher. 2005. Inducing head-driven PCFGs with latent heads: Refining a tree-bank grammar for parsing. In *Proceedings of ECML*.

Y. Sakakibara, M. Brown, R. Hughey, S. Mian, K. Sjölander, R. C. Underwood, and D. Haussler. 1994. Stochastic context-free grammars for tRNA modeling. *Nucleic Acids Research* 22.

A. Saluja, C. Dyer, and S. B. Cohen. 2014. Latent-variable synchronous CFGs for hierarchical translation. In *Proceedings of EMNLP*.

R. Socher, J. Bauer, C. D. Manning, and A. Y. Ng. 2013. Parsing with compositional vector grammars. In *Proceedings of ACL*.

R. Socher, C. D. Manning, and A. Y. Ng. 2010. Learning continuous phrase representations and syntactic parsing with recursive neural networks. In *Proceedings of the NIPS Deep Learning and Unsupervised Feature Learning Workshop*.

M. Stanojević and K. Sima'an. 2015. Reordering grammar induction. In *Proceedings of EMNLP*.

M. Steedman. 2000. *The syntactic process*. MIT Press.

G. Strang. 1993. *Introduction to linear algebra*, volume 3. Wellesley-Cambridge Press Wellesley, MA.

L. Sun, J. Mielens, and J. Baldridge. 2014. Parsing low-resource languages using Gibbs sampling for PCFGs with latent annotations. In *Proceedings of EMNLP*.

L. Tounsi and J. Van Genabith. 2010. Arabic parsing using grammar transforms .

J. Turian, L. Ratinov, and Y. Bengio. 2010. Word representations: a simple and general method for semi-supervised learning. In *Proceedings of ACL*.

X. Zeng, D. F. Wong, L. S. Chao, I. Trancoso, L. He, and Q. Huang. 2014. Lexicon expansion for latent variable grammars. *Pattern Recognition Letters* 42:47–55.

A Proof-Theoretic Semantics for Transitive Verbs with an Implicit Object

Nissim Francez
Computer Science faculty, Technion
Haifa, Israel
`francez@cs.technion.ac.il`

Abstract

The paper presents a proof-theoretic semantics for sentences headed by transitive verbs allowing an unexpressed (implicit) object. Such sentences are shown to have the same (proof-theoretic) meaning as the same sentences with an explicit existentially quantified object something.

This semantics is contrasted with a model-theoretic semantics based on truth-conditions in models. The models used contain in their domain "filler" elements, that have an unclear extra-theoretic significance with an unclear ontological commitments. In contrast, the proof-theoretic meaning is appealing to formal (syntactic) resources that carry no ontological commitment.

Furthermore, the sameness of meaning is based on sameness of deductive role within a meaning-conferring proof-system, based on use.

1   Introduction

The purpose of this paper is to provide yet another argument for the benefits of *proof-theoretic semantics (PTS)*, when applied to natural language (NL), over the traditional *model-theoretic semantics (MTS)*, by focusing on what came be known as *unexpressed objects* of transitive verbs (also referred to as *implicit arguments*) as manifested, for example, by an intransitive use of a transitive verb, as in[1] (1(i)) and(1(ii)) below.

$$(i) \text{ John ate an apple} \quad (ii) \text{ John ate} \quad (1)$$

---

[1] All natural language expressions used in examples are displayed in the San Serif font, and are always mentioned, never used.

PTS is a theory of meaning serving as an alternative to the more traditional Model-Theoretic Semantics (MTS). While the latter identifies meaning with truth-conditions (in arbitrary models of a suitable form), the former identifies meaning with *canonical derivability conditions* in a meaning-conferring proof system. Those canonical derivability conditions provide *grounds for assertion*. For a presentation of the motivation for PTS and a discussion of its advantages over MTS, the reader is referred to the Introduction sections of Francez and Dyckhoff (2010) or Francez and Ben-Avi (2015), which present a PTS for extensive fragments of English. A full presentation can be found in Francez (2015b).

Turning to the unexpressed arguments, the essence of the MTS-based meaning is to interpret such arguments as *existentially quantified*. Hence, (1(ii)) above is seen as equivalent to

$$\text{John ate something} \quad (2)$$

with meaning expressed as usual by a (simply-typed) λ-term, here expressible in 1st-order logic (using simple types on variables and constants):

$$\exists x^e . \mathbf{eat}^{(e,(e,t))}(x, \mathbf{john}) \quad (3)$$

In Dowty (1982), (1(ii)) is seen as resulting from (1(i)) by a general procedure of *argument reduction*.

The structure of the rest of the paper is as follows. In Section 2 I briefly survey two recent approaches to obtaining the above mentioned truth-conditions (MTS-meaning). In Section 3 I briefly review the PTS of a minimal fragment of English in which the issue of unexpressed arguments arises. Fuller details about the PTS (for larger fragments) can be found in Francez and Dyckhoff (2010) or Francez and Ben-Avi (2015) and in part II of Francez (2015b). Section 4 presents the the

59

Proceedings of the 15th Meeting on the Mathematics of Language, pages 59–67,
London, UK, July 13–14, 2017. ©2017 Association for Computational Linguistics

PTS account of the unexpressed argument issue, and compares it to the reviewed MTS accounts. Section 5 ends with some conclusions.

2 A brief review of some model-theoretic approaches

The topic of unexpressed argument has a long history. I will mainly relate to Blom et al. (2012) (see there for references to previous work), and to Giorgolo and Asudeh (2012).

The common feature of those treatments of the unexpressed argument is imposing the sameness of meaning of an intransitive verb used with an implicit argument with the meaning of that verb used transitively with an existentially quantified object. For example,

$$[\![\text{Mary ate}]\!] = \exists x. [\![\text{Mary ate } x]\!] \qquad (4)$$

The difference between Blom et al. (2012) and Giorgolo and Asudeh (2012) is the formalism used for composing meanings, a difference immaterial for my purpose. I consider here only the issue of meaning of such constructs, and not with the syntax-semantics interface used to derive those meanings compositionally, so I will take Blom et al. (2012) as the representative of the MTS to implicit arguments.. A completely different model-theoretic approach is presented in Carlson (1984), appealing to models with (Neo Davidsonian) events and event modifiers. See Blom et al. (2012) for a discussion of this kind of MTS.

The main features of the proposed analysis are the following (expressed in a somewhat different notation).

1. The standard domains D_τ of interpretation of types τ in (simply-typed) λ-calculi are extended with extra elements, designated generically as '$*_\tau$', called the *filler value* of the domain of τ. The type-system is extended so that for every type τ there is a corresponding extended type $\tau^\emptyset$ accommodating the default value so that $D_{\tau^\emptyset} = D_\tau \cup \{*_\tau\}$ (where the union is disjoint).

   See below the qualms about this solution.

2. The lexical meaning of verbs licensing unexpressed objects is expressed via a function of

the (notationally slightly modified) form

$$\mathbf{option}(x_{\tau_1}^o, F_{(\tau_1, \tau_2)}, d_{\tau_2})$$
$$=_{df.} \begin{cases} d & x = *_{\tau_1} \\ F(x) & \text{otherwise} \end{cases} \qquad (5)$$

Here d_{τ_2} is the *default value* of type τ_2, chosen according to a global strategy: an existential closure of the word in the lexical semantic value of which this particular use of **option** is embedded. The **option** operator is of type $\tau_1^o \times (\tau_1, \tau_2) \times \tau_2$. The function $\lambda x.\mathbf{option}(x, F, d)$ is therefore of type (τ_1^o, τ_2).

Example 2.1 *Consider the lexical semantic value associated[2] with the verb* ate, *assuming it licensed an implicit object.*

$$\lambda y_{e^o} \lambda x_e. \mathbf{option}(\; \begin{array}{c} y_{e^o}, \\ \lambda u_e. ate_{(e,(e,t))}(u_e)(y_e), \\ \exists x_e. ate(x)(y) \end{array}) \qquad (6)$$

Now consider the two cases for ate.

explicit object: Mary ate an apple

   See Figure 1:

implicit object: Mary ate

   See Figure 2:

The extra-theoretic interpretation of such "filler elements" $*_\tau \in D_{\tau^o}$ is non-obvious, as is the presence of such elements in lexical meaning assignments of NL words. They cary an ontological commitment which is left unexplained.

As an example of one problematic issue involving filler elements, consider *selectional restrictions*. Typically, the are ignored, but there are ways of incorporating them into the type-based analysis above. Suppose one wants to admit

      Mary ate an apple

but to exclude

      Mary ate a chair

Both are admissible by the above analysis. One way of imposing selectional restrictions is by refining the type e, to include a sub-type $e_{addible}$,

---

[2]I ignore here the finer points involved in implementing this construction in *abstract categorial grammar*, orthogonal to my main concern.

$$\lambda y_{e^o} \lambda x_e.\mathbf{option}(y_{e^o}, \lambda u_e.ate_{(e,(e,t))}(u_e)(y_e), \exists x_e.ate_{(e,(e,t)}(x)(y))([\![an\ apple]\!])([\![Mary]\!])$$
$$= \lambda x_e.\mathbf{option}([\![an\ apple]\!], \lambda u_e.ate_{(e,(e,t))}(u_e)(y_e)([\![an\ apple]\!])([\![Mary]\!]), \exists x_e.ate_{(e,(e,t))}(x)(y))$$
$$= \lambda u.ate(u)([\![an\ apple]\!])([\![Mary]\!])$$
$$= ate([\![an\ apple]\!])([\![Mary]\!])$$
(7)

The second step is due to $[\![an\ apple]\!] \neq *_e$.

Figure 1: $[\![$Mary ate an apple$]\!]$

$$\lambda y_{e^o} \lambda x_e.\mathbf{option}(y_{e^o}, \lambda u_e.ate_{(e,(e,t))}(u_e)(y_e), \exists x_e.ate_{(e,(e,t)}(x)(y)))(*_e)([\![Mary]\!])$$
$$= \lambda x_e.\mathbf{option}(*_e, \lambda u_e.ate_{(e,(e,t))}(u_e)(*_e)([\![Mary]\!]), \exists x_e.ate(x)(y))([\![Mary]\!])$$
$$= \exists x.ate(x)(y)([\![Mary]\!])$$
$$= \exists x.ate(x)([\![Mary]\!])$$
(8)

The second step is due to $*_e = *_e$.

Figure 2: $[\![$Mary ate$]\!]$

---

and requiring the object of **ate** to be of that sub-type (see Ben-Avi and Francez (2004)).

Is $*_e$ of type $e_{addible}$? One way out is to include separate filler values for each sub-type of e, e.g., $*_{e_{addible}}$. But this cause a proliferation of those intelligible values as more selectional restrictions are imposed, for example, admitting

every girl smiled

but excluding

every chair smiled

by introducing a sub-type $e_{animate}$ of e and requiring the subject of **smile** to be of this sub-type..

Populating models with unintelligible entities is typical to MTS in general. One of the main advantages of PTS is avoiding models an unintelligible entities with unclear ontological commitments, dealing only with syntactic entities, artefacts of meaning-conferring proof-systems.

3 A Proof-Theoretic Semantics for a fragment of English

3.1 The language fragment

I briefly present in this section the core of a fragment E_0^+ of English. The core fragment consists of sentences with intransitive and transitive verbs, and determiner phrases with a (count, singular) noun and a determiner. For simplicity, I consider here only the positive determiners **every** and **some**. The inclusion of negative determiners such as **no** or **at most three** involve a certain complication (see Francez and Ben-Avi (2015)) that is orthogonal to the current issue. In addition, there is the copula **isa**. For technical reasons (made clear in Francez and Dyckhoff (2010)) I avoid the use of proper names, and use quantified subjects instead. Some typical sentences are listed below.

every/some girl smiles   (9)

every/some girl loves every/some boy   (10)

A typical sentence with an unexpressed argument to be considered is

every girl ate   (11)

I refer to expressions such as **every girl**, **some boy** as *determiner phrases* (*dps*).

3.2 The meaning-conferring proof-system

The *PTS* is based on a meaning-conferring natural-deduction-proof system N_0^+ with introduction rules and elimination rules (I/E-rules) (see Figure 3). The proof-system is formulated over the language L_0^+, slightly extending E_0^+. Meta-variables

61

X schematise nouns, P – intransitive verbs and R - transitive verbs. Meta-variable S ranges over sentences, and boldface lower-case **j**, **k**, etc., range over $\mathcal{P}$, a denumerable set of *(individual) parameters*, artefacts of the proof-system (not used to make assertions). Syntactically, a parameter in L_0^+ is also regarded as a *dp*. If a parameter occurs in S in some position, I refer to S as a *pseudo-sentence*, and if *all dp*s in S are parameters, the pseudo-sentence S is *ground*. The ground pseudo-sentences play the role of atomic sentences, and their meaning is assumed *given*, externally to the ND proof-system.

The original full system takes into account also *relative clauses*, due to presence of which some scope-related indexing is essential, as there may be more than just two scope levels. For the purpose of dealing with unexpressed arguments, it is enough to have simple sentences, with just two scope-levels, for the subject and object *dp*s. Furthermore, the general sentential variable S can be instantiated by $P(-)$, $R(-,-)$ and $R^u(-)$ for simple sentences headed by intransitive, transitive verbs and transitive verbs allowing intransitive use (lexically specified), respectively. Accordingly, I consider here a simplification of the original proof-system, not dealing with quantifier scope ambiguity. For any *dp*-expression D having a quantifier, I use the notation $S[(D)]$ to refer to a sentence S having a designated position filled by D.

For example, $S[(\text{every } X)]$ refers to a sentence S with a designated occurrence of every X. I use the conventions that within a rule, both $S[D_1], S[D_2]$ refer to the *same* designated position in S. I use Γ, S for the context extending Γ with sentence S.

Derivations (tree shaped), ranged over by $\mathcal{D}$ (possibly indexed) are defined recursively as usual. When presenting example derivations, the context Γ is left implicit, and the notation $[\cdots]_i$ indicates an assumption *discharged* by an application of a rule.

The following is a convenient *derived E*-rule, that can be used to shorten derivations.

$$\frac{\Gamma : S[(\text{every } X)] \quad \Gamma : \mathbf{j} \text{ isa } X}{\Gamma : S[\mathbf{j}]} \ (e\hat{E}) \qquad (12)$$

For example,

$$\frac{\Gamma : \text{every girl smiled} \quad \Gamma : \mathbf{j} \text{ isa girl}}{\Gamma : \mathbf{j} \text{ smiled}} \ (e\hat{E})$$

3.3 Canonical derivations and meaning

Consider an arbitrary natural-deduction proof-system intended as being meaning-conferring for the operators in some object language.

Definition 3.1 *An $\mathcal{N}$-derivation $\mathcal{D}$ for $\vdash_{\mathcal{N}} \Gamma : \psi$ (for a compound ψ) is* canonical *iff it satisfies one of the following two conditions.*

- *The last rule applied in $\mathcal{D}$ is an I-rule (for the main operator of ψ).*

- *The last rule applied in $\mathcal{D}$ is an assumption-discharging E-rule[3] the major premise of which is some φ in Γ, and its encompassed sub-derivations $\mathcal{D}_1, \cdots, \mathcal{D}_n$ (of the minor premises) are all canonical derivations of ψ.*

- *Denote by $[\![S]\!]_\Gamma^c$ the (possibly empty) collection of canonical derivations of S from Γ.*

For Γ empty, the definition reduces to that of a traditional canonical proof. Note the recursion involved in this definition. The important observation regarding this recursion is that it always terminates via the first clause, namely by an application of an I-rule. I refer to such an application of an I-rule (ending a recursive path in the definition) as an *essential* application, the outcome of which is *propagated* throughout the canonical derivation by applications of the assumption-discharging E-rules. Note that each sub-derivation $\mathcal{D}_i$ may start with an essential application of an I-rule, thus having "parallel" essential applications of that rule. I refer to a sequence of occurrences of ψ in a canonical $\mathcal{D}$ starting in the conclusion of an essential application of an I-rule and ending in the conclusion of $\mathcal{D}$ as a *propagation chain* of ψ in $\mathcal{D}$.

The major observation here is that the conclusion ψ of a canonical derivation $\mathcal{D}$ from open assumptions, ending a propagation chain, was:

1. inferred (possibly more than once) by the *essential applications* of the I-rule of its main operator, and

2. *propagated* through applications of assumption-discharging E-rules[4] to its position as the final conclusion of $\mathcal{D}$.

---

[3] General elimination rules (GE) have this structure, hence their association with permuting conversions during normalisation. See more on this in Negri and von Plato (2001).

[4] In propositional intuitionistic logic, say with Gentzen's NJ ND-system, this happens with disjunction.

$$\overline{\Gamma, S : S} \ (Ax)$$

$$\frac{\Gamma, \mathbf{j} \text{ isa } X : S[\mathbf{j}]}{\Gamma : S[(\text{every } X)]} \ (eI) \qquad \frac{\Gamma : \mathbf{j} \text{ isa } X \quad \Gamma : S[\mathbf{j}]}{\Gamma \vdash S[(\text{some } X)]} \ (sI)$$

$$\frac{\Gamma : S[(\text{every } X)] \quad \Gamma : \mathbf{j} \text{ isa } X \quad \Gamma, S[\mathbf{j}] : S'}{\Gamma : S'} \ (eE) \qquad \frac{\Gamma : S[(\text{some } X)] \quad \Gamma, \mathbf{j} \text{ isa } X, S[\mathbf{j}] : S'}{\Gamma : S'} \ (sE)$$

where $\mathbf{j}$ is *fresh* in (eI), and (sE).

Figure 3: The simplified rules for N_0^+

The *(reified) meaning* of a sentence S (in the considered object language), denoted by $[\![S]\!]$, is defined by

Definition 3.2 (reified meaning)

$$[\![S]\!] =^{df.} \lambda \Gamma . [\![S]\!]_\Gamma^c \tag{13}$$

This is a very fine-granularity notion of meaning, see Francez (2014). A natural coarsening is obtained by considering *grounds of assertion.*

Definition 3.3 (grounds for assertion)

$$GA[\![S]\!] = \{\Gamma \mid \vdash^c \Gamma : S\} \tag{14}$$

Thus, any Γ that canonically derives S serves as grounds for assertion of S. This is naturally extended to $GA[\![\Gamma]\!]$, the grounds of asserting all $S' \in \Gamma$. For the methodological role of this concept in the theory of meaning adhered to by PTS, see Dummett (1993).

By appealing to equality of grounds, we get an equivalence relation coarser than meaning as defined above, inducing a natural 'sameness of meaning' relation.

Definition 3.4 (equi-groundedness) S_1 *and* S_2 *are* equi-grounded*, denoted by* $S_1 \equiv_{GA} S_2$*, iff* $GA[\![S_1]\!] = GA[\![S_2]\!]$.

Grounds of assertion lead (see Francez (2015a)) to the following notion of *proof-theoretic consequence*, that captures the pre-theoretic entailment.

Definition 3.5 (proof-theoretic consequences)
S *is a* proof-theoretic consequence *of* Γ $(\Gamma \Vdash S)$ *iff* $GA[\![\Gamma]\!] \subseteq GA[\![S]\!]$.
Thus, proof-theoretic consequences is based on *grounds propagation*: every grounds for collectively asserting all of Γ are already grounds for asserting S.

4 Proof-theoretic semantics for unexpressed arguments

In this section, I extend the proof-system with I/E-rules for the sentences allowing (lexically licensed) intransitive uses of transitive verbs, underlying the proof-theoretic account of their meaning.

The main ideas on which the PTS for implicit arguments is based is the following:

- *Sameness of meaning is based on sameness of I-rules, implying sameness of canonical derivations.*

- *The reduction of an argument is reflected proof-theoretically by omission of a premise (in I-rules).*

4.1 Adding something

As the alert reader might have noticed, all the *nps* in E_0^+ contain an explicit noun-component; for example, some girl, every boy. I will refer to such *nps* as *nominally qualified*. Indeed, typical NL-quantification is what is known as *restricted quantification*. However, for the sake of the specification of the meaning of sentences with an implicit argument, what is needed[5] is to augment the language fragment with an *np* expressing *unrestricted existential quantification* in the form of *something*, lacking a nominal qualification.

The nominally-unqualified existential quantification in object position is governed by the fol-

---

[5]Another alternative is to consider a *universal noun* thing, and use it as the noun-qualifier. I consider this less elegant in the context of PTS.

lowing I/E-rules[6].

$$\frac{\Gamma : \mathbf{j}\, R\, \mathbf{k}}{\Gamma : \mathbf{j}\, R\, \mathsf{something}}\ (usI)$$

$$\frac{\Gamma : \mathbf{j}\, R\, \mathsf{something} \quad \Gamma, \mathbf{j}\, R\, \mathbf{k} : S'}{\Gamma : S'}\ (usE) \quad \mathbf{k}\ \mathsf{fresh.}$$

$$(15)$$

The difference between those two I/E-rules and the I/E-rules for nominally-qualified *nps* ((sI) and (sE) below) is the absence of a nominally-qualifying premise $\mathbf{k}$ isa X (for some noun X) in the I-rule, and the absence of a corresponding discharged assumption in the E-rule.

For example,

$$\frac{\Gamma : \mathbf{j}\, \mathsf{ate}\, \mathbf{k}}{\Gamma : \mathbf{j}\, \mathsf{ate}\, \mathsf{something}}\ (usI)$$

$$\frac{\Gamma : \mathbf{j}\, \mathsf{ate}\, \mathsf{something} \quad \Gamma, \mathbf{j}\, \mathsf{ate}\, \mathbf{k} : S'}{\Gamma : S'}\ (usE) \quad \mathbf{k}\ \mathsf{fresh}$$

4.2 Sentential meanings

In contrast to the view of argument reduction (mentioned above), I present *separate* I-rules for the transitive and intransitive uses of sentences with R^u-verbs; furthermore, the *same I-rule* serves *both* for the introduction for sentences with R^u-verbs with an unexpressed object *and for* the corresponding R with an explicit object *something*. This establishes the *sameness of meaning* of pairs of sentences like (1(ii)) and (2), where this sameness of meaning is based on both sentences having the same grounds of assertion.

Transitive verb with an expressed object:
These are the usual rules (sI) and (sE), simplified to objects only, of simple sentences.

$$\frac{\Gamma : \mathbf{k}\ \mathsf{isa}\ X \quad \Gamma : \mathbf{j}\, R\, \mathbf{k}}{\Gamma : \mathbf{j}\, R\, \mathsf{some}\ X}\ (sI)$$

$$\frac{\Gamma : \mathbf{j}\, R\, \mathsf{some}\ X \quad \Gamma, \mathbf{k}\ \mathsf{isa}\ X, \mathbf{j}\, R\, \mathbf{k} : S'}{\Gamma : S'}\ (sE)$$
$$\mathbf{k}\ \mathsf{fresh}$$

$$(16)$$

---

[6]Similar rules govern $\mathsf{something}$ in a subject position, but they do not matter here and are omitted.

Transitive verb with an unexpressed object:

$$\frac{\Gamma : \mathbf{j}\, R^u\, \mathbf{k}}{\Gamma : \mathbf{j}\, R^u}\ (uI)$$

$$\frac{\Gamma : \mathbf{j}\, R^u \quad \Gamma, \mathbf{j}\, R^u\, \mathbf{k} : S'}{\Gamma : S'}\ (uE) \quad \mathbf{k}\ \mathsf{fresh}$$

$$(17)$$

Here only a sentence with an R^u-verb can be used. Note the "missing" premise $\mathbf{k}$ isa X (for some X), leaving the object nominally-unqualified, hence omit-able. It is crucial to note that the subject of R^u *has to be a parameter*, assuring the meaning-identity with the explicit existentially case below. The parameter $\mathbf{j}$ is a basis for introduction of a universally quantified subject with a higher scoping quantifier.

For example, with $R^u = $ ate (with contexts omitted):

$$\frac{\mathbf{j}\ \mathsf{ate}\ \mathbf{k}}{\mathbf{j}\ \mathsf{ate}}\ (uI) \qquad \frac{\mathbf{j}\ \mathsf{ate} \quad \mathbf{j}\ \mathsf{ate}\ \mathbf{k} : S'}{\Gamma : S'}\ (uE) \quad \mathbf{k}\ \mathsf{fresh}$$

$$(18)$$

Note that no similar (uI)-application can be used to derive, for example, $\mathbf{j}$ smiled, under the assumption that smiled is a proper intransitive verb.

Example 4.2 *As an example, I show that*

$$\vdash \mathsf{every\ girl\ ate\ some\ apple}\ : \mathsf{every\ girl\ ate} \quad (19)$$

The derivation is shown in Figure 4. So, the "essence" of the derivation is not using the premise $\mathbf{k}$ isa apple, later discharged by the application of (sE), in deriving $\mathbf{j}$ ate.

4.3 Sameness of meaning

The following proposition, expressing the equality of meaning between intransitive R^u-headed pseudo-sentence and explicitly existentially-quantified R-headed pseudo-sentences is a simple consequence of those rules.

Proposition 4.1

$$[\![\mathbf{j}\, R^u]\!] = [\![\mathbf{j}\, R\, \mathsf{something}]\!] \quad (21)$$

and in particular

$$\mathbf{j}\, R^u \equiv_{GA} \mathbf{j}\, R\, \mathsf{something} \quad (22)$$

Proof: immediate, as there is a one-one correspondence between canonical derivations on both

$$\frac{[\mathbf{j}\ \text{isa girl}]_1 \quad \text{every girl ate some apple}}{\mathbf{j}\ \text{ate some apple}}\ (e\hat{E}) \qquad \frac{[\mathbf{j}\ \text{ate k}]_2, [\mathbf{k}\ \text{isa apple}]_3}{\mathbf{j}\ \text{ate}}\ (uI)$$

$$\frac{\mathbf{j}\ \text{ate}}{\text{every girl ate}}\ (eI^1) \qquad\qquad (sE^{2,3}) \qquad\qquad (20)$$

Figure 4: Derivation of **every girl ate**

sides of the equality, because the premises of the respective *I*-rules are the same.

A consequence of proposition 4.1 is the following theorem[7], establishing that the sameness of meaning for ground pseudo-sentences with R^u-verbs extends to a congruence of sameness of meaning for arbitrary sentences with R^u-verbs in the fragment (namely, arbitrary quantified *nps* as subject). As we have here only two possible subjects, we state the theorem for both, explicitly.

Theorem 4.1

$$[\![\text{every}\ X\ R^u]\!] = [\![\text{every}\ X\ R\ \text{something}]\!] \quad (23)$$

$$[\![\text{some}\ X\ R^u]\!] = [\![\text{some}\ X\ R\ \text{something}]\!] \quad (24)$$

Proof: Suppose $\mathcal{D}$ is a canonical derivation of every $X\ R^u$. Then, omitting contexts, $\mathcal{D}$ has to have a subderivation of the following form.

$$\frac{\begin{array}{c}[\mathbf{j}\ \text{isa}\ X]_i \\ \mathcal{D}' \\ \mathbf{j}\ R^u\end{array}}{\text{every}\ X\ R^u}\ (eI^i) \qquad (25)$$

There are two sub-cases to consider, depending whether $\mathcal{D}'$ itself is canonical or not.

1. First, assume $\mathcal{D}'$ is canonical. By proposition 4.1, $\mathcal{D}'$ is also a canonical derivation of

$$\begin{array}{c}[\mathbf{j}\ \text{isa}\ X]^i \\ \mathcal{D}' \\ \mathbf{j}\ R\ \text{something}\end{array}$$

And, hence,

$$\frac{\begin{array}{c}[\mathbf{j}\ \text{isa}\ X]_i \\ \mathcal{D}' \\ \mathbf{j}\ R\ \text{something}\end{array}}{\text{every}\ X\ R\ \text{something}}\ (eI^i) \qquad (26)$$

is a canonical derivation, establishing

$$[\![\text{every}\ X\ R^u]\!] \subseteq [\![\text{every}\ X\ R\ \text{something}]\!] \qquad (27)$$

Showing

$$[\![\text{every}\ X\ R\ \text{something}]\!] \subseteq [\![\text{every}\ X\ R^u]\!] \qquad (28)$$

is similar.

2. Next, suppose $\mathcal{D}'$ is not canonical. Thus, the conclusion $\mathbf{j}\ R^u$ was obtained by an application of an *E*-rule. In the small fragment considered here, there are two possible *E*-rules that could yield this consequence.

   (a) The one possibility is the $(e\hat{E})$-rule. This means that $\mathcal{D}'$ must have the following form.

$$\frac{\begin{array}{cc}\mathcal{D}'_1 & \mathcal{D}'_2 \\ \text{every}\ Y\ R^u & \mathbf{j}\ \text{isa}\ Y\end{array}}{\mathbf{j}\ R^u}\ (e\hat{E}^j)$$

   In the subderivation $\mathcal{D}'_1$, the conclusion must have been introduced by the (eI)-rule. So this premise looks like

$$\frac{\begin{array}{c}[\mathbf{k}\ \text{isa}\ Y]_j \\ \mathcal{D}''_1 \\ \mathbf{k}\ R^u\end{array}}{\text{every}\ Y\ R^u}\ (eI^j)$$

   Since $\mathcal{D}''_1$ is smaller than $\mathcal{D}$, by induction it follows that

$$\frac{\begin{array}{c}[\mathbf{k}\ \text{isa}\ Y]_j \\ \mathcal{D}''_1 \\ \mathbf{k}\ R\ \text{something}\end{array}}{\text{every}\ Y\ R\ \text{something}}\ (eI^j)$$

   from which we get the required derivation

$$\frac{\begin{array}{cc}\mathcal{D}'_1 & \mathcal{D}'_2 \\ \text{every}\ Y\ R\ \text{something} & \mathbf{j}\ \text{isa}\ Y\end{array}}{\mathbf{j}\ R\ \text{something}}\ (e\hat{E}^j)$$

   The other direction is similar.

---

[7]Note that neither Blom et al. (2012) nor Giorgolo and Asudeh (2012) relate to this issue, discussing only sentences with proper names as subjects.

(b) The other possibility is that the relevant E-rules are (sE) and (uE) (for a subject some X). The details are similar to the previous case and omitted.

This proof could be extended for subjects with other determiners, e.g. at least seven girls ate, having the same meaning as at least seven girls ate something, not in the current fragment, though.

There is an interesting observation arising from the proof-theoretic meaning device above. Consider sentences with *intensional* transitive verb (ITVs) with an indefinite object np, like

$$\text{Mary seeks/looks for a secretary} \qquad (29)$$

Such sentences are known to have a *notional* reading (unspecific) of their object. Thus, (29) has a reading by which Mary looks for *any* secretary.

Such sentences have no equivalent counterpart with an unexpressed object.

$$(*) \text{ Mary seeks/looks for} \qquad (30)$$

The explanation for the impossibility of the object omission can be found in the proof-theoretic semantics for ITVs provided in Francez (2016). Under this semantics, the objects of ITVs are *different kind of parameters* than the individual parameters, for which the omission rules present here do not apply. This shows the power of syntactic methods, introducing formal distinctions not carrying any ontological commitments, unlike the need to populate models with entities of unclear nature as in model-theoretic semantics.

5 Conclusions

In this paper I have presented a proof-theoretic semantics for sentences headed by transitive verbs allowing an unexpressed (implicit) object. Such sentences are shown to have the same (proof-theoretic) meaning as the same sentences with an explicit existentially quantified object something.

This semantics is contrasted with a model-theoretic semantics based on truth-conditions in models. The models used contain in their domain "filler" elements, that have an unclear extra-theoretic significance with an unclear ontological commitments. In contrast, the proof-theoretic meaning is appealing to formal (syntactic) resources that carry no ontological commitment.

Furthermore, the sameness of meaning is based on sameness of deductive role within a meaning-conferring proof-system, based on use.

References

Gilad Ben-Avi and Nissim Francez. 2004. Categorial grammar with ontology-refined types. In *Categorial Grammars 2004: An efficient tool for Natural Language Processing*. Montpellier, France.

Chris Blom, Philippe de Groote, Yoad Winter, and Joost Zwarts. 2012. Implicit arguments: event modification or option type categories. In Maria Aloni, Vadim Kimmelman, Floris Roelofsen, Galit W. Sasson, Katrin Schulz, and Matthijs Westera, editors, *Proceedings of the 2011 Amsterdam Colloquium*. Springer, LNCS 7218, pages 240–2501.

Greg N. Carlson. 1984. Thematic roles and their role in semantic interpretation. *Linguistics* 22:259 – 279.

David Dowty. 1982. Grammatical relations and montague grammar. In Pauline I. Jacobson and Geoffrey K. Pullum, editors, *The Nature of Syntactic Representation*, Reidel, Dordrecht, pages 79–130.

Michael Dummett. 1993. *The Logical Basis of Metaphysics*. Harvard University Press, Cambridge, MA., paperback edition. Hardcover 1991.

Nissim Francez. 2014. The granularity of meaning in proof-theoretic semantics. In Nicholas Asher and Sergei Soloview, editors, *Proceedings of the 8th International Conference on Logical Aspects of Computational Linguistics (LACL), Toulouse, France, June 2014*. Springer Verlag, LNCS 8535, Berlin, Heidelberg, pages 96–106.

Nissim Francez. 2015a. On distinguishing proof-theoretic consequence from derivability. *Logique et Analysis* (to appear).

Nissim Francez. 2015b. *Proof-theoretic Semantics*. College Publications, London.

Nissim Francez. 2016. Proof-theoretic semantics for intensional transitive verbs. *Journal of Semantics* 33(4):803–826. Doi: 10.1093/jos/ffv013.

Nissim Francez and Gilad Ben-Avi. 2015. A proof-theoretic reconstruction of generalized quantifiers. *Journal of Semantics* 32(3):313–371. Doi:10.1093/jos/ffu001.

Nissim Francez and Roy Dyckhoff. 2010. Proof-theoretic semantics for a natural language fragment. *Linguistics and Philosophy* 33(6):447–477.

Gianluca Giorgolo and Ash Asudeh. 2012. Missing resources in a resource-sensitive semantics. In *Proceedings of the 17th Lexical Functional Grammar (LFG) conference, Bali, June-July 2012*. CSLI online publications.

Sara Negri and Jan von Plato. 2001. *Structural Proof Theory*. Cambridge University Press, Cambridge, UK.

Why We Speak

Rohit Parikh
City University of New York
Brooklyn College and
CUNY Graduate Center
rparikh@gc.cuny.edu

Abstract

We explain the relevance of Nash, Hoare and others in explaining Gricean implicature and cheap talk. We also develop a general model to address cases where communication is not cooperative, i..e cases of deception as well as cases where there is common knowledge of different interests in speaker and hearer. Tow models, one qualitative and one quantitative are introduced.

1   Introduction

In his book *Making the Social World*, John Searle indicates that speech acts have two directions. There is the world to word direction, as in the statement "the cat is on the chair" which is true if the world is as the sentence says it is, i.e., the cat is indeed on the chair. There is also a word to world direction as in "Please shut the door" where the relevant proposition becomes true when the door is closed. The speech act seeks to change the world rather than represent it. However, an utterance even of "The cat is on the chair" can be seen as having a word to world direction in that it could be a way to prevent someone from sitting in the chair, and hence on the cat. An utterance is a social act and seeks to change the mental states of listeners and perhaps, eventually, change their actions. So it has the same word to world direction as a request or a command. This fact was noticed originally by Wittgenstein with his language games, but he characteristically refused to give a more formal account. But he was followed by Austin, by Searle himself and various game theorists like Crawford, Farrell and Sobel. Clearly if an utterance has a word to world direction, then it is a move in a game and can be of interest to game theorists. Thus, paradoxically, a sentence and an utterance are of opposite types in Searle's terminology

2   A Taxonomy

The simplest case is where an utterance merely seeks to inform someone of something. Suppose a sentence A is uttered, the listener already believes T and A is consistent with T. Then the listener moves from T to T + A, the logical closure of T $\cup$ {A}. The listener has been informed of A.[1]

If T is not consistent with A then the listener may apply a revision operator a la AGM (Alchourron et al, 1985) and go from T to T – A (which reduces T to a subtheory not containing ¬A and then adds A to that subtheory) Note that in the second case, some formula B which was in T will no longer be in T – A and it could have been the aim of the speaker to achieve the deletion of B more than the addition of A in the listener. To a man expecting his wife any moment, one could say "Her flight has been delayed by 3 hours," the intention being to remove from him the belief that he will see her at 4 PM.

As we see, even the base case has complexities which Frege did not anticipate. But there is more, namely the phenomenon of implicature noticed and investigated by Paul Grice. Since the utterance of A is a move, the listener not only knows that A has been said, but that it has been said under the particular circumstances of the conversation and can typically infer more than just A and it logical consequences. A motorist says to a pedestrian, "My car is out of gas" and the pedestrian responds, "There is a gas station around the corner."

---

[1]My usage here is the reverse of Searle's own usage but will seem more natural to those used to functions. In the one case the world is operating on the word to give it a truth value and in the other case the word is operating on the world to make it as desired. Some of the material in this paper has been presented in various venues but is unpublished. Some other material, especially the model is new.

Proceedings of the 15th Meeting on the Mathematics of Language, pages 68–74,
London, UK, July 13–14, 2017. ©2017 Association for Computational Linguistics

Grice points out that the implicature is that the gas station is open as far as the pedestrian knows, for otherwise why say what she has said. An example from Pinker et al is where a representative of the mob comes around to a store owner and says, "Nice store you have here. It would be a pity if something were to happen to it." The store owner already knows that it would be a pity and does not need to be told. Here the message is in the fact that it is said at all and by whom. The implicature is likely to be something like "Give us \$500 every month and you will be fine."

Deborah Tannen in her You Just Don't Understand gives an example of a husband and wife travelling along a highway when they are nearing a restaurant. "Are you hungry?" says the wife. "No," says the husband and drives on. Later on he is puzzled why she is angry. Here it is clear that the signaling mechanism has not been sufficiently established between husband and wife. He should have said, "No, I am not hungry, but maybe you are?" Should we call this also an example of an implicature? If it is then it is one which did not succeed for Grice asks that implicatures should be calculable. Clearly the husband failed in his calculation, or else it was not a genuine implicature in Grice's sense.

A particularly interesting but also difficult case is where the utterance takes advantage of something like a Nash equilibrium of communication and understanding.

Prashant Parikh gives an example of a man saying, "I am going to the bank" where bank could mean a financial institution or a river bank.

There are four possible situations here. It is good if the speaker means financial institution and the listener understands financial institution. A bad one is where the speaker means financial institution and the listener understands river bank. There are two more cases, one good and one bad. It requires cooperation on the part of the speaker and listener to agree to a good interpretation and not a bad one. Perhaps common knowledge is involved, perhaps something else.

A final case is one discussed by both Austin and Searle. A minister says to a couple, "I now pronounce you man and wife." Here the minister is not reporting on the world, nor is he suggesting a future action. His statement is the action which changes the world in the required way.

Can there be a general theory which deals with all these cases? Perhaps there is no general theory, but we want to suggest that there might well be a general framework of which these are all special cases. The general framework is one where there is an existing social situation and the speaker is using the utterance as a way of altering the situation in some way. One way the utterance can be seen is as an operator applying to a situation S and creating a new situation S'.

However, the "bank" example and the Deborah Tannen example show that the situation contemplated by the speaker might not be the one which does arise. So then we should think of an utterance as non-deterministic operator which converts the existing situation S into one of (a finite number of) situations $\{S_1; S_2; ... S_n\}$. Perhaps the speaker intends S_1 and what comes about is S_2. The speaker might then have to repair the damage, or it may happen that the damage is not easy to repair.

In the movie *When Harry Met Sally*, Harry tells his friend Sally that she is a very attractive person. "You are coming on to me!" says Sally. And when Harry offers to take it back she says, "You can't take it back; it is already out there now!" Harry does manage a repair but it takes him the rest of the movie.

The notion of situation needed here would necessarily be rich to handle all these cases. Something which would work in most cases would be a contemplated game (or action) with a state of knowledge (belief?) for the various actors, their preferences (which might remain constant) and a set of permitted sequences of action. What a particular actor would do (after an utterance) then would depend on his perception of the (new) situation. As Parikh and Tasdemir point out, the action of an agent can be manipulated by changing her state of information, a point made also by Shakespeare in his *Much Ado about Nothing*, where the actions of Benedick and Beatrice are changed by changing their beliefs.

3 Information versus Action

In this paper, we shall consider the cases where the first speaker **A** has some goal in mind which is common knowledge, and the second speaker **B** makes a statement which is relevant to this goal. In that case the (common) knowledge of the goal is part of the context and is typically used to calculate the implicature.

Prashant Parikh and Benz and van Rooij do

point out that some implicatures might *help* someone to *make a decision* even though other implicatures might just be informational.

It will turn out that some contributions made by the computer scientist Tony Hoare and by the economist John Nash will be relevant. The connection of implicatures with these two eminent scholars seems to have been overlooked thus far in the literature.

3.1 Hoare Semantics

Here is how Hoare semantics enters. A Hoare assertion takes the form

$$\{X\}\alpha\{G\}$$

where X, G are propositions, X is the precondition, G is the goal and α is a contemplated action (actions are *programs* for Hoare).

Formally, S is a state space, X, G are subsets of S, and α becomes a relation R_α on S.

Then the Hoare condition is
$$(\forall s)(\forall t)(s \in X \ \& \ (s,t) \in R_\alpha \to t \in G)$$
If X holds when the program starts, then G will hold when it finishes.

Now **A** wants to reach G using α. We will assume that G is common knowledge between **A** and **B**, and **B** has some information which would affect the possibility of reaching G.

This could happen in four ways.

1. **B** has *part of* the information which implies a suitable X. Thus **B** is supporting both the goal and the action.

2. **B** has information which would cause **A** to modify α in some way (e.g. to replace it by a more specific action, or, technically, a sub-action.)

3. **B** has information which *suggests* a particular action α to **A**.

4. **B** has information which would cause **A** to *abandon* the method α. (Which may mean abandoning the goal altogether or using some completely unrelated method.)

B volunteers information which indicates whether 1) or 2) or 3) or 4) is the case and leaves **A** to

1. Either supply some *other information* which will complete the process of deducing X or

2. modify α in some way, or

3. Conclude that $\neg\{X\}\alpha\{G\}$.

This *other information* or the modification, or $\neg\{X\}\alpha\{G\}$ is the implicature.

And note that the modification is not in itself a *proposition*, it may be an imperative. Thus part of our thesis is that while an implicature may be a proposition, it might well be something else which affects **A** in some way.

4 The Cooperative Case

Here we assume, as Grice does, that the utilities of **A** and **B** are in accord. We do not assume that they are the same since what **A** gains from receiving the information is likely to be much more than the pleasure that **B** gets from helping out.

Consider the case of the motorist. The goal of the motorist[2] is clear. She wants to fill her tank with gas. The statement which **B** makes points to an algorithm, *go around the corner and get gas there*. Here it is unclear whether **A** should *walk* there carrying a gas can, or has at least enough gas to *drive* around the corner. But in either case, *going around the corner* is indicated (by **B**) as the α.

Our Hoare assertion is

$$\{X_1 \& X_2 \& X_3\}\alpha\{G\}$$

Where X_1 is that there is a gas station around the corner, X_2 is that the gas station is open and X_3 is that **A** has enough money to pay for the gas.

Only **A** knows whether X_3 is true and this is not **B**'s business. **B** has volunteered X_1. It is obvious that α is useless unless X_2 is true. Since **B** has indicated that he supports the action α it follows that X_2 is true as far as **B** knows. If he is not supporting the action then he should not have said X_1. So in this case X_2 is the implicature.

4.1 The Hiring Problem

In this example **A** is the chair of the hiring committee for some college, **B** is a professor and **C** is the professor's student who is an applicant for a position at **A**'s college. **B** writes about **C**

He has excellent handwriting and he always came to class on time.

---

[2] In Grice's original version

Here A's goal G is to have a colleague who will be a good teacher and researcher, α is the action of hiring C, and the precondition is that C *is a good philosopher.*

B's statement does *not* give A the information which A needs, or even supports it in some relevant way, and the implicature is that the precondition is false. A should abandon α. The implicature is $\neg\{X\}\alpha\{G\}$ where X is any true condition.

But why doesn't B simply say, "He is not a good philosopher"? Clearly because B is C's teacher and professional ethics preclude him from saying something negative about C. What he does instead is to say something positive which is not good enough.

Here is a joke which makes a similar point.

> A tired and depressed looking man walks into a restaurant and sits down. A waiter comes over and asks what the man wants.
>
> "Two scrambled eggs with rye toast, and a kind word," says the man.
>
> After a while the waiter comes back and puts an order of eggs and rye toast before the man.
>
> As the waiter is walking away, the man says, "What about the kind word?"
>
> "Don't eat them eggs," says the waiter.

Just like the professor, the waiter is forbidden to say something negative about the restaurant. but his "don't eat them eggs" carries the implicature that the eggs are not good, may even cause illness. And that indeed is a kind word.

4.2 Modifying α

We note that in many cases actions are not disjoint from each other. If we think of a (nondeterministic) action as a binary relation on the state space, then two actions may be disjoint, may overlap, or one may be included in the other.

One Hoare-like rule is

$$\frac{\{X\}\alpha\{G\}, \beta \subseteq \alpha}{\{X\}\beta\{G\}}$$

If a correctness condition is satisfied by α then it is also satisfied by a subaction β but not necessarily vice versa.

Here is an example.

A to B, "I am thinking of going to Times square by public transport."

B, "Buses will be very slow during the rush hour."

A likely implicature is "Take the subway."

Here action α is the action of taking *some* public transport, β is the action of taking the subway and γ is the action of taking a bus.[3] α is the union of β and γ.

Let X be the current situation, and G be the goal of getting to Times square on time. Then $\{X\}\beta\{G\}$ is true, but $\{X\}\alpha\{G\}$ is not[4]. B is suggesting that the action be changed from α to β by eliminating γ.

When an action satisfies a Hoare condition then so does a sub-action. But this is not the case if we are trying to maximize expected utility. It is quite possible that the expected utility of α is higher than that of β even though $\beta \subseteq \alpha$.

For example, if I am betting on a horse, then it is better to choose a horse at random than to choose a specific horse which is well known to be a nag.

However, if *satisficing* is our desired condition then subactions would be at least as good as an action. If all outcomes of α are satisfactory, and β is a subaction of α then all outcomes of β are also going to be satisfactory.

5 Nash Bargaining

In Grice's treatment of implicature, he assumes a principle of cooperation. Thus for one of his first examples, when a motorist says, "My car is out of gas" and the pedestrian replies "There is a gas station around the corner," there is an implicature that the station is open. And this follows from the presumption that the pedestrian's desires are the same as those of the motorist, although perhaps less intense and so the pedestrian *wants* the motorist to get gas for his car. This tradition has been followed in much of the subsequent literature.

However, there are exceptions. The economics literature on cheap talk no longer assumes that the utilities are aligned. What the speaker wants and what the listener wants need no longer be fully aligned, although some overlap is necessary for communication to take place at all. Stalnaker in

---

[3]We assume it to be common knowledge that a taxicab is out of the question given the traffic.

[4]This is a consequence of the nondeterminism of the two actions. β is *guaranteed* to achieve the goal whereas α might but is not guaranteed to do so.

his paper Cheap talk and credibility, follows this tradition as well.

Yet as we noted, some degree of cooperation is requisite, for otherwise why communicate at all?

We would like to suggest a slight generalization of the Grice principle which looks like it might bridge the gap between cooperation and strategizing. This principle was originally formulated by John Nash in his paper "The Bargaining problem."

In Nash's framework two players **A** and **B** are trying to decide on a point in two space. There is a convex set S of possible solutions and each point p in S yields utilities u(p), v(p) to **A** and **B** respectively. Nash presumes that the actual bargain, i.e. the point p which is finally chosen will be Pareto optimal. That is to say, Nash assumes that there is no q in S such that $u(q) \geq u(p)$ and $v(q) > v(p)$ or that $u(q) > u(p)$ and $v(q) \geq v(p)$. There is no way to make one person better off without making the other person worse off.

Nash assumes moreover that the space S is convex.[5]

Using very natural axioms on the solution concept Nash proves that the final bargain will be the unique point p such that $u(p) \times v(p)$ is maximum.

It is obvious that assuming that the players are choosing a Pareto optimal point, and there are at least two such, then there is a conflict. Neither can gain without the other losing.

The element of cooperation enters through Nash's notion of a *fallback point*. The fallback point F is the point to which they "fall back" in case they cannot arrive at a bargain, and this point is worse (for both) than any other point in S.

Thus cooperation arises through the fact that both players want to avoid the fallback point and each needs the help of the other to achieve this.

Grice's cooperative principle is a special case of Nash's. For suppose the utilities *are* aligned. I.e., if for any two points p and q we have $u(q) > u(p)$ iff $v(q) > v(p)$, then the Nash bargaining point which maximizes the product $u(p) \times v(p)$ is also

the point which maximizes $u(p)$. **B** gains by helping **A** to gain. The pedestrian helps the motorist to get gas for the sake of the small pleasure of helping another[6].

But as we noted this is not the only case. The mere fact that the players both want to avoid X does not imply that their utilities are fully aligned.

We now offer an example of how the Nash principle works.

Suppose that an American tourist is in India and wants to buy a carved wooden elephant. He has already seen such an elephant in a store for Rs. 500 but sees a hawker selling the identical elephant for Rs. 400.

It is customary to bargain with hawkers but what should the tourist offer?[7]

In this situation, the fallback situation is that the tourist abandons the hawker and buys his elephant in the store. But the hawker himself has bought the elephant for Rs. 40 and so any price paid from 41 rupees to 499 rupees would be better for both than the fallback situation, which is no sale for the hawker and a cost of Rs. 500 for the tourist.

The element of cooperation arises because both parties want to avoid the fallback situation, but given this fact there is an element of conflict in that the hawker wants to charge more and the tourist wants to pay less.

Here we assume that the utilities of **A** and **B** are not aligned although there must be some concord for communication to take place at all.[8]

6 A Model

In the following I am going to make two assumptions and offer a caveat.

1. Each party in a two way conversation expects to benefit from the conversation.

2. This *expectation* is common knowledge.

3. But the *benefit* might not be common knowledge and might not even be true.

Thus suppose Ann says something, A, to Bob in response to a query. Then Bob believes that he has benefited by hearing A (which, in normal cases

---

[5]To take an example rather like that of Nash's original example. Suppose that the two are restricted to a point in the set $\{(x, y)|2x + y \leq 3\}$. The utilities are x for **A** and y for **B**. The fallback point is (0,0). Then the Pareto optimal points will be all the points on the line $2x + y = 3$. But which particular point should be chosen? The product of the utilities is maximized at the point (.75, 1.5).

But Nash does not speak about communication and there is no guarantee even that a Pareto optimal point will be reached, let alone Nash's "ideal" point. To take a real life example, it seems highly unlikely that a Pareto optimal point will be reached in Ukraine.

[6]See for instance Tomasello, (2009)

[7]In a similar situation, Aumann offered 200, the offer was accepted and Aumann bought the elephant, only to find that the proper price would have been Rs. 50.

[8]We do not consider the important and interesting case where **A** *thinks* they are aligned but they are not.

would also mean that Bob believes he benefits by believing A).

And Ann, even if she does not *benefit* as much expects to suffer no loss.

Thus suppose that the two of them were in states (S, T) before A was said, and are now in states (S',T'), then Bob believes that T' is better for him than T and this is common knowledge. And Ann believes that S' is not worse for her than S, and this too is common knowledge.

In a cooperative dialogue, and if Ann is well informed, then both these beliefs will be true. They could be false if Ann was mistaken about A, and Bob believed her, or if Ann was trying to mislead Bob by saying A and he was not aware that she was doing so.

In order to define notions of better and worse we need a notion of pragmatic belief.

A *pragmatic belief* for Bob is a formula of the form X(a) = u which means that performing action a will yield utility u to Bob. If Bob has n actions $(a_1, a_2, , a_n)$ available then a *pragmatic state of belief* is a map X from $\{a_1, ..., a_n\}$ into the real numbers.

A statement A made by Ann will cause a change from one state S of pragmatic belief to another state S'.

"But doesn't the statement A cause a change in Bob's beliefs about the world?" To be sure it does. But it might not mean the addition of A to Bob's beliefs if Bob does not believe Ann. Or it might mean the addition of more than A, A+A' where A' is the implicature. And finally, the change in Bob's state of beliefs will eventuate in a change in what he would do and why.

If Ann tells Bob that the bridge is closed, he does normally come to believe that the bridge is closed and hence not plan to take the bridge to the other side. But it is his plan not to take the bridge which will be our central concern. If Bob had had no intention of going to the other side, he would wonder why Ann had said that the bridge was closed.

So facts result in actions, and it is actions that will be our main concern.

So let us suppose that Bob's initial pragmatic state of beliefs was S, Ann's saying A changes it to S' and the true state is S". So let

$$S = \{(a_1, u_1), , (a_n, u_n)\}$$
$$S' = \{(a_1, u_1'), , (a_n, u_n')\}$$

$$S" = \{(a_1, u"_1), , (a_n, u"_n)\}$$

Let m be Bob's best action according to S, m' be the best action according to S'.

Thus suppose that u_3 is the largest of $u_1, .., u_n$, and u_5' is the largest of $u_1', , u_n'$ then m is a_3 and m' is a_5.

Then Bob's gain in utility is

S"(m') – S"(m).

Bob was thinking according to S and would have done m. After hearing A he will now do m. But his real utilities should be evaluated according to S".

7 Applications of the Model

Consider the case of the motorist and the pedestrian. Initially the motorist had two options. To keep driving hoping to come to a gas station and go around the corner. Let these actions be a and b. Then a has the higher utility for him since he has no reason to think that b would do him any good. After hearing Ann say, "There is a gas station around the corner," presumably his new best action would be b. If it would still be a then why should Ann bother to speak? Ergo b has a higher utility. But, that is so only if the gas station is open.

Consider the case of the person who wants to go to Times Square. She has two options. To take the subway and take the bus. Perhaps they are equally viable or at least both have a high enough utility. But now consider the remark "buses will be slow during the rush hour." That lowers the utility of the bus and makes the subway the decisive choice.

This model does not yet accommodate cases like that of the tourist in India since bargaining goes beyond mere converation. But we hope to deal with it in the next version of the paper.

The other cases are similar and can be dealt with by looking at utilities and actions.

8 A Comparison

We have offered two models to understand the phenomena of implicature and cheap talk. The first model was based on Hoare and was qualitative. We do not need to assign numbers to the value of getting faster to Times Square. The second model makes shameless use of numerical utilities. Of course the first model is more general and would apply also to situations where we have no idea of utilities or probabilities. But the second model has the potential for clean mathematical results.

Acknowledgements We thank Nicholas Allott, Anton Benz, Luciana Belotti, Michael Devitt, Stephen Neale, Prashant Parikh, Steven Pinker, Adriana Renero, Robert Stainton and Cagil Tasdemir for comments.

References

Nicholas Allott. 2006. Game theory and communication. In Benz, Jäger, and van Rooij, editors, *Game Theory and Pragmatics*, pages 123–152. Palgrave Macmillan, London.

Carlos E. Alchourrn, Peter Grdenfors, and David Makinson. 1985. On the logic of theory change: Partial meet contraction and revision functions. *Journal of Symbolic Logic* 50(2): 510-530.

Anton Benz and Robert van Rooij. 2007. Optimal assertions and what they implicate. *Topoi* 26(1), 63–78.

Joseph Farrell and Matthew Rabin, 1996, Cheap talk, *Journal of Economic Perspectives* 10(3) , 103-118.

Paul Grice,1989, *Studies in the Way of Words*, Harvard U. Press, Cambridge, Mass

C. A. R. Hoare, An axiomatic basis for computer programming,1969, *Communications of the Association for Computing Machinery*, 12(10), 576–580.

John Nash,1950, The bargaining problem, *Econometrica*, 18(2), 155-162.

Rohit Parikh, Cagil Tademir, and Andreas Witzel, 2013,The power of knowledge in games, 2013, *International Game Theory Review* 15(4), 1-28.

Steven Pinker, MA Nowak and JJ Lee, 2008 The logic of indirect speech, *Proceedings of the National Academy of Sciences*, 105(3), 833-838.

Prashant Parikh,2001 *The Use of Language*, CSLI, Stanford.

John Searle, 2010, *Making the social world: The structure of human civilization*, Oxford University Press, Oxford

Robert Stalnaker, 2006, Saying and meaning, cheap talk and credibility, In Benz, Jäger and van Rooij, editors, *Game Theory and Pragmatics*, pages 83-100, Palgrave Macmillan, London

Micchal Tomasello, 2009, *Why we cooperate.* MIT press, Cambridge, MA

A Monotonicity Calculus and Its Completeness

Thomas F. Icard
Stanford University
Stanford, CA, USA
icard@stanford.edu

Lawrence S. Moss
Indiana University
Bloomington, IN, USA
lsm@cs.indiana.edu

William Tune
Motlow State Community College
Smyrna, TN, USA
wtune@mscc.edu

Abstract

One of the prominent mathematical features of natural language is the prevalence of "upward" and "downward" inferences involving determiners and other functional expressions. These inferences are associated with negative and positive polarity positions in syntax, and they also feature in computer implementations of textual entailment. Formal treatments of these phenomena began in the 1980's and have been refined and expanded in the last 10 years. This paper takes a large step in the area by extending typed lambda calculus to the *ordered setting*. Not only does this provide a formal tool for reasoning about upward and downward inferences in natural language, it also applies to the analysis of monotonicity arguments in mathematics more generally.

1 Introduction

Monotonicity reasoning is pervasive across many domains, from mathematics to natural language, indeed in any setting that deals with functions of ordered sets. A function f is *monotone* if it preserves order, that is, if $x \leq y$ implies $f(x) \leq f(y)$. Anti-monotone (or *antitone*) functions f are those that reserve order, that is, for which $x \leq y$ implies $f(y) \leq f(x)$. Natural language constructions that exhibit these patterns are ubiquitous, spanning semantic and grammatical categories. Algorithms have been devised and studied for deriving monotonicity patterns in complex expressions composed of simpler functional expressions (van Benthem, 1986; Sánchez-Valencia, 1991; van Eijck, 2007). For instance, the interaction of quantifier and temporal expressions, together with the fact that $2 \leq 5$, guarantee that *Any play that lasts*

more than 2 hours is too long entails (is "less than" in a sense to be made precise) *Any play that lasts more than 5 hours is too long*. There has been recent theoretical work on monotonicity reasoning as part of a general interest in "natural logic" (Bernardi, 2002; Zamansky et al., 2006; MacCartney and Manning, 2009; Muskens, 2010; Icard, 2012; Moss, 2012; Icard and Moss, 2013; Tune, 2016), and much of this work has made its way into psycholinguists (Geurts, 2003; Geurts and van der Slik, 2005) and natural language processing (MacCartney and Manning, 2007; Angeli and Manning, 2014; Bowman et al., 2015; Abzianidze, 2015). (For review see Icard and Moss 2014.)

Whereas monotonicity reasoning in natural language is often seen as comprising a fragment of higher-order logic, we can also construe it as encoding a logical system in its own right relative to a suitably coarsened model-theoretic interpretation. In this context standard metalogical questions such as *completeness* can be raised. A completeness result would tell us that a proof system is sufficient to derive everything that follows on the intended model of monotonicity reasoning.

Though our primary interest here is natural language, it bears mention that such reasoning in higher-order settings is also ubiquitous in other areas, e.g., in mathematics. Consider the convergence test for improper integrals, which states that if $0 \leq f(x) \leq g(x)$ on an interval $[a, \infty)$, then $\int_a^\infty f(x)dx$ converges if $\int_a^\infty g(x)dx$ does. As an example of this, note that knowing $\int_1^\infty e^{-x}dx = e^{-1}$ converges allows, by monotonicity reasoning alone, to infer that $\int_1^\infty e^{-x^2}$ also converges:

$$\frac{\dfrac{\dfrac{\dfrac{1 \leq x}{x \leq x^2}}{-x^2 \leq -x}}{e^{-x^2} \leq e^{-x}}}{\int_1^\infty e^{-x^2}dx \leq \int_1^\infty e^{-x}dx}$$

Proceedings of the 15th Meeting on the Mathematics of Language, pages 75–87,
London, UK, July 13–14, 2017. ©2017 Association for Computational Linguistics

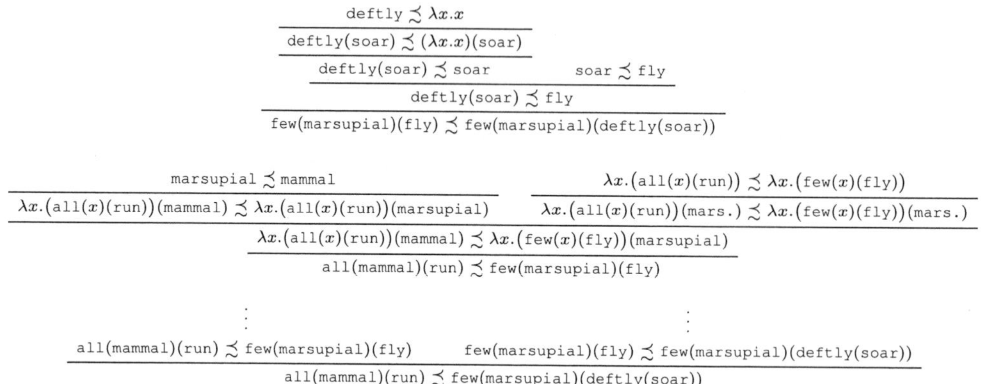

Figure 1: Example proof that *All mammals run* implies *Few marsupials deftly soar* (in three parts).

This argument, similar to those we will be considering, depends only on the *monotonicity profiles* of the relevant functions (multiplication, exponentiation, etc.) on the relevant domains.

The aim of the present contribution is to formulate a suitable system for monotonicity reasoning in a higher-order setting, appropriate to the task of capturing common entailment patterns in natural language in particular, and to prove a completeness result for an associated proof system. Along the way we also prove an analogue of Lyndon's (1959) Theorem for first order logic, showing exactly when, in our general setting, a subterm occurrence stands in a monotone or antitone position. For reasons of space, we skip some of the less central proofs.

2   Motivating Example

To motivate the specific formal apparatus that we will employ, consider the following small fragment. As in previous work (Icard and Moss, 2013), we will be considering an extended simply typed lambda calculus where the functional types can be "marked" with monotonicity information, $+$ for monotone, $-$ for antitone, and $\cdot$ for neither (or unknown). Suppose we have two base types t and p, corresponding to truth values and predicates (more commonly, functions from entity type to truth value type), and the following typed terms:

$$
\begin{array}{lcl}
\texttt{all} & : & p \xrightarrow{-} (p \xrightarrow{+} t) \\
\texttt{few} & : & p \xrightarrow{-} (p \xrightarrow{-} t) \\
\texttt{mammal, marsupial} & : & p \\
\texttt{run, fly, soar} & : & p \\
\texttt{deftly} & : & p \xrightarrow{\cdot} p
\end{array}
$$

Let us furthermore assume the following background entailment facts Γ, where $M \precsim N$ is understood as entailment generalized to all types:

$$
\begin{array}{rcl}
\lambda x.\big(\texttt{all}(x)(\texttt{run})\big) & \precsim & \lambda x.\big(\texttt{few}(x)(\texttt{fly})\big) \\
\texttt{deftly} & \precsim & \lambda x.x \\
\texttt{marsupial} & \precsim & \texttt{mammal} \\
\texttt{soar} & \precsim & \texttt{fly}
\end{array}
$$

The first statement encodes the assumption that if all members of a given category run, it can be inferred that few members of that category fly. The second statement essentially says that *deftly* is subsective (Kamp and Partee, 1995): deftly v'ing involves v'ing (see Figure 1). The third and fourth capture basic lexical entailments. We can then use these assumptions to derive *Few marsupials deftly soar* from *All mammals run*. A proof using our monotonicity calculus appears in Figure 1.

There are several important points to notice about this example. First, in order to state and use assumptions such as the first two above, we make crucial use of lambda abstraction and β-reduction.

Second, note that we can state entailment facts between terms even when those terms have different (marked) types, as in the first two statements. For example, though in our typing system $\lambda x.x$ will be of type $p \xrightarrow{+} p$, it can nonetheless be compared with $\texttt{deftly}$ because (denotations of) terms of type $p \xrightarrow{+} p$ can be semantically "coerced" to type $p \xrightarrow{\cdot} p$. This is simply because the domain $D_{p \xrightarrow{\cdot} p}$ for terms of type $p \xrightarrow{\cdot} p$ will be the class of all functions from D_p to D_p, which certainly includes all the monotone functions.

Third, it can be useful to derive monotonicity information for complex terms, e.g., so that we can derive $\lambda x.\big(\texttt{all}(x)(\texttt{run})\big)(\texttt{mammal}) \precsim \lambda x.\big(\texttt{all}(x)(\texttt{run})\big)(\texttt{marsupial})$ in one step. Theorem 8.2 below guarantees that the way we type lambda abstractions is in a sense optimal.

The framework developed in this paper is motivated by the desire to capture patterns like these. Such patterns could be derived in an inequational system of full higher-order logic: given a constant $\vee$ for disjunction at a given type, it is easy to see that a term f will be a monotone function just in case we have $\lambda x.\lambda y.f(x) \precsim \lambda x.\lambda y.f(x \vee y)$. Proofs of facts like that above might then be derived in a higher order logic proof system with monotonicity declarations as additional premises. We of course could not have completeness in this setting, but more importantly, we would rather like to isolate and understand what is characteristic of monotonicity reasoning as such.

There are many instances of this kinds of reasoning outside of natural language. As a simple illustration of the main concepts and definitions, throughout the paper we will be considering a running example of elementary mathematical reasoning about real number functions.

3 Types and Domains

Our set $\mathcal{T}$ of types is defined inductively from a set $\mathcal{B}$ of base types b:

$$\tau \ ::= \ b \mid \tau \xrightarrow{\cdot} \tau \mid \tau \xrightarrow{+} \tau \mid \tau \xrightarrow{-} \tau$$

Definition 3.1 (Markings and types). The set Mar of *markings* is $\{+, -, \cdot\}$. We use m and m' to denote markings. We always take Mar to be ordered with $+ \sqsubseteq \cdot$, $- \sqsubseteq \cdot$, and $m \sqsubseteq m$ for all m. We also define a binary operation $\circ$ on Mar by $+ \circ + = +$, $+ \circ - = -$, $- \circ + = -$, $- \circ - = +$; and otherwise $m \circ m' = \cdot$. Notice that $\circ$ is associative.

We have a natural ordering $\precsim$ on types, where $\sigma \precsim \tau$ can be read as: any term of type σ could also be considered of type τ (cf. Def. 3.5 below).

Definition 3.2 ($\precsim$ on types). Define $\precsim \ \subseteq \mathcal{T} \times \mathcal{T}$ to be the least preorder with the property that whenever $\sigma' \precsim \sigma$ and $\tau \precsim \tau'$, and $m \sqsubseteq m'$, we have $\sigma \xrightarrow{m} \tau \precsim \sigma' \xrightarrow{m'} \tau'$.

Definition 3.3 (the functions $\uparrow$, $\vee$, and $\sigma \mapsto \hat{\sigma}$ on types). $(\mathrm{Mar}, \sqsubseteq)$ is an upper semilattice. So we have an operation $\vee$ on it. Explicitly, $m \vee m = m$ for all m, and for $m \neq m'$, $m \vee m' = \cdot$. We also define $\uparrow$ to be the smallest relation on types, and $\vee$ to be the smallest partial function on types, with the properties that for all σ, τ_1, and τ_2:

1. $\sigma \uparrow \sigma$, and $\sigma \vee \sigma = \sigma$.

2. If $\tau_1 \uparrow \tau_2$, then $(\sigma \xrightarrow{m_1} \tau_1) \uparrow (\sigma \xrightarrow{m_2} \tau_2)$ for all $m_1, m_2 \in \mathrm{Mar}$, $(\sigma \xrightarrow{m_1} \tau_1) \vee (\sigma \xrightarrow{m_2} \tau_2) = \sigma \xrightarrow{m_1 \vee m_2} (\tau_1 \vee \tau_2)$.

Finally, we define $\sigma \mapsto \hat{\sigma}$ on $\mathcal{T}$ by $\hat{\sigma} = \sigma$ for σ basic, and $(\sigma \xrightarrow{m} \tau)\hat{\ } = \sigma \xrightarrow{\cdot} \hat{\tau}$.

We also note the following characterization of the order $\precsim$. In it, we use the height function defined by: $\mathrm{ht}(\sigma) = 0$ for σ basic, and $\mathrm{ht}(\sigma \xrightarrow{m} \tau) = 1 + \max(\mathrm{ht}(\sigma), \mathrm{ht}(\tau))$.

Proposition 3.4. Let R_0 be the identity relation on the set $\mathcal{T}$ of types. Given R_n, let

$$R_{n+1} = \ \{(\sigma \xrightarrow{m} \tau, \sigma' \xrightarrow{m'} \tau') : \sigma, \sigma', \tau, \tau' \\ \text{have height} \leq n; m \sqsubseteq m'; \text{and} \\ \text{both } (\sigma', \sigma), (\tau, \tau') \text{ belong to } R_n\}$$

Then $\bigcup_n R_n$ is the order $\precsim$.

Definition 3.5. A *pre-structure* $\mathbb{D} = \{\mathbb{D}_\tau\}_{\tau \in \mathcal{T}}$ is given by a class of preorders $\mathbb{D}_\tau = (D_\tau, \leq_\tau)$ for each type $\tau \in \mathcal{T}$, and a family of maps

$$\pi_{\sigma,\tau} : \mathbb{D}_\sigma \to \mathbb{D}_\tau$$

when $\sigma \precsim \tau$, subject to the following constraints:

1. $D_{\sigma \xrightarrow{+} \tau} \cup D_{\sigma \xrightarrow{-} \tau} \subseteq D_{\sigma \xrightarrow{\cdot} \tau} \subseteq D_\tau^{D_\sigma}$.

2. $f \in D_{\sigma \xrightarrow{+} \tau}$ and $a \leq_\sigma b$ imply $f(a) \leq_\tau f(b)$.

3. $f \in D_{\sigma \xrightarrow{-} \tau}$ and $a \leq_\sigma b$ imply $f(b) \leq_\tau f(a)$.

4. $f \leq_{\sigma \xrightarrow{m} \tau} g$ iff for all $a \in D_\sigma : f(a) \leq_\tau g(a)$.

5. $\pi_{\sigma,\sigma}$ is the identity on $\mathbb{D}_\sigma$.

6. If $\sigma \precsim \tau \precsim \mu$, then $\pi_{\sigma,\mu} = \pi_{\tau,\mu} \circ \pi_{\sigma,\tau}$.

7. Each map $\pi_{\sigma,\tau}$ is order-preserving.

Definition 3.6. Here is a family of pre-structures called the *full pre-structures* based on an assignment of preorders $\mathbb{D}_\sigma$ to base types σ. Then one defines $\mathbb{D}_\sigma$ by recursion on the height of σ:

$$\begin{aligned} \mathbb{D}_{\sigma \xrightarrow{\cdot} \tau} &= (D_\tau)^{D_\sigma}, \text{ all functions from } D_\sigma \text{ to } D_\tau \\ \mathbb{D}_{\sigma \xrightarrow{+} \tau} &= \{f \in \mathbb{D}_{\sigma \xrightarrow{\cdot} \tau} : f \text{ is monotone}\} \\ \mathbb{D}_{\sigma \xrightarrow{-} \tau} &= \{f \in \mathbb{D}_{\sigma \xrightarrow{\cdot} \tau} : f \text{ is antitone}\} \end{aligned}$$

The order in all cases is the pointwise order. We define the maps $\pi_{\sigma,\tau}$ in terms of the characterization in Proposition 3.4. For $n = 0$, the only time we have $(\sigma, \tau) \in R_0$ is when $\sigma = \tau$; in this case, we set $\pi_{\sigma,\sigma}$ to be the identity on $\mathbb{D}_\sigma$. Notice that each $\pi_{\sigma,\tau}$ is order preserving (since τ must be σ when $n = 0$), and also that $\pi_{\sigma,\mu} = \pi_{\tau,\mu} \circ \pi_{\sigma,\tau}$.

Given $\pi_{\sigma,\tau}$ for all pairs $(\sigma, \tau) \in R_n$, here is how we extend the definition to $R_{n+1} \setminus R_n$. Given $\sigma' \preceq \sigma$ and $\tau \preceq \tau'$, $m \sqsubseteq m'$, and also $\pi_{\sigma,\sigma'}$ and $\pi_{\tau,\tau'}$, we define $\pi_{\sigma \xrightarrow{m} \tau, \sigma' \xrightarrow{m'} \tau'}$ to be

$$k \in D_{\sigma \xrightarrow{m} \tau} \quad \mapsto \quad \pi_{\tau,\tau'} \circ k \circ \pi_{\sigma',\sigma}. \qquad (1)$$

It is easy to verify the properties in Def. 3.5.

Example 3.7. Take $\mathcal{B} = \{r\}$, with r intuitively standing for *real numbers*. Then we will have types in $\mathcal{T}$ such as those shown below. As an example of $\uparrow$, $(r \xrightarrow{-} r) \uparrow (r \xrightarrow{+} r)$, while $(r \xrightarrow{-} r) \vee (r \xrightarrow{+} r) = r \xrightarrow{\cdot} r$. We build the full pre-structure using $\mathbb{D}_r = \mathbb{R}$, the reals with the usual order $\leq$. Then we have, e.g.,

σ	$\mathbb{D}_\sigma$
$r \xrightarrow{\cdot} r$	functions from $\mathbb{R}$ to $\mathbb{R}$
$r \xrightarrow{+} r$	monotone functions from $\mathbb{R}$ to $\mathbb{R}$
$r \xrightarrow{-} r$	antitone functions from $\mathbb{R}$ to $\mathbb{R}$
$r \xrightarrow{+} (r \xrightarrow{-} r)$	monotone functions from $\mathbb{R}$ to $\mathbb{D}_{r \xrightarrow{-} r}$

4 The Language $\mathcal{L}_\lambda$ of Terms

Our language $\mathcal{L}_\lambda$ is a variant of the typed λ-calculus which makes use of the marked types that we saw in Section 3.

We begin with a set $\mathcal{C}$ of constants, each coming with a unique type, and a set $\mathcal{V}$ of variables, also with their types. We define the language $\mathcal{L}_\lambda$ of all terms using a typing calculus. Beginning with a set of typing statements determined from $\mathcal{C}$ and $\mathcal{V}$, we define several things simultaneously: *terms with their types* (denoted $M : \sigma$, $N : \tau$, etc.), *occurrences of free variables in terms*, and *the valence of each free variable occurrence in M*.

1. For a variable $x : \tau$ in $\mathcal{V}$, $x : \tau$ is a term. Further, x occurs free in itself in the evident way, and x is the only variable that occurs free in itself. The valence is $+$.

2. Each constant $c : \sigma$ is a term, so there are no free occurrences of any variables in c.

3. Let $m \in$ Mar. We have the following rule:

$$\frac{M : (\sigma \xrightarrow{m} \tau) \qquad N : \sigma'}{M(N) : \tau} \; \sigma' \preceq \sigma$$

The free occurrences of x in $M(N)$ are the free occurrences of x in M together with the free occurrences of x in N.

Any free occurrence of x in $M(N)$ is either an occurrence in M, or an occurrence in N:

(a) For a free occurrence of x in M, the valence in $M(N)$ is that in M.

(b) For a free occurrence of x in N, with valence m', the valence of x in $M(N)$ is $m \circ m'$, where $M : \sigma \xrightarrow{m} \tau$.

4. Finally, if x is a variable,

$$\frac{x : \sigma \qquad M : \tau}{\lambda x.M : \sigma \xrightarrow{m} \tau}$$

If all free occurrences of x in M are $+$, and if there is at least one free occurrence of x in M, then $m = +$. If all free occurrences of x in M are $-$, and if there is at least one free occurrence of x in M, then $m = -$. If there are either no free occurrences of x in M, or if there are free occurrences but they are not all $+$ and also not all $-$, then $m = \cdot$.

There are no free occurrences of x in $\lambda x.M$. The free occurrences of variables $y \neq x$ in $\lambda x.M$ are the free occurrences of y in M. Those occurrences have the same valence in $\lambda x.M$ as they have in M.

We define $FV(M)$ to be the variables with free occurrences in M. We define $BV(M)$ to be the variables with bound occurrences in M. (We have not defined these, but they are defined as usual.) The main point about the valences of variable occurrences will come shortly, in Lemma 5.2.

Remark 4.1. Note that every term M has a unique type. For this reason, we often omit the type when it is not pertinent to the discussion.

Example 4.2. We build on Example 3.7. Let us take the set $\mathcal{C}$ of constants to be given as follows:

constant c	type σ	standard interpretation
0	r	0
1	r	1
$+$	$r \xrightarrow{+} r \xrightarrow{+} r$	$a \mapsto (b \mapsto a + b)$
$-$	$r \xrightarrow{+} r \xrightarrow{-} r$	$a \mapsto (b \mapsto a - b)$

We shall present the semantics of terms in Section 5 below. The "standard interpretation" is not quite the semantics $[\![\,]\!]$ in our sense because $[\![\,]\!]$ is defined in terms of valuations. The difference is

term M	type σ	$[\![M]\!]_\phi$	term M	type σ	$[\![M]\!]_\phi$
$+(0)(1)$	r	1	$-(1)(x)$	r	$1 - \phi(x)$
$+(1)(1)$	r	2	$-(x)(1)$	r	$\phi(x) - 1$
$-(0)(1)$	r	-1	$-(x)(y)$	r	$\phi(x) - \phi(y)$
x	r	$\phi(x)$	$\lambda x. - (1)(x)$	$r \xrightarrow{-} r$	$a \mapsto (1 - a)$
$-(x)$	$r \xrightarrow{-} r$	$b \mapsto (\phi(x) - b)$	$\lambda x. - (x)(1)$	$r \xrightarrow{+} r$	$a \mapsto (a - 1)$
$-(0)$	$r \xrightarrow{-} r$	$b \mapsto -b$	$\lambda f.\lambda x. - (0)(f(x))$	$(r \xrightarrow{\cdot} r) \to (r \xrightarrow{\cdot} r)$	$f \mapsto -f$

Figure 2: Examples of terms, types, and interpretations, under an arbitrary valuation ϕ. See Example 3.7.

harmless. Further, we take variables $x, y, z : r$, and $f, g : r \xrightarrow{\cdot} r$. Figure 2 has examples of terms, again with types and semantics under a valuation ϕ. We assume that the semantics interprets the constants as above. Of course, it would be more sensible to write $1 + 1$ instead of $+(1)(1)$.

5   Semantics of $\mathcal{L}_\lambda$: Structures

At this point, we turn to the semantics of our language. We interpret $\mathcal{L}_\lambda$ in what we call *structures*. These are pre-structures together with additional information needed to interpret variables and constant symbols.

Definition 5.1. Let $\mathbb{D}$ be a pre-structure. We let $\Phi = \Phi(\mathbb{D})$ be the set of functions ϕ whose domain is the set of (typed) variables, with the property that if $x : \sigma$, then $\phi(x) \in D_\sigma$. We call such functions ϕ *valuations in* $\mathbb{D}$.

An *interpretation function in* $\mathbb{D}$ is a function

$$[\![\,]\!] : \mathcal{L}_\lambda \times \Phi \to \mathbb{D},$$

mapping the terms M of the language $\mathcal{L}_\lambda$ together with valuations to elements of $\mathbb{D}$. As before, we require that if $M : \sigma$, then $[\![M]\!]_\phi \in D_\sigma$.

A *structure* is a pair

$$\mathbb{S} \; = \; (\mathbb{D}, [\![\,]\!]),$$

where $\mathbb{D}$ is a pre-structure, and $[\![\,]\!]$ is an interpretation in $\mathbb{D}$ such that the following conditions hold:

1. For $M : \sigma \xrightarrow{m} \tau$, and $N : \sigma' \preceq \sigma$,
   $[\![M(N)]\!]_\phi = [\![M]\!]_\phi \big(\pi_{\sigma', \sigma}[\![N]\!]_\phi\big)$.

2. For $M : \sigma$, $x : \tau$, and $a \in \mathbb{D}_\tau$,
   $[\![\lambda x.M]\!]_\phi(a) = [\![M]\!]_{\phi_x^a}$.

In the last point, we use our notation for modifying functions when we write

$$\phi_x^a(y) \; = \; \begin{cases} \phi(y) & \text{if } y \neq x \\ a & \text{if } y = x \end{cases}$$

Again, see Figure 2 for examples.

5.1   Positivity Entails Monotonicity; Negativity Entails Antitonicity

Recall that positive or negative occurrences of variables are syntactic notions, whereas monotonicity and antitonicity are semantic notions. One of the contributions of this paper is to explore the connection between these notions.

Lemma 5.2. Let $M : \tau$, and let $x : \sigma$ be a variable. Let $\mathbb{S} = (\mathbb{D}, [\![\,]\!])$ be any structure.

1. If all free occurrences of x in M are $+$, then for all ϕ, $a \mapsto [\![M]\!]_{\phi_x^a}$ is monotone.

2. If all free occurrences of x in M are $-$, then for all ϕ, $a \mapsto [\![M]\!]_{\phi_x^a}$ is antitone.

Proof. By induction on M. We prove both parts simultaneously.

Let M be a variable. We have two cases, depending on whether $M = x$ or not. If $M = x$, then all occurrences of x in x are $+$. Moreover, $a \mapsto [\![M]\!]_{\phi_x^a}$ is the identity and hence monotone in a. (Also, it is not the case that all occurrences of x in x are $-$.) If M is a variable $y \neq x$, then all occurrences of x in M (there are none) are both $+$ and $-$. And in this case, $[\![M]\!]_{\phi_x^a} = \phi(y)$. So $a \mapsto [\![M]\!]_{\phi_x^a}$ is a constant function. As such, it is both monotone and antitone.

Now suppose (1) and (2) for M and for N, and consider $M(N)$. First, suppose that all free occurrences of x in $M(N)$ are $+$. Then all free occurrences of x in M are $+$. We have two cases.

First, we consider the case when M is of functional type $\sigma \xrightarrow{+} \tau$. In this case, all occurrences of x in N must be $+$. By induction hypothesis, $a \mapsto [\![M]\!]_{\phi_x^a}$ is monotone, and so is $a \mapsto [\![N]\!]_{\phi_x^a}$. Hence so is $a \mapsto [\![M(N)]\!]_{\phi_x^a}$. In more detail, let $a \leq b$. Then

$$
\begin{aligned}
[\![M(N)]\!]_{\phi_x^a} &= [\![M]\!]_{\phi_x^a}\big([\![N]\!]_{\phi_x^a}\big) \\
&\leq [\![M]\!]_{\phi_x^a}\big([\![N]\!]_{\phi_x^b}\big) \\
&\leq [\![M]\!]_{\phi_x^b}\big([\![N]\!]_{\phi_x^b}\big) \\
&= [\![M(N)]\!]_{\phi_x^b}.
\end{aligned}
$$

We have suppressed the type information on the inequality signs $\leq$.

The case when M is of negative functional type is similar. This concludes our (abridged) discussion of point (1).

We turn to (2). Suppose that all free occurrences of x in $M(N)$ are $-$. Then all free occurrences of x in M are $-$. We again have two cases.

First, we consider the case when M is $+$. So all free occurrences of x in N are $-$. Thus $a \mapsto [\![M]\!]_{\phi_x^a}$ is monotone, and $a \mapsto [\![N]\!]_{\phi_x^a}$ is antitone. Each $[\![M]\!]_{\phi_x^a}$ is antitone. Now let $a \leq b$. Then

$$
\begin{aligned}
[\![M(N)]\!]_{\phi_x^b} &= [\![M]\!]_{\phi_x^b}([\![N]\!]_{\phi_x^b}) \\
&\leq [\![M]\!]_{\phi_x^a}([\![N]\!]_{\phi_x^b}) \\
&\leq [\![M]\!]_{\phi_x^a}([\![N]\!]_{\phi_x^a}) \\
&= [\![M(N)]\!]_{\phi_x^a}.
\end{aligned}
$$

Finally, we have the case that M is $-$. Each $[\![M]\!]_\phi$ is antitone, and $a \mapsto [\![M]\!]_{\phi_x^a}$ is antitone. Further, all free occurrences of x in N are $+$, so $a \mapsto [\![N]\!]_{\phi_x^a}$ is monotone. And for $a \leq b$ we have

$$
\begin{aligned}
[\![M(N)]\!]_{\phi_x^b} &= [\![M]\!]_{\phi_x^b}([\![N]\!]_{\phi_x^b}) \\
&\leq [\![M]\!]_{\phi_x^b}([\![N]\!]_{\phi_x^a}) \\
&\leq [\![M]\!]_{\phi_x^a}([\![N]\!]_{\phi_x^a}) \\
&= [\![M(N)]\!]_{\phi_x^a}.
\end{aligned}
$$

This concludes our work on application terms.

We conclude the overall induction by considering abstraction terms $\lambda y.M$. Let $a \leq b$. To see that $[\![\lambda y.M]\!]_{\phi_x^a} \leq_{\tau \overset{m}{\to} \sigma} [\![\lambda y.M]\!]_{\phi_x^b}$, let $d \in \mathbb{D}_\sigma$:

$$
\begin{aligned}
[\![\lambda y.M]\!]_{\phi_x^a}(d) &= [\![M]\!]_{(\phi_x^a)_y^d} \\
&= [\![M]\!]_{(\phi_y^d)_x^a} \\
&\leq [\![M]\!]_{(\phi_y^d)_x^b} \qquad \text{ind. hyp.} \\
&= [\![M]\!]_{(\phi_x^b)_y^d} \\
&= [\![\lambda y.M]\!]_{\phi_x^b}(d)
\end{aligned}
$$

Note that we apply the induction hypothesis to ϕ_y^d, not to ϕ. This for all d shows (1) for $\lambda y.M$. (Recall that we are using the pointwise order for functional types $\tau \overset{m}{\to} \sigma$.) The same reasoning applies to (2) for the term $\lambda y.M$.

This concludes the proof. $\dashv$

Example 5.3. The full pre-structures introduced in Definition 3.6 give structures in the following way. We define $[\![M]\!]_\phi$ by recursion on M, simultaneously for all ϕ, and we also at the same time verify that for $M : \sigma$, $[\![M]\!]_\phi \in D_\sigma$. The definitions are related to (but not identical to) what we saw in the definition of a structure:

1. For $M : \sigma \overset{m}{\to} \tau$, and $N : \sigma' \preceq \sigma$,
$$[\![M(N)]\!]_\phi = [\![M]\!]_\phi\big(\pi_{\sigma',\sigma}[\![N]\!]_\phi\big).$$

2. For $M : \sigma$, and $x : \tau$, $[\![\lambda x.M]\!]_\phi$ is the function $a \mapsto [\![M]\!]_{\phi_x^a}$.

Now along with the definition, we carry along the proof of Lemma 5.2. We do this in order to know that the typings of the abstractions $\lambda x.M$ are correct. For example, suppose that our typing has $\lambda x.M : \sigma \overset{+}{\to} \tau$. According to our definition in Section 4, we know that all free occurrences of x in M are $+$. So by Lemma 5.2, we know that $[\![\lambda x.M]\!]_\phi$ as defined above really is a monotone function; that is, it really belongs to $D_{\sigma \overset{+}{\to} \tau}$.

6 Term Substitution and Reduction

A *substitution* is a function s from variables to terms, sending $x : \sigma$ to some $s(x) : \sigma' \preceq \sigma$.

One example is the identity substitution Id. For any substitution s, and any variable $x : \sigma$ and $M : \sigma' \preceq \sigma$, we get a new substitution s_x^M, defined by $s_x^M(x) = M$, and for $y \neq M$, $s_x^M(y) = s(y)$. When the subscript/superscript notation becomes cumbersome, we might change it. For example, we usually write Id_x^M as $[M/x]$.

The notion of capture-avoiding substitution is something of a challenge to get correct. We adopt the definitions of Stoughton (1988) and then quote the results from this paper, adapted to our setting.

Given a term M and a substitution s, we define $M[s]$ by induction on s. It represents the result of substituting, for each x, $s(x)$ for every free occurrence of x in s. We only use the notation $M[s]$ when no variable occurs bound in M and free in any $s(x)$. (That is, we insist that no variable free in any $s(x)$ has bound occurrences in M.)

$$
\begin{aligned}
x[s] &= s(x) \\
M(N)[s] &= M[s](N[s]) \qquad (2) \\
(\lambda x.M)[s] &= \lambda y.(M[s_x^y])
\end{aligned}
$$

In the last line, y is the least variable in some pre-set list such that y is not free in M, nor in any $s(z)$ for z free in M. Also s_x^y is just like s except that $s_x^y(x) = y$. But y can be any variable z with those properties; by Corollary 3.11 of Stoughton (1988), the result $\lambda z.(M[s_x^z])$ will be α-equivalent to $\lambda y.(M[s_x^y])$. (We define α-equivalence below.)

Lemma 6.1. For all terms M and substitutions s, $M[s]$ is a proper term, and the type of $M[s]$ is $\preceq$ the type of M.

Lemma 6.2. Let M be a term, and consider a free occurrence of a variable x in M with valence m. Let s be a substitution, and let y be a variable which occurs free in $s(x)$ with valence m'. Then in $M[s]$, the free occurrences of y which arise as substitutions for the given occurrence of x all have valence $m \circ m'$.

Proof. By induction on M. When M is a variable, this variable must be x. Since the valence of x in itself is $+$, and since $+$ is a neutral element for $\circ$, our result follows.

When M is a constant, our result is vacuous.

Consider an application term $M(N)$, and assume our lemma for M and for N. Recall that $M(N)[s] = (M[s])(N[s])$. Consider a free occurrence of x in $M(N)$.

First, we consider the case when our free occurrence of x in $M(N)$ is actually a free occurrence in M. In this case, the valence of all the corresponding occurrences of y in $M(N)[s]$ is the same as the valence of those occurrences in $M[s]$. And so the result in this case follows easily from the induction hypothesis.

Second, we have the case when our free occurrence of x in $M(N)$ is a free occurrence of x in N. Let m_1 and m_2 be such that $M : \sigma \overset{m_1}{\to} \tau$ and the valence of our occurrence in N is m_2. Then m, the valence of x in $M(N)$, is $m_1 \circ m_2$. The corresponding occurrences of y in $(M(N))[s]$ are free occurrences of y in $N[s]$, and by induction hypothesis, their valences there are $m_2 \circ m'$. So their valences in $(M(N))[s]$ are $m_1 \circ (m_2 \circ m') = (m_1 \circ m_2) \circ m' = m \circ m'$. (We have used the associativity of $\circ$.) This is as desired.

We conclude with the induction step for abstraction. Let M be $\lambda z.N$ with $z \neq x$. We assume our lemma for N, and we have an occurrence of x in M; its valence there is the same as the valence of the corresponding occurrence in N. Recall that $M[s]$ is $\lambda w.N[s]$, with w suitably fresh. A free occurrence of y in $M[s]$ corresponds to a free occurrence in $N[s]$, and the valence is the same. Our result follows from the induction hypothesis. $\dashv$

The next two results will guarantee that the usual reduction rules of lambda calculus involve well-defined operations on our set of terms.

Theorem 6.3 (Subject Reduction Theorem). The type of $M[^N\!/_x]$ is $\preceq$ the type of $(\lambda x.M)N$.

Proof. The type of $(\lambda x.M)N$ is the type of M, so the result follows from Lemma 6.1. $\dashv$

Theorem 6.4 (Subject Reduction Theorem for valences). Consider a free occurrence occ of y in $(\lambda x.M)N$ with valence m, either $+$ or $-$. Also, consider the term that results from (β) reduction, $M[^N\!/_x]$. Then the occurrences of y in $M[^N\!/_x]$ which correspond to occ also have valence m.

Proof. If the free occurrence of y is in $\lambda x.M$, then our result is easy. So we focus on the case when it is in N. Now $m = m_1 \circ m_2$, where m_1 is such that $\lambda x.M : \sigma \overset{m_1}{\to} \tau$, and m_2 is the valence of occ in N. We are assuming that m_1 is either $+$ or $-$. By the way we type abstractions, all free occurrences of x in M have valence m_1. By Lemma 6.2, the occurrences of y which correspond to occ also have valence $m_1 \circ m_2$. $\dashv$

Definition 6.5. Define $\approx$ to be the least equivalence relation between $\mathcal{L}_\lambda$-terms closed under:

$$(\alpha) \ \frac{}{\lambda x.M \approx \lambda y.M[^y\!/_x]} \ {}^{[y \notin FV(M) \cup BV(M)]}$$

$$(\beta) \ \frac{}{(\lambda x.M)N \approx M[^N\!/_x]} \ {}^{[BV(M) \cap FV(N) = \emptyset]}$$

$$(\eta) \ \frac{}{\lambda x.Mx \approx M} \qquad (\xi) \ \frac{M \approx N}{\lambda x.M \approx \lambda x.N}$$

$$(\text{Cong}) \ \frac{M \approx M' \qquad N \approx N'}{M(N) \approx M'(N')}$$

In the (η) rule we also assume $x \notin FV(M)$.

The following proposition guarantees that equivalent terms are assigned the same meaning.

Proposition 6.6. If $M \approx N$, then for all $\mathbb{S}$ and ϕ, $\pi_{\sigma_1,\tau}[\![M]\!]_\phi = \pi_{\sigma_2,\tau}[\![N]\!]_\phi$, where $\tau = \sigma_1 \vee \sigma_2$.

We say that a term M is in *normal form* if it has no β- or η-redexes, those defined in the usual way.

7 Term Structures

In this section, we outline a method to define a prestructure from a preorder on terms of the language. Given a term M, we denote its $\approx$-equivalence class by $\langle M \rangle$. When we define a function ι on the $\approx$-equivalence classes, we generally write $\iota\langle M \rangle$ rather than $\iota(\langle M \rangle)$. Let

$$L_\tau = \{\langle M \rangle : M \text{ is an } \mathcal{L}_\lambda\text{-term of type } \tau\}$$
$$T_\tau = \bigcup\{L_\sigma : \sigma \preceq \tau\}$$

We have *inclusion maps* $i_{\sigma,\tau} : T_\sigma \to T_\tau$.

Proposition 7.1. The family $i_{\sigma,\tau}$ has the following functoriality properties: $i_{\sigma,\sigma}$ is the identity on T_σ, and if $\sigma \preceq \tau \preceq \mu$, then $i_{\sigma,\mu} = i_{\tau,\mu} \circ i_{\sigma,\tau}$.

Definition 7.2. A *term structure* $\mathbb{T}$ is a family $\{T_\tau, \sqsubseteq_\tau\}$ of preorders, subject to the following:

1. If $\langle M \rangle \in T_{\sigma \xrightarrow{+} \tau}$ and $\langle N \rangle \sqsubseteq_\sigma \langle O \rangle$, then $\langle M(N) \rangle \sqsubseteq_\tau \langle M(O) \rangle$.

2. If $\langle M \rangle \in T_{\sigma \xrightarrow{-} \tau}$ and $\langle N \rangle \sqsubseteq_\sigma \langle O \rangle$, then $\langle M(O) \rangle \sqsubseteq_\tau \langle M(N) \rangle$.

3. $\langle M \rangle \sqsubseteq_{\sigma \xrightarrow{m} \tau} \langle N \rangle$ iff $\langle M(O) \rangle \sqsubseteq_\tau \langle N(O) \rangle$ for all $\langle O \rangle \in T_\sigma$.

Lemma 7.3. For any term structure and any type σ, if $\langle M \rangle \sqsubseteq_\sigma \langle N \rangle$ and $\sigma \preceq \tau$, then also $\langle M \rangle \sqsubseteq_\tau \langle N \rangle$. In other words, the inclusion maps $i_{\sigma,\tau}$ are order-preserving.

Proof. By induction on types. For basic types the order is trivial, so suppose that $\langle M \rangle \sqsubseteq_{\sigma \xrightarrow{m} \tau} \langle N \rangle$, and that $\sigma \xrightarrow{m} \tau \preceq \sigma' \xrightarrow{m'} \tau'$, so that $\sigma' \preceq \sigma$, $\tau \preceq \tau'$, and $m \sqsubseteq m'$. Then:

$$
\begin{aligned}
& \langle M \rangle \sqsubseteq_{\sigma \xrightarrow{m} \tau} \langle N \rangle \\
\Leftrightarrow \quad & \text{for all } \langle O \rangle \in T_\sigma : \langle M(O) \rangle \sqsubseteq_\tau \langle N(O) \rangle \\
\Rightarrow \quad & \text{for all } \langle O \rangle \in T_{\sigma'} : \langle M(O) \rangle \sqsubseteq_\tau \langle N(O) \rangle \\
\Rightarrow \quad & \text{for all } \langle O \rangle \in T_{\sigma'} : \langle M(O) \rangle \sqsubseteq_{\tau'} \langle N(O) \rangle \\
\Leftrightarrow \quad & \langle M \rangle \sqsubseteq_{\sigma' \xrightarrow{m'} \tau'} \langle N \rangle
\end{aligned}
$$

The second implication is because $T_{\sigma'} \subseteq T_\sigma$. The third implication is by induction hypothesis. $\dashv$

Proposition 7.4. For any term structure $\{\mathbb{T}_\tau\}_{\tau \in \mathcal{T}}$ there is an associated pre-structure $\{\mathbb{D}_\tau\}_{\tau \in \mathcal{T}}$ with order-isomorphisms $\iota_\tau : \mathbb{T}_\tau \to \mathbb{D}_\tau$, such that:

$$
\iota_{\sigma \xrightarrow{m} \tau}\langle M \rangle \big(\pi_{\sigma',\sigma}(\iota_{\sigma'}\langle N \rangle)\big) = \iota_\tau \langle M(N) \rangle. \tag{3}
$$

Proof. We build preorders $\{\mathbb{D}_\tau\}_{\tau \in \mathcal{T}}$ and order-isomorphisms $\iota_\tau : \mathbb{T}_\tau \to \mathbb{D}_\tau$ using recursion on the set of types. For base types $b \in \mathcal{B}$ we simply take $\mathbb{D}_b = \mathbb{T}_b$, and ι_b is the identity.

Suppose we have already defined $\mathbb{D}_\sigma$ and $\mathbb{D}_\tau$, and we have isomorphisms $\iota_\sigma : \mathbb{T}_\sigma \to \mathbb{D}_\sigma$ and $\iota_\tau : \mathbb{T}_\tau \to \mathbb{D}_\tau$. For $\mathbb{D}_{\sigma \xrightarrow{m} \tau}$, we use

$$
D_{\sigma \xrightarrow{m} \tau} = \{ M^* : \langle M \rangle \in T_{\sigma \xrightarrow{m} \tau} \}
$$

where

$$
\begin{aligned}
M^*\big(\iota_\sigma \langle N \rangle\big) &= \iota_\tau \langle M(N) \rangle \\
M^* \leq_{\sigma \xrightarrow{m} \tau} N^* &\text{ iff } M \sqsubseteq_{\sigma \xrightarrow{m} \tau} N \text{ in } \mathbb{T}_{\sigma \xrightarrow{m} \tau}
\end{aligned}
$$

In other words, we define M^* exactly so that (3) is satisfied. The map M^* is well-defined because $\approx$ respects term application. The order-embedding

$\iota_{\sigma \xrightarrow{m} \tau} : \mathbb{T}_{\sigma \xrightarrow{m} \tau} \to \mathbb{D}_{\sigma \xrightarrow{m} \tau}$ is obviously given by $\iota_{\sigma \xrightarrow{m} \tau}(\langle M \rangle) = M^*$. We show this map is 1-1.

Suppose $\iota_{\sigma \xrightarrow{m} \tau}\langle M \rangle = \iota_{\sigma \xrightarrow{m} \tau}\langle N \rangle$. Choose some variable $v \notin FV(M) \cup FV(N)$. Then by definition of $\iota_{\sigma \xrightarrow{m} \tau}$ we have $\iota_\tau \langle M(v) \rangle = \iota_\tau \langle N(v) \rangle$, which means by induction hypothesis that $\langle M(v) \rangle = \langle N(v) \rangle$, i.e, that $M(v) \approx N(v)$. By rule (ξ) we also have $\lambda v.M(v) \approx \lambda v.N(v)$, and by two applications of (η) and transitivity we have $M \approx N$, whence $\langle M \rangle = \langle N \rangle$.

It remains only to show that $\{\mathbb{D}_\tau\}_{\tau \in \mathcal{T}}$ is a well defined pre-structure with maps $\pi_{\sigma,\tau}$ given by

$$
\pi_{\sigma,\tau} = \iota_\tau \circ i_{\sigma,\tau} \circ \iota_\sigma^{-1}.
$$

Condition 1 in Definition 3.5 holds trivially. Condition 2 comes from condition 1 on the term structure, condition 3 from point 2, condition 4 from point 3, and condition 7 from Lemma 7.3. The functoriality properties 5 and 6 come from Proposition 7.1. $\dashv$

Lemma 7.5 is the main construction of semantic models for our calculus besides the full structures which we saw in Definition 3.6 and Example 5.3. In it, note that if ψ is an assignment function (a map from variables to terms), then composing with the natural map (taking terms to $\approx$-classes) gives a map into the term structure. So further composing with ι gives a valuation into a pre-structure. We thus define $\langle \psi \rangle$ to be the valuation function given by $\langle \psi \rangle(x) = \iota \langle \psi(x) \rangle$ for all x. What is more, every valuation function into a model of this type is of the form $\langle \psi \rangle$, and ψ is determined uniquely up to $\approx$.

Lemma 7.5. Let $\mathbb{T}$ be a term structure, and let $\mathbb{D}$ be its associated pre-structure from Proposition 7.4. Define an interpretation function in $\mathbb{D}$:

$$
[\![M]\!]_{\langle \psi \rangle} = \iota \langle M[\psi] \rangle, \tag{4}
$$

for all terms M and term substitutions ψ. Let $\mathbb{S} = (\mathbb{D}, [\![\ ]\!])$. Then $\mathbb{S}$ is a structure.

Proof. We check the two requirements on the interpretation function. The first requirement concerns applications. Let $M : \sigma \xrightarrow{m} \tau$, and $N : \sigma' \preceq \sigma$. Fix a substitution ψ. Then

$$
\begin{aligned}
& [\![M(N)]\!]_{\langle \psi \rangle} \\
= \quad & \iota \langle M(N)[\psi] \rangle && \text{by (4)} \\
= \quad & \iota \langle M[\psi](N[\psi]) \rangle && \text{by def. of } [\psi] \\
= \quad & \iota \langle M[\psi] \rangle (\pi_{\sigma',\sigma}\langle \iota(N[\psi]) \rangle) && \text{by (3)} \\
= \quad & [\![M]\!]_\psi \big(\pi_{\sigma',\sigma}[\![N]\!]_\psi\big) && \text{by (4), twice}
\end{aligned}
$$

Finally, consider a term $\lambda x.M$, and fix a substitution ψ. Let $a \in D_\sigma$, and let A be a term such that $\iota\langle A \rangle = a$. Let y be a variable which is not free in A, and also not free in M or any $\psi(z)$ for z a free variable of M. Then

$$
\begin{aligned}
& (\iota\langle(\lambda x.M)[\psi]\rangle)(a) \\
=\ & (\iota\langle\lambda y.M[\psi_x^y]\rangle)(\iota\langle A\rangle) && \text{by def. of } A \text{ and } y \\
=\ & (\lambda y.M[\psi_x^y])^*(\iota\langle A\rangle) && \text{by definition of } \iota \\
=\ & \iota((\lambda y.M[\psi_x^y])(A)) && \text{by definition of } * \\
=\ & \iota(M[\psi_x^y][\mathrm{id}_y^A]) && \text{by } \beta\text{-equivalence} \\
=\ & \iota\langle M[\psi_x^A]\rangle && \text{by choice of } y \\
=\ & [\![M]\!]_{\langle\psi_x^A\rangle} && \text{by the def. in (4)} \\
=\ & [\![M]\!]_{\langle\psi\rangle_x^a} && \text{because } \iota\langle A\rangle = a \\
=\ & [\![\lambda x.M]\!]_{\langle\psi\rangle}(a) && \text{semantics of } \lambda x.M
\end{aligned}
$$

This completes the proof. $\dashv$

8 Monotonicity Entails Positivity; Antitonicity Entails Negativity

The main result of this section is Theorem 8.2, a converse (of sorts) to Lemma 5.2.

8.1 A Term Structure Built "Freely" from an Inequality

The proof of Theorem 8.2 employs a specific term structure. Let σ be a type, and let x, y, and z be distinct variables of type σ. We take $\mathbb{T} = \mathbb{T}(x, y, z)$ to be the term structure obtained by defining for each type ρ, $\langle P \rangle \sqsubseteq_\rho \langle Q \rangle$ if and only if the following holds: P and Q are in normal form, there is a term S with no occurrences of y or z, and there are pairwise disjoint sets of occurrences A, B, Y, and Z of x in S such that all occurrences in A are positive, all occurrences in B are negative, and

$$
\begin{aligned}
P &= S[y/x_A, z/x_B, y/x_Y, z/x_Z] \\
&= S[y/x_{A\cup Y}, z_{B\cup Z}] \\
Q &= S[z/x_A, y/x_B, y/x_Y, z/x_Z] \\
&= S[y/x_{B\cup Y}, z_{A\cup Z}]
\end{aligned} \quad (5)
$$

In other words, if $\langle P \rangle \sqsubseteq_\rho \langle Q \rangle$, then we can obtain Q from P, assuming these are in normal form, by "increasing" some positive occurrences y to z (those occurrences in A) and "decreasing" some negative occurrences of z to y (those in B). The sets Y and Z are needed in order to make the whole construction work. More specifically, it follows from (5) that

$$
P[x/y, x/z] = S = Q[x/y, x/z].
$$

Lemma 8.1. $\mathbb{T}$ is a term structure.

Theorem 8.2 (Lyndon Theorem). Suppose $M : \tau$ is a typed term in normal form, and let $x : \sigma$ be a variable. Then the following are equivalent:

a. All free occurrences of x in M are $+$ $(-)$.

b. For all structures $\mathbb{S} = (\mathbb{D}, [\![\ ]\!])$ and assignments ϕ, $a \mapsto [\![M]\!]_{\phi_x^a}$ is monotone (antitone).

Proof. The (a) $\Rightarrow$ (b) directions follow from Lemma 5.2. We show (b) $\Rightarrow$ (a). We only argue that "monotone implies positive", as the argument that "antitone implies negative" is similar. Let $M : \tau$ be a term, and suppose $y : \sigma$ and $z : \sigma$ are distinct variables not appearing in M.

Fix a normal form M and a variable $x : \sigma$ that occurs freely in it. Take $\mathbb{T}$ to be the term structure $\mathbb{T}(x, y, z)$ studied in Lemma 8.1. Take ϕ to be the assignment generated by the identity substitution, $\langle\phi\rangle(w) = \langle\mathrm{id}(w)\rangle = \langle w\rangle$.

Let $(\mathbb{D}, [\![\ ]\!])$ be the structure obtained from $\mathbb{T}$ using Lemma 7.5. We apply (1b) to this structure. By monotonicity of ι, $\iota\langle y\rangle \leq_\sigma \iota\langle z\rangle$ in $\mathbb{D}$. We thus see that in $\mathbb{D}_\tau$,

$$
\begin{aligned}
& \iota\langle M[y/x]\rangle \\
=\ & [\![M]\!]_{\langle\phi_x^y\rangle} && \text{by (4)} \\
=\ & [\![M]\!]_{\langle\phi\rangle_x^{\iota\langle y\rangle}} && \text{since } \langle\phi_x^y\rangle(x) = \iota\langle y\rangle \\
\leq_\tau\ & [\![M]\!]_{\langle\phi\rangle_x^{\iota\langle z\rangle}} && \text{by hypothesis on } M \\
=\ & [\![M]\!]_{\langle\phi_x^z\rangle} && \text{since } \langle\phi_x^z\rangle(x) = \iota\langle z\rangle \\
=\ & \iota\langle M[z/x]\rangle && \text{by (4)}
\end{aligned}
$$

Since ι reflects order, $\langle M[y/x]\rangle \sqsubseteq_\tau \langle M[z/x]\rangle$.

Notice that $M[y/x]$ and $M[z/x]$ are β-normal forms, since M is a β-normal form. By definition of the order in $\mathbb{T}$, there is a term S and sets of free occurrences of x in S, say A, B, Y, and Z, such that all occurrences in A are positive, all occurrences in B are negative, and

$$
\begin{aligned}
S &= M[y/x][x/y, x/z] \\
&= M[z/x][x/y, x/z] \\
M[y/x] &= S[y/x_{A\cup Y}, z/x_{B\cup Z}] && (6) \\
M[z/x] &= S[y/x_{B\cup Y}, z/x_{A\cup Z}] && (7)
\end{aligned}
$$

However, z does not occur in the term on the left of (6), since z does not occur in M. And so $B = Z = \emptyset$. Similarly, y does not occur in the term on the right of (7), so $Y = \emptyset$ also. Thus $M[y/x] = S[y/x_A]$. And as x is not free in $M[y/x]$, it is not free in $S[y/x_A]$. This means that A is the set of

$$
\text{(Ref)}\ \frac{}{M \precsim_\sigma M}
\qquad
\text{(Trans)}\ \frac{M \precsim_\sigma N \qquad N \precsim_\sigma O}{M \precsim_\sigma O}
\qquad
\text{(Point)}\ \frac{M \precsim_{\sigma \xrightarrow{m} \tau} N}{M(O) \precsim_\tau N(O)}
$$

$$
\text{(Mono)}\ \frac{N \precsim_\sigma O}{M(N) \precsim_\tau M(O)}\ {}_{[M : \sigma \xrightarrow{+} \tau]}
\qquad
\text{(Anti)}\ \frac{N \precsim_\sigma O}{M(O) \precsim_\tau M(N)}\ {}_{[M : \sigma \xrightarrow{-} \tau]}
$$

$$
\text{(Pres)}\ \frac{M \precsim_\sigma N}{M \precsim_\tau N}\ {}_{\sigma \preceq \tau}
\qquad
\text{(Equiv)}\ \frac{}{M \precsim_\tau N}\ {}_{M \approx N}
\qquad
\text{(Func)}\ \frac{M \precsim_\tau N}{\lambda x.M \precsim_{\sigma \xrightarrow{m} \tau} \lambda x.N}
$$

Figure 3: Rules of the Monotonicity Calculus. We omit the types on the terms, except for the side conditions in the (Mono) and (Anti) rules.

all free occurrences of x in S. As a result, all free occurrences of x in S are $+$. Furthermore,

$$ S \;=\; M[y/x][x/y, x/z] \;=\; M. $$

(The last equality holds because $M[y/x][x/y, x/z]$ takes M, then changes all free occurrences of x in M to y, and then changes y and z back to x. Since y and z do not occur free in M, this is M itself.) Again, $S = M$. Thus all free occurrences of x in M are $+$, as desired. $\dashv$

9 A Complete Proof System

We come to the centerpiece of this work, the Monotonicity Calculus given by the rules of inference in Figure 3.

Syntax Our setting is similar to equational reasoning in simply typed lambda calculus (Friedman, 1975); however, our calculus deals with *inequality assertions* $M \precsim_\sigma N$. We make such assertions when the types of M and N are both $\preceq \sigma$. We use Γ for a set of inequality assertions. We write $\Gamma \vdash M \precsim_\sigma N$ if there is a proof of $M \precsim_\sigma N$ from Γ, that is, if there is a finite tree with root $M \precsim_\sigma N$, and each node either a leaf from Γ, or an application of one of the rules in Figure 3.

Example 9.1. In Figure 4 we give a derivation using our ongoing example of real functions. The derivation is similar to the one depicted in Figure 1; we only include this one in full for reasons of space. The proof of "$1 - 1 \le 2 - 0$" uses two basic assumptions: "$0 \le 1$" and "$x \le x + 0$ for any x." Note in particular the use of Lemma 9.5. (We could alternatively have assumed $\lambda x.x \precsim \lambda x. + (x)(0)$ in the same way we used the assumption `deftly` $\precsim \lambda x.x$ in Figure 1.)

Semantics Given a structure $\mathbb{S} = (\mathbb{D}, [\![\,]\!])$, we write $\mathbb{S} \vDash_\phi M \precsim_\sigma N$ if the following hold:

1. The types of M and N are both $\preceq \sigma$.

2. $\pi_{\sigma_1, \sigma}([\![M]\!]_\phi) \le \pi_{\sigma_2, \sigma}([\![N]\!]_\phi)$ in $\mathbb{D}_\sigma$.

Frequently we leave off the type σ in assertions $\mathbb{S} \vDash_\phi M \precsim_\sigma N$. We also write $\Gamma \vDash M \precsim N$ if for all structures $\mathbb{S}$ such that $\mathbb{S} \vDash_\phi G \precsim H$ for all $G \precsim H \in \Gamma$ and all assignments ϕ, we also have $\mathbb{S} \vDash_\phi M \precsim N$ for all assignments ϕ.

9.1 Basic Properties of the System

Proposition 9.2. If $\Gamma \vdash M \precsim_\sigma N$, and also $M \approx M'$ and $N \approx N'$, then $\Gamma \vdash M' \precsim_\sigma N'$

The next result is a key fact about our system. It emphasizes the fact that we take open assertions in hypothesis sets Γ to be "universally quantified."

Proposition 9.3. Let $M : \sigma_1 \to \tau_1$ and $N : \sigma_2 \to \tau_2$ be terms, let $m_1, m_2 \sqsubseteq m$, and let σ and τ be types with $\sigma \preceq \sigma_1, \sigma_2$ and $\tau_1, \tau_2 \preceq \tau$. Thus, $\sigma_1 \xrightarrow{m_1} \tau_1, \sigma_2 \xrightarrow{m_2} \tau_2 \preceq \sigma \xrightarrow{m} \tau$. Suppose that for all terms $O : \sigma$, $\Gamma \vdash M(O) \precsim_\tau N(O)$. Then $\Gamma \vdash M \precsim_{\sigma \xrightarrow{m} \tau} N$.

Proof. Let x be a variable of type σ which does not occur in M or N. Our hypotheses tell us that $\Gamma \vdash M(x) \precsim_\tau N(x)$. By (Func), $\Gamma \vdash \lambda x.M(x) \precsim_{\sigma \xrightarrow{m} \tau} \lambda x.N(x)$. By ($\eta$), (Equiv), and (Trans), $\Gamma \vdash M \precsim_\tau N$. $\dashv$

Remark 9.4. We do not know whether Proposition 9.3 holds without (Func). This would be important if one were to revise our meaning of the calculus. Currently $\Gamma \vDash M \precsim N$ means that for all structures $\mathbb{S}$ such that $\mathbb{S} \vDash_\phi G \precsim H$ for all $G \precsim H \in \Gamma$ and all assignments ϕ, we also have $\mathbb{S} \vDash_\phi M \precsim N$ for all assignments ϕ. Suppose we wish to change this to mean: for all $\mathbb{S}$ and ϕ such that $\mathbb{S} \vDash_\phi G \precsim H$ for all $G \precsim H \in \Gamma$, $\mathbb{S} \vDash_\phi M \precsim N$ for the same ϕ. Then our rules (ξ) and (Func) are no longer sound. We conjecture that dropping (ξ) and (Func) results in a complete system with the revised semantic interpretation.

$$\dfrac{\dfrac{x \precsim +(x)(0)}{1 \precsim +(1)(0)} \text{ Lemma 9.5} \quad \dfrac{0 \precsim 1}{+(1)(0) \precsim +(1)(1)} \text{ (Mono)}}{\dfrac{\dfrac{1 \precsim 2}{-(1) \precsim -(2)} \text{ (Mono)}}{-(1)(1) \precsim -(2)(1)} \text{ (Point)} \qquad \dfrac{0 \precsim 1}{-(2)(1) \precsim -(2)(0)} \text{ (Anti)}}{-(1)(1) \precsim -(2)(0)}} \text{ (Trans)}$$

Figure 4: Example of elementary monotonicity reasoning with real numbers and functions.

Lemma 9.5. For all terms M and N, and all term substitutions ψ,

$$M \precsim N \vdash M[\psi] \precsim N[\psi].$$

Proof. Let $x_1, \ldots x_n$ be the free variables of M and N. By repeated use of (Func), we have

$$M \precsim N \vdash \lambda x_1 \ldots \lambda x_n.M \precsim \lambda x_1 \ldots \lambda x_n.N$$

Now, because of (Equiv) and α equivalence, we may change each x_i into a variable y_i with $y_i \notin FV(\psi(x_j))$ for each $i, j = 1, \ldots n$. Then by repeated use of the (Point), we have

$$\begin{aligned} M \precsim N \ \vdash \ & (\lambda y_1 \ldots \lambda y_n.M')\psi(x_1) \ldots \psi(x_n) \\ \precsim \ & (\lambda y_1 \ldots \lambda y_n.N')\psi(x_1) \ldots \psi(x_n) \end{aligned}$$

where

$$\begin{aligned} M' &= M[y_1/x_1] \ldots [y_n/x_n] \\ N' &= N[y_1/x_1] \ldots [y_n/x_n] \end{aligned}$$

Then by repeated use of (Equiv) and β reductions:

$$\begin{aligned} M \precsim N \ \vdash \ & M'[\psi(x_1)/y_1] \ldots [\psi(x_n)/y_n] \precsim \\ & N'[\psi(x_1)/y_1] \ldots [\psi(x_n)/y_n] \end{aligned}$$

Now because each y_i does not appear in $\psi(x_j)$ for $j \leq i$, then these substitutions can be done simultaneously, i.e.,

$$\begin{aligned} & M'[\psi(x_1)/y_1] \ldots [\psi(x_n)/y_n] \\ &\quad = M'[\psi(x_1)/y_1, \ldots, \psi(x_n)/y_n] \\ & N'[\psi(x_1)/y_1] \ldots [\psi(x_n)/y_n] \\ &\quad = N'[\psi(x_1)/y_1, \ldots, \psi(x_n)/y_n] \end{aligned}$$

And since

$$\begin{aligned} & M'[\psi(x_1)/y_1, \ldots, \psi(x_n)/y_n] \\ &\quad = M[\psi(x_1)/x_1, \ldots, \psi(x_n)/x_n] \\ & N'[\psi(x_1)/y_1, \ldots, \psi(x_n)/y_n] \\ &\quad = N[\psi(x_1)/x_1, \ldots, \psi(x_n)/x_n] \end{aligned}$$

then $M \precsim N \vdash M[\psi(x_1)/x_1, \ldots, \psi(x_n)/x_n] \precsim N[\psi(x_1)/x_1, \ldots \psi(x_n)/x_n]$, or in other words, we have $M \precsim N \vdash M[\psi] \precsim N[\psi]$. $\dashv$

Lemma 9.6. If $\Gamma \vdash M \precsim N$, then for all term substitutions ψ, $\Gamma \vdash M[\psi] \precsim N[\psi]$.

Proof. This is an easy induction on the proof system whose base case is Lemma 9.5. $\dashv$

Theorem 9.7 (Completeness of the Monotonicity Calculus). $\Gamma \vdash M \precsim_\sigma N$ iff $\Gamma \vDash M \precsim_\sigma N$.

9.2 Soundness

We fix a structure $\mathbb{S}$ and a valuation ϕ making all sentences in Γ true in $\mathbb{S}$. We show by induction on derivations from Γ that if $\Gamma \vdash M \precsim_\sigma N$, with the types of M and N being $\sigma_1 \precsim \sigma$ and $\sigma_2 \precsim \sigma$, then $\pi_{\sigma_1,\sigma}(\llbracket M \rrbracket_\phi) \leq \pi_{\sigma_2,\sigma}(\llbracket N \rrbracket_\phi)$ in $\mathbb{S}_\sigma$. We use the properties of pre-structures in Definition 3.5.

The most basic derivations from Γ are the elements of Γ itself. This case is trivial.

We begin with (Pres). Assume we have $\Gamma \vdash M \precsim_\tau N$ via proof whose last line is an application of (Pres), with $\sigma \precsim \tau$. So $\Gamma \vdash M \precsim_\sigma N$. Our induction hypothesis tells us, $\mathbb{S} \vDash_\phi M \precsim_\sigma N_\phi$. Suppose that the types of M and N are σ_1 and σ_2, respectively. Let us write m for $\llbracket M \rrbracket_\phi$ and n for $\llbracket N \rrbracket_\phi$. Then we have $\pi_{\sigma_1,\sigma}m \leq \pi_{\sigma_2,\sigma}n$ in $\mathbb{D}_\sigma$. And now we calculate easily that $\pi_{\sigma_1,\tau}(m) \leq \pi_{\sigma_1,\tau}(n)$. As a result, $\mathbb{S} \vDash_\phi M \precsim_\tau N$.

For (Ref), we use the fact that each relation $\sqsubseteq_\sigma$ is reflexive. We omit the easy details on (Trans).

In the rest of this proof, we shall deal with assertions $\Gamma \vdash M \precsim N$ without any notation for the overall type; that is, we shall assume that the types of M and N are exactly σ.

For (Point), assume that $\llbracket M \rrbracket_\phi \leq_\sigma \llbracket N \rrbracket_\phi$ in $\mathbb{D}_\sigma$. Then by the "pointwise property" (Definition 3.5 part 4) of $\mathbb{S}$,

$$\begin{aligned} \llbracket M(O) \rrbracket_\phi \\ = \ & \llbracket M \rrbracket_\phi(\llbracket O \rrbracket_\phi) \\ \leq_\sigma \ & \llbracket N \rrbracket_\phi(\llbracket O \rrbracket_\phi) \\ = \ & \llbracket N(O) \rrbracket_\phi. \end{aligned}$$

For (Mono), let $M : \sigma \xrightarrow{+} \tau$. Assume that $\llbracket N \rrbracket_\phi \leq_\sigma \llbracket O \rrbracket_\phi$ in $\mathbb{D}_\sigma$. By Lemma 5.2,

85

$[\![M]\!]$ is a monotone function $\mathbb{D}_\sigma \to \mathbb{D}_\tau$. Hence $[\![M(N)]\!]_\phi \leq_\sigma [\![M(O)]\!]_\phi$. The (Anti) rule is treated similarly. The soundness of the (Equiv) rule follows easily from Proposition 6.6.

9.3 Completeness

We next show that the monotonicity calculus is complete. Assume that $\Gamma \models M^\star \precsim_\sigma N^\star$, with the types of $M^\star$ and $N^\star$ being σ_1 and σ_2. We shall show that $\Gamma \vdash M^\star \precsim_\sigma N^\star$ using a term structure $\mathbb{T}$ whose associated structure $\mathbb{S}$ from Lemma 7.5 is called the *canonical model of* Γ. We recall Definition 7.2 and the notation there, especially the fact that each T_σ is the set of $\approx$-equivalence classes of terms of type $\preceq \sigma$. We order T_σ by

$$\langle M \rangle \sqsubseteq_\sigma \langle N \rangle \quad \text{iff} \quad \Gamma \vdash M \precsim_\sigma N.$$

This relation is well-defined by Proposition 9.2.

Claim 9.8. $\mathbb{T}$ is a term structure.

Proof. Let $\langle M \rangle \in T_{\sigma \overset{+}{\to} \tau}$ and $\langle N \rangle \sqsubseteq_\sigma \langle O \rangle$. Then $M : \sigma' \overset{+}{\to} \tau'$, for some types $\sigma \preceq \sigma'$ and $\tau' \preceq \tau$. We thus have a derivation from Γ:

$$
\cfrac{\cfrac{\cfrac{\vdots}{N \precsim_\sigma O}}{\cfrac{N \precsim_\sigma O}{N \precsim_{\sigma'} O} \text{ (Prec)}}{M(N) \precsim_{\tau'} M(O)} \text{ (Mono)}}{M(N) \precsim_\tau M(O)} \text{ (Prec)}
$$

Thus, $\langle M(N) \rangle \sqsubseteq_\tau \langle M(O) \rangle$. For a similar reason, Property 2 also holds, only (Anti) is used instead. One direction of Property 3 is easy, using (Point), and so we omit it. For the reverse direction, suppose $\langle M \rangle(\langle O \rangle) \sqsubseteq_\tau \langle N \rangle(\langle O \rangle)$ for all $\langle O \rangle \in T_\sigma$. Using Proposition 9.3, we see that $\Gamma \vdash M \precsim_{\sigma \overset{m}{\to} \tau} N$. $\quad\dashv$

Claim 9.8 proved, we appeal to Proposition 7.4 and Lemma 7.5 to obtain an associated structure which we call $\mathbb{S}$. As we know, for any assignment ϕ on $\mathbb{D}$, there is a substitution ψ such that $\langle \psi \rangle = \phi$. To find such a ψ, note that for every variable x, $\phi(x) = \iota(\langle A \rangle)$ for some term A, so we can define $\psi(x) = A$. Then $\langle \psi \rangle(x) = \iota(\langle \psi(x) \rangle) = \iota(\langle A \rangle)$.

Lemma 9.9. Let $G : \sigma_1$ and $H : \sigma_2$, with $\sigma_1, \sigma_2 \preceq \sigma$. If $\Gamma \vdash G \precsim_\sigma H$, then $\mathbb{S} \models G \precsim_\sigma H$. In other words, for every assignment ϕ,

$$\pi_{\sigma_1,\sigma}([\![G]\!]_\phi) \leq_\sigma \pi_{\sigma_2,\sigma}([\![H]\!]_\phi).$$

In particular, for every assertion $G \precsim_\sigma H$ in Γ, $\mathbb{S} \models G \precsim_\sigma H$.

Proof. Suppose $\Gamma \vdash G \precsim_\sigma H$, and fix an assignment ϕ, and let ψ be a substitution such that $\langle \psi \rangle = \phi$. By Lemma 7.5,

$$[\![G]\!]_\phi = [\![G]\!]_{\langle\psi\rangle} = \iota(\langle G[\psi] \rangle), \qquad \text{and}$$
$$[\![H]\!]_\phi = [\![H]\!]_{\langle\psi\rangle} = \iota(\langle H[\psi] \rangle).$$

As $\pi_{\sigma_1,\sigma}$, $\pi_{\sigma_2,\sigma}$, and ι are all order preserving, to prove our lemma we only need to show that $\langle G[\psi] \rangle \sqsubseteq_\sigma \langle H[\psi] \rangle$: in other words, $\Gamma \vdash G[\psi] \precsim_\sigma H[\psi]$. And this was shown in Lemma 9.6. $\quad\dashv$

We conclude the proof of completeness. We started with $\Gamma \models M^\star \precsim N^\star$, and now we show that $\Gamma \vdash M^\star \precsim N^\star$. By Lemma 9.9, the canonical model $\mathbb{S}$ satisfies Γ under all assignments. We apply this with the assignment ϕ given by $\phi(x) = \langle x \rangle$. Therefore $\mathbb{S} \models_\phi M^\star \precsim_\sigma N^\star$. So $\langle M^\star[\phi] \rangle \sqsubseteq_\sigma \langle N^\star[\phi] \rangle$. However, because $\langle M^\star[\phi] \rangle = \langle M^\star \rangle$ and $\langle N^\star[\phi] \rangle = \langle N^\star \rangle$, we see from the ordering on $\mathbb{S}$ that $\Gamma \vdash M^\star \precsim_\sigma N^\star$.

10 Conclusion

We have presented a calculus extending the simply typed lambda calculus with enough order-theoretic infrastructure to represent arguments about increasing and decreasing functions. The calculus provides a mathematical foundation for a style of monotonicity reasoning that is often implicit in practical NLP work (e.g., MacCartney and Manning 2007; Angeli and Manning 2014; Angeli et al. 2016). Typically this research draws upon lexical resources such as WordNet and entailment relations learned from data, and uses these assumptions as input for proof search over derivations similar to those we considered here. Functional expressions will be marked as monotone or antitone either "by hand" or in an automated way.

In addition to formalizing the proof procedures used in existing work, we believe the present study also suggests further possibilities for applied work. For instance, we have shown how more complex entailment assumptions can be stated and used in a flexible way, e.g., allowing comparison between functions of different polarity types (recall the example in Section 2).

From a technical point of view, our completeness result is analogous to a standard result for simply typed lambda calculus due to Friedman (1975). While our setting is considerably more complex, there still remain open issues here that were settled in the simpler setting: e.g., how to obtain completeness for "full" structures (Ex. 5.3). We leave such questions for future work.

References

Lasha Abzianidze. 2015. A tableau prover for natural logic and language. In *EMNLP*. pages 2492–2502.

Gabor Angeli and Christopher Manning. 2014. NaturalLI: Natural logic inference for common sense reasoning. In *EMNLP*. pages 534–545.

Gabor Angeli, Neha Nayak, and Christopher Manning. 2016. Combining natural logic and shallow reasoning for question answering. In *ACL*. pages 442–452.

Raffaella Bernardi. 2002. *Reasoning with Polarity in Categorial Type Logic*. Ph.D. thesis, University of Utrecht.

Samuel R. Bowman, Christopher Potts, and Christopher D. Manning. 2015. Learning distributed word representations for natural logic reasoning. In *AAAI Spring Symposium on Knowledge Representation and Reasoning*. pages 10–13.

Harvey Friedman. 1975. Equality between functionals. In Rohit Parikh, editor, *Proceedings of Logic Colloquium '73*. volume 53 of *Lecture Notes in Mathematics*, pages 22–37.

Bart Geurts. 2003. Reasoning with quantifiers. *Cognition* 86(3):223–251.

Bart Geurts and Frans van der Slik. 2005. Monotonicity and processing load. *Journal of Semantics* 22:97–117.

Thomas F. Icard. 2012. Inclusion and exclusion in natural language. *Studia Logica* 100(4):705–725.

Thomas F. Icard and Lawrence S. Moss. 2013. A complete calculus of monotone and antitone higher-order functions. In *Proceedings, Topology, Algebra, and Categories in Logic 2013*, Vanderbilt University, volume 23 of *EPiC Series*, pages 96–99.

Thomas F. Icard and Lawrence S. Moss. 2014. Recent progress on monotonicity. *Linguistic Issues in Language Technology* 9(7):167–194.

Hans Kamp and Barbara Hall Partee. 1995. Prototype theory and compositionality. *Cognition* 57(2):129–191.

Roger C. Lyndon. 1959. An interpolation theorem in the predicate calculus. *Pacific Journal of Mathematics* 9(1):129–142.

Bill MacCartney and Christopher D. Manning. 2007. Natural logic for textual inference. In *ACL Workshop on Textual Entailment and Paraphrasing*. pages 193–200.

Bill MacCartney and Christopher D. Manning. 2009. An extended model of natural logic. In *IWCS-8, Proceedings of the Eighth International Conference on Computational Semantics*. pages 140–156.

Lawrence S. Moss. 2012. The soundness of internalized polarity marking. *Studia Logica* 100(4):683–704.

Reinhard Muskens. 2010. An analytic tableau system for natural logic. In Maria Aloni, Haarold Bastiaanse, Tikitu de Jager, and Katrin Schulz, editors, *Logic, Language and Meaning*, Lecture Notes in Computer Science, volume 6042, pages 104–113.

Victor Sánchez-Valencia. 1991. *Studies on Natural Logic and Categorial Grammar*. Ph.D. thesis, Universiteit van Amsterdam.

Allen Stoughton. 1988. Substitution revisited. *Theoretical Computer Science* 59(3):317–325.

William Tune. 2016. *A Lambda Calculus for Monotonicity Reasoning*. Ph.D. thesis, Indiana University.

Johan van Benthem. 1986. *Essays in Logical Semantics*. Reidel, Dordrecht.

Jan van Eijck. 2007. Natural logic for natural language. In Balder ten Cate and Henk Zeevat, editors, *6th International Tbilisi Symposium on Logic, Language, and Computation*. Springer, pages 216–230.

A. Zamansky, N. Francez, and Y. Winter. 2006. A 'natural logic' inference system using the Lambek calculus. *Journal of Logic, Language, and Information* 15(3):273–295.

DAG Automata for Meaning Representation

Frank Drewes
Umeå University, Sweden
drewes@cs.umu.se

Abstract

Languages of directed acyclic graphs (DAGs) are of interest in Natural Language Processing because they can be used to capture the structure of semantic graphs like those of Abstract Meaning Representation. This paper gives an overview of recent results on a family of automata recognizing such DAG languages.

1 Introduction

This paper attempts to survey, motivate, and explain recent work on automata that recognize sets of directed acyclic graphs (DAGs). These automata are called DAG automata and the languages they recognize regular DAG languages. While DAG automata are of interest in various areas of computer science, different application areas place different requirements on what constitutes a good model of such automata. Here we are interested in DAG automata that are suitable for capturing the structure of semantic graphs in Natural Language Processing, and in particular meaning representations such as Abstract Meaning Representation (AMR).

AMR was introduced by Banarescu et al. (2013) as a domain-independent graph notation for the semantics of meanings in natural language, and since then a quickly growing AMR bank for English has been built.[1] The purpose of such graphbanks is to enable research towards domain-independent semantic language processing, thus mirroring the advances in syntactic processing that have been made during the past 20 years thanks to the existence of large syntactic treebanks.

AMR serves a similar purpose in the realm of semantic representation as the well-known con-stituent tree does for syntactic representation. For the latter, we have a great variety of well-studied formal models for capturing the structure of correct representations, distinguishing them from faulty ones, and efficiently processing these formal objects. One of the simplest and at the same time most useful models is the finite-state tree automaton or, equivalently, the regular tree grammar (see Gécseg and Steinby (1984, 1997) and Drewes (2006, Appendix A)). Simplicity, though resulting in limited expressive power, is an is an asset in this context. It is to their simplicity that finite-state tree automata owe their usefulness. One of the advantages of the model is that it can easily be extended by weights (Fülöp and Vogler, 2009), then associating with every tree a value that indicates how "good" or perhaps probable the tree is. Such weighted automata are especially useful in Natural Language Processing where one rarely finds a clear dividing line between correct and wrong representations, and where one furthermore has to find ways to resolve ambiguities.

An AMR[2] is usually not a tree but a DAG such as the (somewhat simplified and abstracted) AMR in Figure 1. Its vertices are mostly PropBank concepts (Kingsbury and Palmer, 2002) connected by edges which are labelled by role labels, intuitively supplying the concepts with their semantic arguments. Readers familiar with dependency trees probably notice the similarity, but also the difference: if several concepts share a semantic argument the latter is still represented only once, being pointed to by several edges. This is what turns AMRs into DAGs, thus calling for formal models that allow to specify DAG languages or even weighted DAG languages. Unsurprisingly, it turns

---

[1] See http://amr.isi.edu.

[2] We use the abbreviation AMR to refer to the general concept of Abstract Meaning Representation as defined by Banarescu et al. (2013), but also to refer to individual graphs that follow the AMR specification.

Proceedings of the 15th Meeting on the Mathematics of Language, pages 88–99,
London, UK, July 13–14, 2017. ©2017 Association for Computational Linguistics

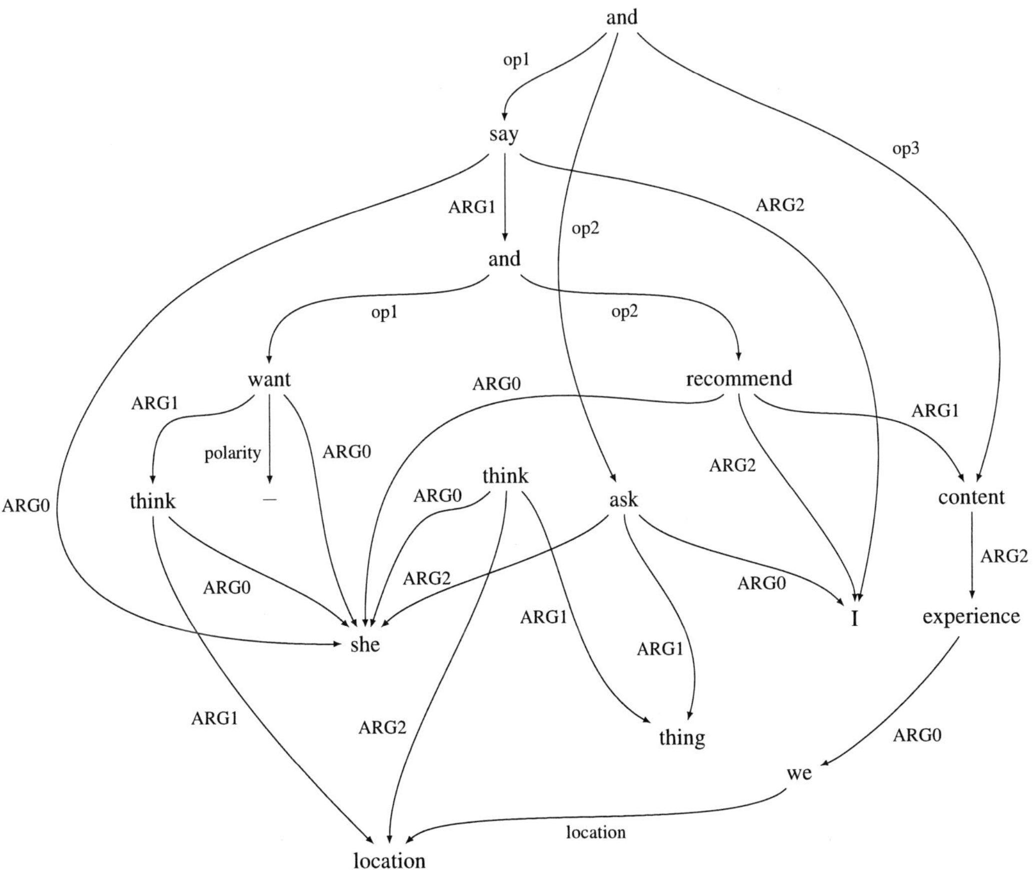

Figure 1: An AMR from the AMR Bank: "I asked her what she thought about where we'd be and she said she doesn't want to think about that, and that I should be happy about the experiences we've had (which I am)." The PropBank frame names (the vertex labels) have been simplified for the sake of readability. They can be found in Chiang et al. (2016), which also uses this example. As a side note, one may note that the AMR does not specify whether her saying was the answer to my asking, the other way around, or the two were independent. This, however, would contribute to a discussion about the limits of AMR rather than about formal automata models for AMR.

out that the greater complexity and expressivity of DAGs makes it all too easy to develop natural types of DAG automata that are surprisingly powerful, thus lacking the simplicity that is the great advantage of finite-state tree automata.

The amount of research that has been devoted to DAG automata models that extend finite-state tree automata to the realm of DAGs is rather limited. Moreover, researchers have come up with various *different* models that, owing to different intended areas of application, work on different types of DAGs and exhibit different properties. Most appear to be more powerful than what appears to be reasonable from a linguistic point of view, thus making the model more complex than

desirable. One of the potential problems that comes with greater complexity is computational inefficiency, another one is that such models, when being trained, are prone to overfitting.

2 What Is a "Good" Automaton Model for Meaning Representation?

Let us have a look at a few aspects that come to mind when thinking about an appropriate automaton model for meaning representation such as AMR.

2.1 Types of DAGs

A typical meaning representation, such as the one in Figure 1 carries labels on both vertices and edges. However, as an edge labelled a may easily be replaced by a sequence of two unlabelled edges with a vertex labelled a in between, one may prefer to view edge labels as "syntactic sugar" which, in the name of simplicity, can be removed from the core formalism.

As in the case of tree languages, DAG languages may be of bounded or unbounded vertex degree (usually called *ranked* and *unranked* in the context of tree languages). In principle, there is no bound on the number of incoming or outgoing edges of vertices in meaning representations because an arbitrary number of optional modifiers may be attached to a concept, and there may be any number of references to a vertex (cf. the 'she' vertex in Figure 1). Thus, the vertex degree is only bounded by the length of the sentence represented, and even this is only true as long as representations of single sentences are considered. In the future, one may very well want to consider AMRs that represent an entire text. However, it is certainly meaningful to study DAG languages of bounded degree as an important base case, especially if the two can be linked in some way. Such a link will be discussed in Section 7.

Another question is whether there should be an order on the incoming and outgoing edges of a vertex. Standard edge labels such as `arg0`, `arg1`, etc seem to indicate an order on outgoing edges. However, there can also be other labels, such as `polarity` and `location` in Figure 1, and incoming edges are not naturally equipped with an order in the first place. This indicates that meaning representations should be viewed as unordered DAGs. On the other hand, as an order only adds expressiveness without affecting the formal properties of the resulting model very much, one may wish to study both possibilities, especially as long as DAG languages of bounded degree are considered. One reason why the ordered case is of interest is that it can more readily be related to the case of regular tree languages (which consist of ordered trees). Another one is that the notion of determinism is rather meaningless in the unordered case.

One may consider DAGs with multiple roots or DAGs which are required to have a unique root.[3] As discussed by Chiang et al. (2016) AMRs should preferably be viewed as multiply-rooted DAGs because their standard representation turns them into singly-rooted graphs only by introduc-

ing cycles. Moreover, ensuring single-rootedness often requires the addition of a topmost 'and' vertex. This may become awkward later on if several related sentences shall be viewed as a collection of statements to be represented in a single DAG. As will be discussed below, there are also good formal reasons for not imposing the single-root requirement (and not providing DAG automata with this ability either).

In the literature one also finds DAG automata that work on planar DAGs obeying additional structural conditions (Kamimura and Slutzki, 1981), and finite-state tree automata applied to trees with maximal sharing, meaning that isomorphic subtrees are represented only once and can thus be processed only once even in the nondeterministic case (Charatonik, 1999; Anantharaman et al., 2005). Meaning representations are often non-planar once they get sufficiently complex, and may contain isomorphic substructures that represent distinct instances of otherwise identical concepts and relations.

Several other types of DAG automata are briefly discussed by Chiang et al. (2016). Together with (Quernheim and Knight, 2012; Blum and Drewes, 2016, 2017; Drewes, 2017), the latter represents the line of research discussed in this paper. Despite a rather different formal presentation, its DAG automata are closely related to those by Priese (2007), except for the fact that the latter includes a concept of initial and final states, and can thus in particular restrict the number of roots of accepted DAGs. The importance of this distinction will be discussed in Section 4.

2.2 Expressive Power vs Algorithmic and Language Theoretic Properties

As the expressive power of formal models increases, their algorithmic and language theoretic properties become less favourable. As mentioned initially, for meaning representations expressive power appears to be less important than simplicity and good computational properties. In practice, (weighted) DAG automata will have to be learned from and trained on large semantic graphbanks. For this to be feasible, DAG automata must be implemented as efficiently as possible, it must be possible to check their behaviour, and good closure properties may be required as well.

What makes it reasonable to look for a comparatively weak type of DAG automaton is that mean-

---

[3]We call a vertex a root if it has no incoming edges.

ing representations, viewed as DAG languages, seem to be relatively simple. One aspect that can be used as an indicator of sufficient simplicity is the complexity of the path languages of recognizable DAG languages. Given a DAG language (with labels only on the vertices, say), its path language consists of all strings obtained by reading the symbols on root-to-leaf paths in the DAGs of the language. It seems linguistically reasonable to assume that, for DAG languages formalizing meaning representations such as AMR, these path languages are regular. Hence, DAG automata that give rise to non-regular path languages are suspiciously powerful from the point of view of meaning representation and should be avoided in favour of simpler ones.

3 DAG Languages of Bounded Degree

Let us now give a formal definitions of DAGs and, afterwards, DAG automata on DAGs of bounded vertex degree.

3.1 DAGs

Let Σ be a (finite) alphabet of vertex labels. A *directed graph* over Σ is a tuple $G = (V, E, lab, src, tar)$ consisting of

- finite sets V and E of *vertices* and *edges*,

- mappings $src, tar\colon E \to V$ associating with each edge $e \in E$ a *source* $src(e)$ and a *target* $tar(e)$, and

- a mapping $lab\colon V \to \Sigma$ that assigns a label $lab(v)$ to every vertex $v \in V$.

For every vertex v we let $IN(v) = tar^{-1}(v)$ and $OUT(v) = src^{-1}(v)$ denote its sets of incoming and outgoing edges.

A path from u to v is an alternating sequence $v_0 e_1 \cdots v_{k-1} e_k v_k$ of vertices and edges such that $v_0 = u$, $v_k = v$, and $\{v_{i-1}, v_i\} = \{src(e_i), tar(e_i)\}$ for all $i \in \{1, \ldots, k\}$. The path is empty if $k = 0$, directed if $v_{i-1} = src(e_i)$ for all $i \in \{1, \ldots, k\}$, and a (simple) cycle if $v_1, \ldots, v_k$ are pairwise distinct and $u = v$. G is acyclic, a *DAG* for short, if it does not contain any nonempty directed cycle.

3.2 DAG Automata

Given a finite set Q, let us denote the set of all finite multisets over Q by $\mathcal{M}(Q)$. In other words, $\mathcal{M}(Q)$ is the set of all functions $M\colon Q \to \mathbb{N}$. A *DAG automaton* is a triple $A = (\Sigma, Q, R)$ consisting of

- an alphabet Σ,

- a finite set Q of states, and

- a finite set R of rules $I \xrightarrow{\sigma} O$, where $I, O \in \mathcal{M}(Q)$ and $\sigma \in \Sigma$.

A *run* of A on a DAG $D = (V, E, lab, src, tar)$ is a mapping $\rho\colon E \to Q$. For a set $E' = \{e_1, \ldots, e_n\}$ of edges in E, we let $\rho(E')$ denote the multiset $\{\rho(e_1), \ldots, \rho(e_n)\}$. (Formally, $\rho(E')(q) = |\{i \in \{1, \ldots, n\} \mid \rho(e_i) = q\}|$ is the number of times q occurs in $\rho(e_1), \ldots, \rho(e_n)$.) The run ρ is *accepting* if it is locally consistent with the rules in R, i.e., if

$$\rho(IN(v)) \xrightarrow{lab(v)} \rho(OUT(v))$$

is in R for every vertex $v \in V$. Naturally, a DAG D is *accepted* by A if there exists an accepting run of A on D. The *DAG language accepted by* A is the set $L(A)$ of all connected and nonempty DAGs accepted by A. Following Blum and Drewes (2017) we shall in the following call DAG languages of the form $L(A)$ *regular DAG languages*.

The fact that we restrict $L(A)$ to connected and nonempty DAGs deserves a brief reflexion. By the definition of acceptance, a DAG is accepted by A if and only if each of its connected components is accepted individually. Hence the set of all accepted DAGs is uniquely determined by $L(A)$. In contrast to $L(A)$ it is never empty (it always contains the empty DAG, which is the disjoint union of zero DAGs in $L(A)$), and it is finite if and only if $L(A)$ is empty. This shows that $L(A)$ is a much more meaningful object of study. In particular, with this definition of $L(A)$ it becomes sensible to ask whether emptiness and finiteness are decidable properties.

3.3 Variants

As discussed above, one may sometimes prefer to work with ordered DAGs instead of the unordered variant defined above. In that case, a DAG is conveniently defined to be a tuple (V, E, IN, OUT, lab), where $IN(v)$ and $OUT(v)$ are sequences of edges, and the components I and O of a rule $I \xrightarrow{\sigma} O$ are sequences rather than multisets of states. Similarly to the definition above, a run ρ would be accepting if $\rho(IN(v)) \xrightarrow{lab(v)} \rho(OUT(v))$ is in R for every vertex $v \in V$, with ρ being extended to a function from sequences of edges to sequences of states in the canonical way.

For a semiring $\mathbb{S}$, a *weighted DAG automaton* with weights in $\mathbb{S}$ is obtained by turning R into a function that assigns a weight in $\mathbb{S}$ to every potential rule in such a way that all but a finite number of rules are assigned the weight zero. This works in both the ordered and the unordered case. The weight of a run on a DAG D is then the product of the weights of the rules applied at the individual vertices, and the weight of D is the sum of the weights of all runs on D. As usual, unweighted DAG automata are obtained as a special case by choosing the Boolean semiring as $\mathbb{S}$ and representing the set R of rules by its characteristic function.

3.4 Example

As a simple but instructive example, let $\Sigma = \{a, b, \diamond\}$ and $Q = \{p, p', q, q'\}$ with the following rules (where ε denotes the empty sequence):

$$\varepsilon \xrightarrow{a} pp', \ p \xrightarrow{a} pp',$$
$$p' \xrightarrow{\diamond} q'$$
$$pq' \xrightarrow{b} q, \ qq' \xrightarrow{b} q,$$
$$pq' \xrightarrow{b} \varepsilon, \ qq' \xrightarrow{b} \varepsilon.$$

A run on one of the DAGs accepted by this DAG automaton is shown in Figure 2. Note that the second outgoing edge of every a is assigned the state p', which then becomes a q' by passing through $\diamond$, and every b requires an incoming q'. It follows that all accepted DAGs have equal numbers of as and bs. In the DAG of Figure 2 this makes sure that the path not containing any $\diamond$ (i.e., the one obtained by intersection with the regular language a^*b^*) is of the form $a^n b^n$.

4 What a Difference a Root Makes

The preceding example seems to show that the DAG automata discussed here are more powerful than intended, as the path language of $L(A)$ appears to be non-regular. In fact, path languages even seem to exceed the context-free languages as it is easy to add further letters in the same way, thus obtaining paths like $a^n b^n c^n d^n$. However, this is true only if $L(A)$ is restricted to DAGs with a unique root. As there is no way for A to ensure this, we can take a second accepted DAG, add it

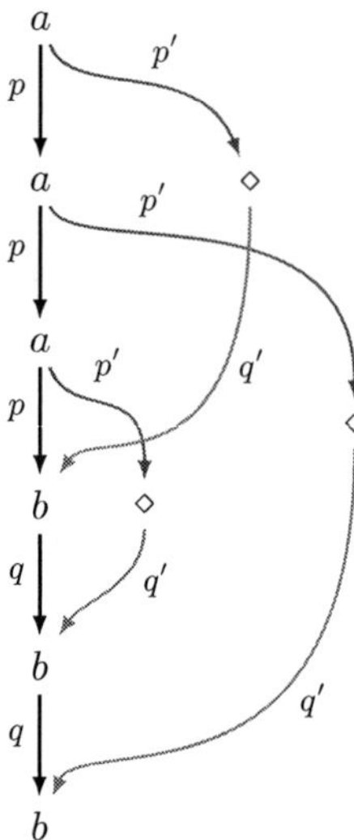

Figure 2: A run of the DAG automaton in Section 3.4; for better visual clarity edges carrying states p' and q' are drawn in blue and red, respectively.

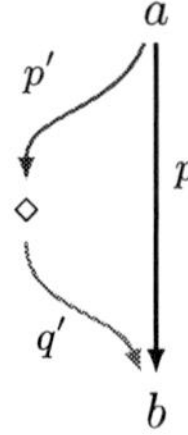

Figure 3: Another run of the DAG automaton in Section 3.4

disjointly to the one in Figure 2, and then connect the two by swapping the targets of two edges with the same state. This swapping operation turns out to be a powerful tool for proofs as it, by the definition of accepting runs, preserves acceptance. For example, by using the DAG in Figure 3 as the second component on can construct the accepting run in Figure 4 on a DAG that contains the paths $a^3 b$ and ab^3.

Thus, if we let $L_u(A)$ denote the set of all DAGs in $L(A)$ having exactly one root, then the path language of $L_u(A)$ may indeed be non-context-free. In contrast, it can be shown that the path language of $L(A)$ is regular for every DAG automaton A. In fact, we can *unfold* a DAG $D = (V, E, lab, src, tar)$ into a set of trees over Σ, as follows. For a vertex $v \in V$ with $lab(v) = \sigma$

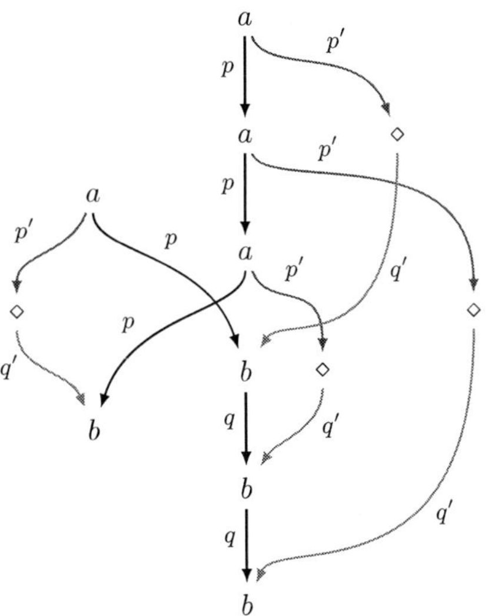

Figure 4: A possible combination of the runs in Figures 2 and 3

and $OUT(v) = \{e_1, \ldots, e_k\}$, let $unfold_D(v)$ be the tree whose root is labelled σ and whose direct subtrees are the trees $unfold_D(tar(e_i))$ for $i = 1, \ldots, k$. Note that this works for ordered DAGs as well as for unordered ones. In the former case $OUT(v) = e_1 \cdots e_k$ is a sequence of edges rather than a set and the direct subtrees are ordered accordingly.

Now the unfolding $unfold(D)$ of D yields the set of all trees $unfold_D(v)$ such that v is a root of D, and the unfolding of a DAG language L is given by $\bigcup_{D \in L} unfold(D)$. Pretending for the moment that we work in the ordered setting, one can then use edge swapping to show that $unfold(L(A))$ is a regular tree language, mainly by turning every rule $p_1 \cdots p_m \xrightarrow{\sigma} q_1 \cdots q_n$ into the set of rules $p_i \to \sigma(q_1, \ldots, q_n)$, obtaining a regular tree grammar. (An initial nonterminal S must be added, with all rules $S \to \sigma(q_1, \ldots, q_n)$ for which A contains the rule $\varepsilon \xrightarrow{\sigma} q_1 \cdots q_n$.)

However, for this argument to work properly one first has to remove useless rules from A because otherwise the regular tree grammar may generate trees that are unfoldings of DAGs not in $L(A)$. Useless rules can be detected by a technique that also allows us to decide whether $L(A)$ is empty. The key observation is that A can be seen as a Petri net whose places are the states of A. Every rule $I \xrightarrow{\sigma} O$ corresponds to a Petri net transition that consumes tokens from the places in

I and produces tokens on the places in O. Again, this works in both the ordered and the unordered case as the Petri net is oblivious to an order on its places. Deciding the emptiness of $L(A)$ boils down to the question whether the Petri net has a *zero cycle*, meaning that it can take the empty configuration back to itself. This is because a run can be viewed as a top-down process that starts with the empty DAG. It then creates a root according to a rule, thus creating some outgoing "dangling" edges with states assigned to them. Then the process continues by applying rules, always taking some of the still unprocessed dangling edges, making them the incoming edges of a vertex and producing new dangling edges with assigned states (unless the vertex created is a leaf). Finally, all dangling edges (and thus "unused" states) must have vanished, corresponding to the null configuration of the Petri net.

Petri nets with a zero cycle are also said to be *structurally cyclic*. Deciding this property is a special case of the Petri net reachability problem which can be solved in polynomial time (Drewes and Leroux, 2015). Consequently, emptiness of regular DAG languages can be decided in polynomial time as well. Moreover, the result in (Drewes and Leroux, 2015) is obtained by presenting a polynomial-time algorithm that computes the set of all transitions of the Petri net that occur in zero cycles. Since these transitions are exactly the useful rules of A, the algorithm detects the latter.

With these pieces, especially the removal of useless rules, in place it can furthermore be shown that the finiteness problem is solvable in polynomial time as well. On the other hand, considering $L_u(A)$ instead of $L(A)$, detection of useless rules, emptiness, and finiteness all become as hard as general Petri net reachability. Though this problem is decidable as well (Mayr, 1984; Kosaraju, 1982), no primitive recursive upper bound on its complexity is known.

In summary:

1. The unfolding of a regular DAG language (of ordered DAGs) yields a regular tree language; in particular, since the path language of a regular DAG language obviously coincides with the path language of its unfolding, it is a regular string language. In contrast, the path language of $L_u(A)$ is not necessarily context-free. Furthermore, the latter shows that the unfolding of $L_u(A)$ is not necessar-

ily a context-free tree language.

2. There is a polynomial algorithm that detects the useless rules of a DAG automaton A. However, one should be aware that some of the remaining rules may be useless for generating DAGs in $L_u(A)$. Detecting those requires to solve the general Petri net reachability problem, which may very well be one of the hardest decidable problems there is.

3. Similarly, the emptiness and finiteness problems can be solved in polynomial time for $L(A)$, but solving the same problems for $L_u(A)$ again requires to solve the general Petri net reachability problem.

These results are especially interesting in the light of the fact that most notions of DAG automata known from the literature can simulate the DAG automata discussed here and are, in addition, able to restrict the number of roots to one.

5 Further Properties of Regular DAG Languages

5.1 Closure under Set Theoretic Operations

Unsurprisingly, the class of regular DAG languages turns out to be closed under union and intersection, using the standard constructions. Slightly less obvious is the fact that that it is *not* closed under complementation or set difference. Consider the language L of all connected nonempty DAGs over $\{a, b\}$ in which every vertex is either a root of out-degree 2 or a leaf of in-degree 2. Thus, the elements of L are simple undirected cycles. Clearly, L is regular, and so is the subset containing only the DAGs over $\{a\}$. Their difference $L_b = L \setminus L'$ is the set of all DAGs in L containing at least one b. Now, assuming that L_b is regular and fixing a DAG automaton that accepts it, consider a run on a DAG $D \in L_b$. If D is large enough, it contains two edges with the same state that can be swapped in such a way that the resulting DAG D' falls apart into two components. As D' is still accepted, each of its components is in L_b. However, if we choose D in such a way that it contains only one b, then only one component of D' can contain that b, a contradiction.

5.2 Pumping

Swapping edges also yields two pumping lemmata. Both work by iterated application of the swapping operation. Given a DAG D, two edges e, e' of D, and some $n \geq 0$, let $D(e \bowtie e')^n$ denote the DAG obtained from $n + 1$ isomorphic copies $D_0, \ldots, D_n$ of D by swapping the copy of e' in D_{i-1} with the copy of e in D_i, for all $i \in \{1, \ldots, n\}$. (As before, swapping two edges f and f' means that $tar(f)$ and $tar(f')$ become the targets of f' and f, respectively.)

Pumping Property 1 For every DAG automaton A there is a constant c such that every DAG $D \in L(A)$ with at least c edges contains edges e, e' such that

(1) A accepts $D(e \bowtie e')^n$ for all $n \geq 0$ and

(2) each $D(e \bowtie e')^n$ contains a connected component of size $\geq n$.

The construction used to prove this property (Blum and Drewes, 2017) actually creates connected components in (b) that grow at a constant rate, which means that regular DAG languages exhibit a linear growth property: if we sort the DAGs in a given regular DAG language by size (number of vertices, say), then there is a global constant d such that the sizes of two consecutive DAGs differ by at most d.

Pumping Property 2 If a regular DAG language L contains a DAG D that has an undirected cycle, then D contains an edge e such that $D(e \bowtie e)^n \in L$ for all $n \geq 0$.

The property follows from the simple fact that swapping an edge e on a cycle in one copy of D with its counterpart in a second copy of D creates a connected DAG. This property entails an interesting consequence, as it is not difficult to show that $\{D\}$ is regular for every connected nonempty DAG D that does *not* contain a cycle.[4] Hence, by the closedness under union, finite sets of DAGs that do not contain cycles are regular. Together with the above property this proves that a finite set of connected nonempty DAGs is regular if and only if none of its DAGs contains a cycle.

6 Recognition

One of the most central computational problems for automata is recognition, also known as the membership problem: given an object D, in this

---

[4]This is a special case of the more general fact that every regular tree language is a regular DAG language (of ordered DAGs).

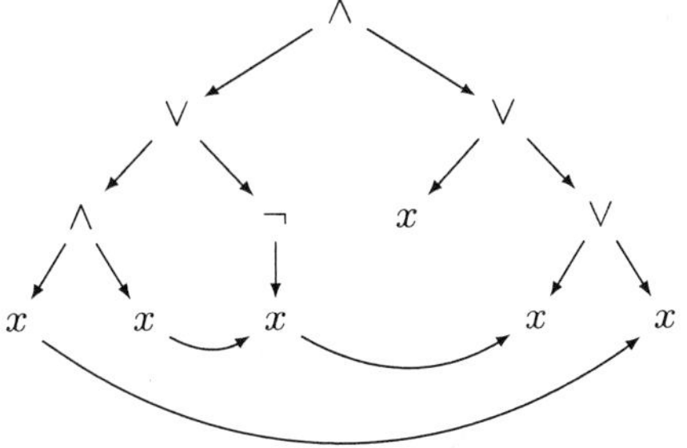

Figure 5: DAG representation of the propositional formula $((x_1 \wedge x_2) \vee \neg x_2) \wedge (x_3 \vee (x_2 \vee x_1))$

case a DAG, is it accepted by the automaton? In the weighted case, the answer is not *yes* or *no*, but the actual weight of D. Recognition comes in two flavors, depending on whether the automaton is part of the input – the *uniform recognition problem* – or not – the *non-uniform recognition problem*. For linguistic applications the potentially more difficult uniform version is the more relevant one, which should preferably be efficiently solvable. Unfortunately, it turns out that even non-uniform recognition is NP-complete, i.e., there exist fixed DAG automata whose accepted language is NP-complete.

6.1 NP-Completeness

An NP-complete regular DAG language is surprisingly easy to construct. Take the regular tree language representing propositional formulas over $\wedge$, $\vee$, $\neg$, and x, where x denotes an (anonymous) occurrence of a propositional variable. Link all occurrences of x that represent the same variable by a linear chain of edges. An example of such a DAG representation of the propositional formula $((x_1 \wedge x_2) \vee \neg x_2) \wedge (x_3 \vee (x_2 \vee x_1))$ is shown in Figure 5. Now a DAG automaton can verify that a formula is satisfiable, as follows: use states t and f representing truth values and rules such as $tt \xrightarrow{x} t$ and $ff \xrightarrow{x} f$, which guess a consistent assignment of truth values to all occurrences of the same variable, and $ff \xrightarrow{\vee} f$, $tf \xrightarrow{\vee} f$, $ft \xrightarrow{\vee} f$, $tt \xrightarrow{\wedge} t$, which implement the operators. Moreover, for every rule $I \xrightarrow{\sigma} t$ there is a rule $I \xrightarrow{\sigma} \varepsilon$. Clearly, such an automaton accepts a DAG representing a propositional formula φ in the way described above if and only if φ is satisfiable.

6.2 Recognition

The NP-completeness result above indicates that, in general, there may be no efficient recognition algorithms for regular DAG languages. However, according to statistics reported by Chiang et al. (2016), DAGs actually arising from real-world AMRs are usually rather benign. In particular, although the treewidth of AMRs is unbounded in principle, real-world AMRs seem to have a small treewidth. Among 20 000 AMRs from the AMR Bank the maximum treewidth turned out to be 4, which was reached by only 31 AMRs, and the average treewidth was 1.55. It thus seems to be a relevant goal to develop recognition algorithms that work well on DAGs of small treewidth.

Let us recall here that a tree decomposition of a graph $G = (V, E, lab, src, tar)$ is a tree whose vertices, called *bags* here, are labelled with subsets of V such that

1. the union of all bags is V,[5]

2. for each edge $e \in E$ there is at least one bag containing both $src(e)$ and $tar(e)$, and

3. for each vertex $v \in V$ the bags containing v form a connected subgraph of the tree.

The treewidth of G is the smallest width of any of its tree decompositions, the width of a tree decomposition being the maximum size of its bags minus one. (A tree has treewidth 1 because it suffices to use a bag of size two for each edge e, namely $\{src(e), tar(e)\}$.)

A basic recognition algorithm generalizes the well-known forward algorithm by Baum (1972). Intuitively, it works as follows: To describe the algorithm, let us call a vertex together with its incident edges a *star*. For each such star in the input DAG, the algorithm records a set of candidate assignments of states to its edges. The initial candidate assignments to be recorded are given by the rules: every rule that could potentially be applied to the star (i.e., has the right vertex label and numbers of states in the left- and right-hand side) yields a candidate assignment to be recorded.

Now, the algorithm repeatedly chooses two stars s_1, s_2 that share an edge.[6] It contracts the

---

[5] For the sake of brevity, we often identify a bag with the set of vertices it is labelled with.

[6] To be precise, one also has to cover the case where more than one edge is shared between s_1 and s_2, but here we disregard this possibility.

edge, merging s_1 and s_2 into a single (possibly larger) star s. The candidate assignments to be recorded in s are obtained as follows: assume that s_1 and s_2 have edges $e_0, \ldots, e_m$ and $f_0, \ldots, f_n$, where $e_0 = f_0$ is the one being contracted. Then s has edges $e_1, \ldots, e_m, f_1, \ldots, f_n$, and for all candidate assignments $p_0, \ldots, p_m$ and $q_1, \ldots, q_n$ recorded in s_1 and s_2 with $p_0 = q_0$ the assignment $p_1, \ldots, p_m, q_1, \ldots, q_n$ is recorded in s. When the process stops, only one star s is left, which consists of a single vertex with no edges. The DAG is accepted if s records the empty candidate assignment, and is rejected if it contains no assignment at all.

With only little modification the algorithm may be used for weighted DAG automata, then computing the weight of the input DAG. In this case, each of the recorded candidate assignments is associated with a weight, which in the initial step is the weight of the rule in question. When two candidate assignments are combined in the process of merging two stars s_1, s_2, their weights are multiplied. If one of the candidate assignments for the combined star emerges in several ways from combinations of candidate assignments recorded in s_1, s_2, the resulting weights are summed up.

6.3 Efficiency of Recognition

The reader may have observed that the order in which edges are contracted may strongly affect the size of stars occurring during the execution of the algorithm, thus making the algorithm more or less efficient. Treating every vertex in the way described means that we are actually working with the *linegraph* $\mathcal{LG}(D)$ of D. It is obtained by turning every edge into a vertex and every vertex v into a clique on the incident edges of v (which are now vertices). The optimal contraction order can be read off an optimal tree decomposition of $\mathcal{LG}(D)$. A closer inspection of these facts (Chiang et al., 2016) revels that the algorithm can be implemented to run in time

$$O(|E| \cdot |Q|^{tw(\mathcal{LG}(D))+1}),$$

where $tw(G)$ denotes the treewidth of a graph G. The exponential dependency on the treewidth of $\mathcal{LG}(D)$ (rather than on the treewidth of D itself) is bad; as soon as D contains a vertex v of degree k the treewidth of $\mathcal{LG}(D)$ is at least $k - 1$ since there must be a bag in the tree decomposition that covers the entire clique of size k that corresponds to v.

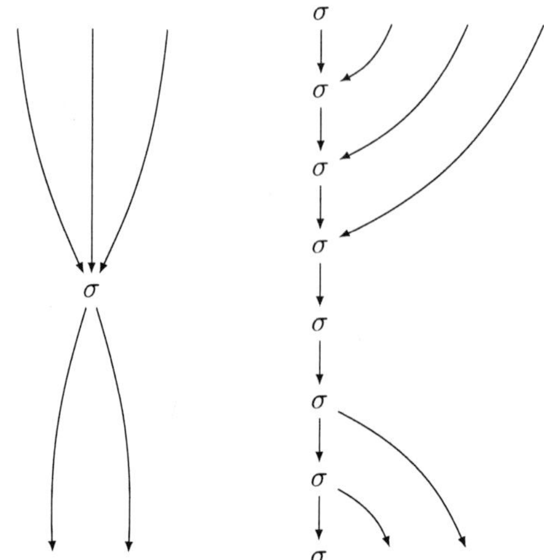

Figure 6: Binarizing a vertex of in-degree 3 and out-degree 2

A potential improvement can be achieved by a method called binarization, which is inspired by the well-known first-sibling next-child encoding of trees of unbounded rank by binary trees. The idea is to replace every vertex by a sub-DAG in which each vertex has in-degree ≤ 2 and out-degree ≤ 1 or else in-degree ≤ 1 and out-degree ≤ 2. A simple binarization scheme is illustrated in Figure 6 for a vertex of in-degree 3 and out-degree 2. Applying this schema to all vertices in a given DAG yields a binary DAG, and it is straightforward to modify the rules of a given DAG automaton in such a way that the automaton accepts the binarized version of the original DAG language.

By turning from an input DAG D to its binarized version D' we have thus reduced the degree of vertices, and hence the size of cliques in $\mathcal{LG}(D')$, to three. However, this is not enough in order to increase the efficiency of the algorithm because any fixed binarization such as the one above may be structurally incompatible with an optimal tree decomposition of D, thus actually making $tw(D')$ and $tw(\mathcal{LG}(D'))$ much larger than $tw(D)$.

The solution to this dilemma is provided by the fact that, for every tree decomposition of a graph G, there is a binary tree decomposition of G (i.e., a tree decomposition which is a binary tree) of the same width. If we, instead of binarizing vertices according to a fixed scheme, base our binarization on an optimal binary tree decomposition of D, it can be shown that a better binarization is obtained.

Here is a rough outline of the method. Consider a binary tree decomposition T of D. It can be shown that we may, without affecting the width of T, build it in such a way that every edge e of D is covered by an explicitly assigned leaf bag of T, these bags being distinct for distinct edges. By the definition of tree decompositions, the bags containing a given vertex v of D form a subtree T_v of T. This tree contains one leaf for each edge e incident on v. Hence we can binarize D by replacing every vertex v by the corresponding T_v, attaching each of the original incident edges of v to the corresponding leaf of T_v. Intuitively, using the subtrees T_v of T to binarize vertices ensures that the structure of the resulting binary DAG D_T is compatible with the tree decomposition T. This makes it possible to "refine" T into a tree decomposition of D_T, in this way showing that, if the width of T is $k \geq 1$, then

$$tw(\mathcal{LG}(D_T)) \leq 2(k+1).$$

In particular, choosing a tree decomposition T of width $tw(D)$, we get

$$tw(\mathcal{LG}(D_T)) \leq 2(tw(D) + 1).$$

As in the case of the specific binarization above, it is not very difficult to turn the original DAG automaton A into a DAG automaton A' that accepts the language of all binarized DAGs D_T such that $D \in L(A)$ and T is a binary tree decomposition of D. The application of a rule of A to a vertex v is simulated by rules that read the sub-DAG T_v, gathering the states on the original incoming and outgoing edges of v one by one. A disadvantage of this method is that it increases the number of states exponentially, but at the same time the inequalities above imply that the exponent $tw(\mathcal{LG}(D))$ is bounded from above by $2tw(D) + 1$. Altogether, running the recognition algorithm above on the binarized versions of D and A yields a running time that is exponential in $|Q|$ and $tw(D)$ rather than in $tw(\mathcal{LG}(D))$. For a detailed discussion as well as formal constructions and proofs see Chiang et al. (2016).

7 DAG Languages of Unbounded Degree

Let us briefly discuss the case of unbounded vertex degree. Regular DAG languages are of bounded

vertex degree simply because the set of rules of a DAG automaton is finite, and every rule applies only to vertices with a fixed in- and out-degree. If we want to change this, we need to represent an infinite set of rules $I \xrightarrow{\sigma} O$ in a finite manner. Similarly to the well-known case of unranked tree languages, we do this by replacing I and O by regular expressions over the alphabet of states. However, unrestricted regular expressions can specify arbitrary semilinear sets, which seems unreasonably powerful from a linguistic perspective: conditions such as "σ has twice as many incoming edges labelled with state q as it has incoming states labelled with state q or q'" lack linguistic motivation and would thus make the model overly powerful.

Extended DAG automata (Chiang et al., 2016), therefore, use Ochmański's c-regular expression (Ochmański, 1985), which are regular expressions in which the Kleene star is only applied to subexpressions over a unary alphabet. For example, $(qq)^*pp^* + qq^*$ is a c-regular expression that specifies the language of all finite multisets $M \in \mathcal{M}(\{p, q\})$ which either contain an even number of qs and at least one p, or otherwise at least one q and no p. (Recall that we are interested in multisets of states rather than sequences, because we are dealing with unordered DAGs. In the ordered case the above c-regular expression would be interpreted as specifying the set of all nonempty strings over $\{p, q\}$ which either consist of an even number of qs followed by at least one p, or contain no p at all.)

In the following, we denote the set of multisets denoted by a c-regular expression α by $[\![\alpha]\!]$. An *extended DAG automaton* is defined like a DAG automaton, except that its set R of rules now consists of extended rules of the form $\alpha \xrightarrow{\sigma} \beta$, where α and β are c-regular expressions over Q. An ordinary rule $I \xrightarrow{\sigma} O$ is an *instance* of $\alpha \xrightarrow{\sigma} \beta$ if $I \in [\![\alpha]\!]$ and $O \in [\![\beta]\!]$. A run ρ on a DAG $D = (V, E, lab, src, tar)$ is accepting if, for every vertex $v \in V$, $\rho(IN(v)) \xrightarrow{\sigma} \rho(OUT(v))$ is an instance of a rule in R.

7.1 The Weighted Case

Extended DAG automata can be made weighted by using weighted c-regular expressions in their rules. The semantics of a weighted c-regular expression α over Q is a function $[\![\alpha]\!] : \mathcal{M}(Q) \to \mathbb{S}$, where $\mathbb{S}$ is the semiring of weights. The weight of an ordinary rule $I \xrightarrow{\sigma} O$ is

$$\sum_{(\alpha \xrightarrow{\sigma} \beta) \in R} [\![\alpha]\!]\,(I) \cdot [\![\beta]\!]\,(O),$$

using the addition and multiplication of $\mathbb{S}$. Now, every run gets as its weight the product of the weights of the rule instances applied in it, and every input DAG gets as its weight the sum of the weights of all its runs.

7.2 Transferring Results by Binarization

Every (weighted) c-regular expression has an equivalent (weighted) finite automaton working on multisets rather than strings, a so-called m-automaton. Essentially, such an m-automaton is an ordinary finite automaton whose behaviour is invariant under reordering the symbols in the input string (which thus, in effect, is treated as a multiset). Using such an m-automaton to implement the rules of an extended DAG automaton, binarization works in essentially the same way as for non-extended DAG automata. In other words, for every extended DAG automaton A there is a binary DAG automaton A' such that $L(A')$ is the set of all binarized representations of DAGs in $L(A)$, and similarly for the weighted case.

It follows immediately that emptiness and finiteness are decidable, and that the recognition algorithm for DAG automata can be used to implement recognition for extended DAG automata, all with essentially the same running times as in the non-extended case.[7] Even without the use of binarization it is clear that the second pumping property of Section 5.2 holds for extended DAG automata as well, since runs are still invariant under edge swapping. The first pumping property, however, does not carry over because its proof relies on the fact that, in the case of DAGs of bounded degree, large DAGs contain long simple paths on which states must eventually repeat. This is not true anymore for DAGs of unbounded degree.

Acknowledgments

I thank everyone who contributed to the work on DAG automata either directly or by discussing ideas and providing opinions at various occasions.

---

[7]Naturally, here the input size must include the m-automata that implement the c-regular expressions appearing in the rules. These m-automata may, however, be compiled into a single one to reduce the input size by representing common parts only once.

In particular, this includes Johannes Blum, David Chiang, Daniel Gildea, Adam Lopez, Giorgio Satta, the members of the research group Foundations of Language Processing at Umeå University, and the participants of the 2014 Johns Hopkins summer workshop in Prague and the Dagstuhl Seminars 15122 and 17142.

References

Siva Anantharaman, Paliath Narendran, and Michaël Rusinowitch. 2005. Closure properties and decision problems of DAG automata. *Information Processing Letters* 94:231–240.

Laura Banarescu, Claire Bonial, Shu Cai, Madalina Georgescu, Kira Griffitt, Ulf Hermjakob, Kevin Knight, Philipp Koehn, Martha Palmer, and Nathan Schneider. 2013. Abstract meaning representation for sembanking. In *Proc. 7th Linguistic Annotation Workshop, ACL 2013 Workshop*.

Leonard E. Baum. 1972. An inequality and associated maximization technique in statistical estimation for probabilistic functions of Markov processes. In Oved Shisha, editor, *Inequalities III: Proceedings of the Third Symposium on Inequalities*. pages 1–8.

Johannes Blum and Frank Drewes. 2016. Properties of regular DAG languages. In A.H. Dediu, J. Janoušek, C. Martín-Vide, and B. Truthe, editors, *Proc. 10th Intl. Conf. on Language and Automata Theory and Applications*. volume 9618 of *Lecture Notes in Computer Science*, pages 427–438.

Johannes Blum and Frank Drewes. 2017. Language theoretic properties of regular DAG languages. Submitted.

Witold Charatonik. 1999. Automata on DAG representations of finite trees. Research Report MPI-I-1999-2-001, Max-Planck-Institut für Informatik, Saarbrücken.

David Chiang, Frank Drewes, Daniel Gildea, Adam Lopez, and Giorgio Satta. 2016. Weighted DAG automata for semantic graphs. Submitted.

Frank Drewes. 2006. *Grammatical Picture Generation – A Tree-Based Approach*. Texts in Theoretical Computer Science. An EATCS Series. Springer.

Frank Drewes. 2017. On DAG languages and DAG transducers. *Bulletin of the European Association for Theoretical Computer Science* 121:142–163.

Frank Drewes and Jérôme Leroux. 2015. Structurally cyclic Petri nets. *Logical Methods in Computer Science* 11(4:15).

Zoltán Fülöp and Heiko Vogler. 2009. Weighted tree automata and tree transducers. In Werner Kuich,

Manfred Droste, and Heiko Vogler, editors, *Handbook of Weighted Automata*, Springer, chapter 9, pages 313–403.

Ferenc Gécseg and Magnus Steinby. 1984. *Tree Automata*. Akadémiai Kiadó, Budapest. Online version available under `https://arxiv.org/abs/1509.06233`.

Ferenc Gécseg and Magnus Steinby. 1997. Tree languages. In G. Rozenberg and A. Salomaa, editors, *Handbook of Formal Languages*. Vol. 3: *Beyond Words*, Springer, chapter 1, pages 1–68.

Tsutomu Kamimura and Giora Slutzki. 1981. Parallel and two-way automata on directed ordered acyclic graphs. *Information and Control* 49:10–51.

Paul Kingsbury and Martha Palmer. 2002. From treebank to propbank. In *Proc. 3rd Intl. Conf. on Language Resources and Evaluation (LREC 2002)*.

S. Rao Kosaraju. 1982. Decidability of reachability in vector addition systems (preliminary version). In *Proceedings of the Fourteenth Annual ACM Symposium on Theory of Computing*. pages 267–281.

Ernst W. Mayr. 1984. An algorithm for the general Petri net reachability problem. *SIAM J. Comput.* 13:441–460.

Edward Ochmański. 1985. Regular behaviour of concurrent systems. *Bulletin of the European Association for Theoretical Computer Science* 27:56–67.

Lutz Priese. 2007. Finite automata on unranked and unordered DAGs. In T. Harju, J. Karhumäki, and A. Lepistö, editors, *Proc. 11th Intl. Conf. on Developments in Language Theory (DLT 2007)*, volume 4588 of Lecture Notes in Computer Science, pages 346–360.

Daniel Quernheim and Kevin Knight. 2012. Towards probabilistic acceptors and transducers for feature structures. In *Proc. 6th Workshop on Syntax, Semantics and Structure in Statistical Translation*. Association for Computational Linguistics, pages 76–85.

(Re)introducing Regular Graph Languages

Sorcha Gilroy
University of Edinburgh
s.gilroy@sms.ed.ac.uk

Adam Lopez
University of Edinburgh
alopez@inf.ed.ac.uk

Sebastian Maneth
Universität Bremen
smaneth@uni-bremen.de

Pijus Simonaitis
Ecole Normale Supérieure de Lyon
pijus.simonaitis@ens-lyon.fr

Abstract

Distributions over strings and trees can be
represented by probabilistic regular languages, which characterise many models in natural language processing. Recently, several datasets have become available which represent natural language phenomena as graphs, so it is natural to ask
whether there is an equivalent of probabilistic regular languages for graphs. This
paper presents **regular graph languages**,
a formalism due to Courcelle (1991) that
has not previously been studied in natural language processing. RGL is crucially a subfamily of both Hyperedge Replacement Languages (HRL), which can
be made probabilistic; and Monadic Second Order Languages (MSOL), which
are closed under intersection. We give
an accessible introduction to Courcelle's
proof that RGLs are in MSOL, providing clues about how RGL may relate to
other recently introduced graph grammar
formalisms.

1 Introduction

NLP systems for machine translation, summarisa-
tion, paraphrasing, and other problems often fail to
preserve the compositional semantics of sentences
and documents because they model language as
bags of words, or at best syntactic trees. To preserve semantics, they must model semantics. In
pursuit of this goal, several datasets have been produced which pair natural language with compositional semantic representations in the form of directed acyclic graphs (DAGs), including the Abstract Meaning Represenation Bank (AMR; Banarescu et al. 2013), the Prague Czech-English
Dependency Treebank (Hajič et al., 2012), Deepbank (Flickinger et al., 2012), and the Universal Conceptual Cognitive Annotation (Abend and

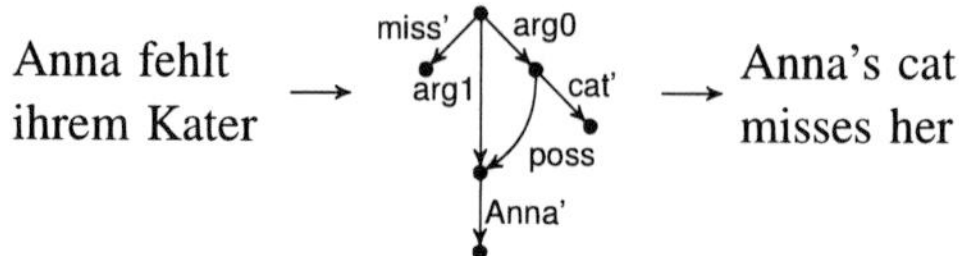

Figure 1: Semantic machine translation using AMR (Jones
et al., 2012). The edge labels identify 'cat' as the subject of
the verb 'miss', 'Anna' as the object of 'miss' and 'Anna' as
the possessor of 'cat'.

Rappoport, 2013). To make use of this data, we
require probabilistic models of graphs.

Consider how we might use compositional
semantic representations in machine translation
(Figure 1). We first parse a source sentence to its
semantic representation, and then generate a target sentence from this representation. To do this
practically, we must be able to compose a string-
to-graph model with a graph-to-string model, and
we must be able to compute the probability of this
composition. To compose the models, we need to
be able to compute the intersection of the graph
domains of each model. Hence, we must be able to
define probability distributions over the graph domains and efficiently compute their intersection.

For NLP problems in which data is in the form
of strings and trees, such distributions can be represented by finite automata (Mohri et al., 2008; Allauzen et al., 2014), which are closed under intersection and can be made probabilistic. It is therefore natural to ask whether there is a family of
graph languages with similar properties to finite
automata. Recent work in NLP has focused pri-
marily on two families of graph languages: **hyperedge replacement languages** (HRL; Drewes
et al. 1997), a context-free graph rewriting formalism that has been studied in an NLP context
by several researchers (Chiang et al., 2013; Peng
et al., 2015; Bauer and Rambow, 2016); and **DAG
automata languages**, (DAGAL; Kamimura and
Slutzki 1981), studied by (Quernheim and Knight,

Proceedings of the 15th Meeting on the Mathematics of Language, pages 100–113,
London, UK, July 13–14, 2017. ©2017 Association for Computational Linguistics

2012). (Thomas, 1991) showed that the latter are a subfamily of the **monadic second order languages** (MSOL), which are of special interest to us, since, when restricted to strings or trees, they exactly characterise the **recognisable**—or regular—languages of each (Büchi, 1960; Büchi and Elgot, 1958; Trakhtenbrot, 1961).

The HRL and MSOL families are incomparable: that is, the context-free graph languages do not contain the recognisable graph languages, as is the case in languages of strings and trees (Courcelle, 1990). So, while each formalism has appealing characteristics, neither appear adequate for the problem outlined above: HRLs can be made probabilistic, but they are not closed under intersection; and while DAGAL and MSOL are closed under intersection, it is unclear how to make them probabilistic (Quernheim and Knight, 2012).[1]

This situation suggests that we might want a family of languages that is a subfamily of both HRL and MSOL. Courcelle (1991) defines all such languages as the family of **strongly context-free languages** (SCFL).[2,3] Unfortunately, SCFLs are defined non-constructively, but Courcelle (1991) exhibits a constructively-defined subfamily: **Regular Graph Languages** (RGL), defined as a restricted form of HRL, which Courcelle demonstrates is also in MSOL.

Recently, two new graph grammar formalisms have been defined which are also restricted forms of HRL: Tree-like Grammars (TLG; Matheja et al. 2015) and Restricted DAG Grammars (RDG; Björklund et al. 2016). TLGs are claimed to be in SCFL, but the relationship of RDG to SCFL is unknown. The grammar restrictions of TLGs, RDGs and RGGs are incomparable, but they share important characteristics, which we discuss in §5.

This paper provides an accessible proof that RGL are a subfamily of MSOL, since only a sketch is provided in Courcelle (1991). Our aim in studying the proof is to gain insights into the re-

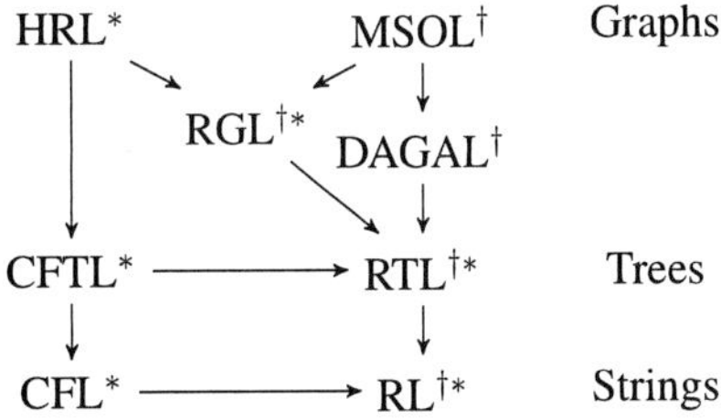

Figure 2: Containment relationships for families of regular and context-free string and tree languages, hyperedge replacement languages (HRL), monadic second order definable graph languages (MSOL), directed acyclic graph automata languages (DAGAL), and the regular graph languages (RGL). ∗ indicates that the family of languages is probabilistic and † indicates that the family of languages is intersectible.

lationship of RGL, TLG, and RDG, which might enable us to define more general classes of graph languages that are also within SCFL. Our discussion emphasises points at which Courcelle's proof relies on particular restrictions of RGL, and is intended to highlight the places where relaxations of these restrictions may be possible.

Figure 2 summarises the relationship of RGL to other formalisms and their properties. The proof of each Lemma, Proposition and Theorem in this paper that does not appear here is provided in full in the supplementary materials.[4]

2 Monadic Second-Order Logic

The regular string and tree languages precisely coincide with the monadic second-order logic (MSO) definable sets of strings and trees, respectively (Büchi, 1960; Büchi and Elgot, 1958; Trakhtenbrot, 1961), so it is natural to look at MSO over graphs.

We use the following notation. If n is an integer, $[n]$ denotes the set $\{1, \ldots, n\}$. If A is a set, $s \in A^*$ denotes that s is a sequence of arbitrary length, each element of which is in A. We denote by $|s|$ the length of s. A **ranked alphabet** is an alphabet A paired with an arity function rank: $A \to \mathbb{N}$.

Definition 1. *A **hypergraph** over a ranked alphabet Γ is a tuple $G = (V_G, E_G, att_G, lab_G, ext_G)$ where V_G is a finite set of nodes; E_G is a finite set of edges (distinct from V_G); $att_G : E_G \to V_G^*$ maps each edge to a sequence of nodes; $lab_G : E_G \to \Gamma$ maps each edge to a label such that $|att_G(e)| = rank(lab_G(e))$; and ext_G is an ordered subset of V_G called the **external nodes** of G.*

---

[1] *Semiring-weighted* MSOLs have been defined, where weights may be in the tropical semiring (Droste and Gastin, 2005). However, for the weights to define a probability distribution, they must meet the stronger condition that the sum of multiplied weights over all definable objects is one. This does not appear to have been demonstrated for DAGAL, which violate the sufficient conditions that (Booth and Thompson, 1973) give for probabilistic languages. We suspect that there are DAGAL (hence MSOL) for which it is not possible.

[2] Courcelle's definition of strongly context-free is unrelated to use of this term in NLP.

[3] The equality of SCFL and MSOL ∩ HRL was recently proved by (Bojanczyk and Pilipczuk, 2016).

---

[4] goo.gl/Z5L2gP

and quantifiers $\exists, \forall$. We allow $\text{vert}(x, i) = y$ to hold only when x is an edge and y is a node. In the case of edge-labelled graphs, the x in $\text{lab}_\gamma(x)$ must be an edge. We define the formula $\text{edg}(x, y_1, \ldots, y_k) : \text{vert}(x, 1) = y_1 \wedge \ldots \text{vert}(x, k) = y_k \wedge \bigwedge_{k' > k} \forall y \neg \text{vert}(x, k') = y$ which expresses $\text{att}(x) = (y_1, \ldots, y_k)$.

We can write down an MSO formula to express that sets $X_1, \ldots, X_n$ partition the domain. $\text{PART}(X_1, \ldots, X_n)$:

$$\forall x[x \in X_1 \cup \cdots \cup X_n \wedge \neg x \in X_1 \cap X_2 \atop \wedge \neg x \in X_1 \cap X_3 \wedge \cdots \wedge \neg x \in X_{n-1} \cap X_n] \quad (1)$$

We use ! to denote unique existential quantification. For any formula R:

$$\exists! x R(x) : \exists x R(x) \wedge \forall y R(y) \to x = y.$$

We can define an MSO statement expressing that the graph is a string by defining an edge labelled graph where the edges have rank 2, there is exactly one node with no incoming edge, there is exactly one node with no outgoing edge, and every node has at most one incoming edge and at most one outgoing edge:

$\text{STRING} : \forall y \exists! x_1 \exists! x_2 \text{edg}(y, x_1, x_2) \wedge$

$\exists! x_1 \forall y \neg \text{vert}(y, 2) = x_1 \wedge \exists! x_2 \forall y \neg \text{vert}(y, 1) = x_2$

$\wedge \forall x \neg x = x_1 \Rightarrow \exists! y \exists! x' \text{edg}(y, x', x)$

$\wedge \forall x \neg x = x_2 \Rightarrow \exists! y \exists! x' \text{edg}(y, x, x')$

Let $\text{First}(x)$ denote that x has no incoming edges and $\text{Last}(x)$ denote that x has no outgoing edges.

Example 2. Let A be the automaton in Figure 4. The corresponding MSO quantifies over a subset X_i for each state q_i in the automaton. The subsets partition the nodes of the string graph to simulate a run of the automaton.

Finally, we encode each transition of the form (q_i, a, q_j) as $x \in X_i \wedge \exists y \exists x' \text{edg}(y, x, x') \wedge \text{lab}_a(y) \Rightarrow x' \in X_j$.

From A, we construct the formula aut_A:

$\text{aut}_A : \text{STRING} \wedge \exists X_0 \exists X_1 \text{PART}(X_0, X_1)$

$\wedge \forall x \text{First}(x) \Rightarrow x \in X_0$

$\wedge \forall x \text{Last}(x) \Rightarrow x \in X_1$

$\wedge \forall x \in X_0 \wedge \exists y \exists x' \text{edg}(y, x, x') \wedge \text{lab}_a(y) \Rightarrow x' \in X_1$

$\wedge \forall x \in X_1 \wedge \exists y \exists x' \text{edg}(y, x, x') \wedge \text{lab}_a(y) \Rightarrow x' \in X_1$

$\wedge \forall x \in X_1 \wedge \exists y \exists x' \text{edg}(y, x, x') \wedge \text{lab}_b(y) \Rightarrow x' \in X_0$

For a graph G and an MSO statement ϕ we say that $G \models \phi$ (or G satisfies ϕ) when there is an assignment of variables of ϕ to nodes and edges of

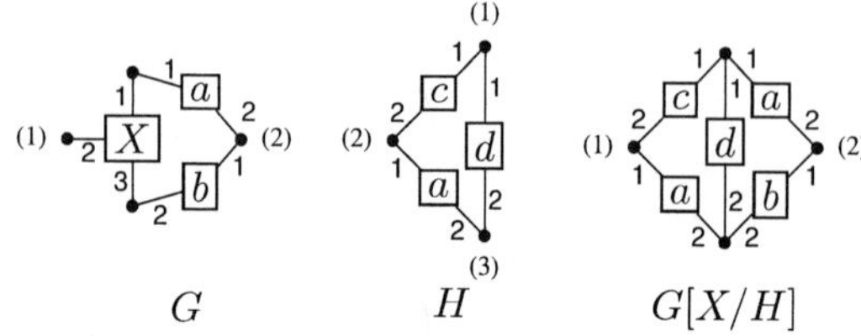

Figure 3: The replacement of the X-labeled edge in G by the graph H.

We assume that both the elements of ext_G and the elements of $\text{att}_G(e)$ for each edge e are pairwise distinct. An edge e is attached to its nodes by **tentacles**, each labeled by an integer indicating the node's position in $\text{att}_G(e) = (v_1, \ldots, v_k)$. The tentacle from e to v_i has label i, so the tentacle labels lie in the set $[k]$ where $k = \text{rank}(e)$. To express that a node v is attached to the i-th tentacle of an edge e we say $\text{vert}(e, i) = v$. The nodes in ext_G are labeled by their position in ext_G. In figures, the i-th external node is labeled (i). The **rank** of an edge e is k if $\text{att}(e) = (v_1, \ldots, v_k)$ (or equivalently, $\text{rank}(\text{lab}(e)) = k$). The **rank** of a hypergraph G is $|\text{ext}_G|$. An **induced subgraph** of a hypergraph G by edges $E' \subseteq E_G$ is the subgraph of G formed by including all edges in E' and their endpoints. Define $\text{HG}_{\Sigma, \Gamma}$ to be the set of all hypergraphs with node labels in Σ and edge labels in Γ (the hypergraphs as defined here have no node labels so are in $\text{HG}_{\emptyset, \Gamma}$).

Example 1. Hypergraph G in Figure 3 has four nodes (shown as black dots) and three hyperedges labeled a, b, and X (shown boxed). The bracketed numbers (1) and (2) denote its external nodes and the numbers between edges and the nodes are tentacle labels. Call the top node v_1 and, proceeding clockwise, call the other nodes v_2, v_3, and v_4. Call its edges e_1, e_2 and e_3. Its definition would state:

$$\begin{aligned} \text{att}_G(e_1) &= (v_1, v_2) & \text{lab}_G(e_1) &= a \\ \text{att}_G(e_2) &= (v_2, v_3) & \text{lab}_G(e_2) &= b \\ \text{att}_G(e_3) &= (v_1, v_4, v_3) & \text{lab}_G(e_3) &= X \\ \text{ext}_G &= (v_4, v_2). \end{aligned}$$

MSO on graphs quantifies over nodes, sets of nodes, edges, and sets of edges.[5] The atomic formulas are $x \in X$, $x = y$, $\text{lab}_\gamma(x)$, and $\text{vert}(x, i) = y$. We construct MSO sentences using the atomic formulas, connectives $\wedge, \vee, \neg, \Rightarrow$,

---

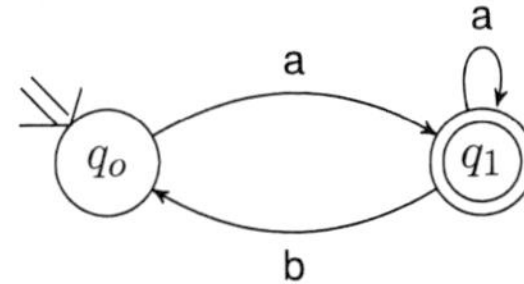

Figure 4: The finite automaton A.

Figure 5: The graph representing the string $aaba$.

G that makes ϕ true.

Example 3. The string graph $G = aaba$ as shown in Figure 5 can be produced by automaton A. The letters are edge labels and call its nodes from left to right v_0, v_1, v_2, v_3, and v_4. If we let $X_0 = \{v_0, v_3\}$ and $X_1 = \{v_1, v_2, v_4\}$ then $G \models \text{aut}_A$.

Let $\text{aut}'_A(X_0, X_1)$ be the MSO formula identical to aut_A with $\exists X_0 \exists X_1$ removed from the beginning of the formula. X_0 and X_1 are free variables of aut'_A, and we refer to the set of free variables of a formula as its **parameters**. Given a graph G and a formula $\phi(\mathcal{W})$ with parameters $\mathcal{W}$, let α be function from $\mathcal{W}$ to subsets of nodes and edges in G. Then we say that $(G, \alpha) \models \phi(\mathcal{W})$ if G and α satisfy $\phi(\mathcal{W})$. We call α a **parameter assignment**. The MSO interpretation of an automaton is satisfied if we can find a parameter assignment that simulates a run of the automaton—more precisely, $G \models \text{aut}_A$ if $(G, \alpha) \models \text{aut}'_A(X_0, X_1)$. In general, there may be more than one such α.

Example 4. Let $\mathcal{W} = \{X_0, X_1\}$ be parameters. If $\alpha(X_0) = \{v_0, v_3\}$ and $\alpha(X_1) = \{v_1, v_2, v_4\}$ then $(G, \alpha) \models \text{aut}'_A(\mathcal{W})$.

We can use an MSO statement ϕ to define a language, $L(\phi) = \{G \mid G \models \phi\}$, and we call the family of languages definable this way as **MSOL**. We define the intersection of two languages $L(\phi_1) \cap L(\phi_2) = \{G \mid G \models \phi_1 \wedge \phi_2\}$. This clearly shows that MSO languages are closed under intersection.

2.1 MSO Transductions

One way to show that a language is MSO definable is to use the backwards translation theorem (Courcelle and Engelfriet, 2011), which depends on **MSO transductions (MSOT)**, a generalisation of finite-state string and tree transductions. The theorem is a generalisation to graphs of the fact that regular string and tree languages are closed

under inverse finite-state transductions (Hopcroft and Ullman, 1979; **?**).

Theorem 1 (Backwards Translation Theorem). *If L is an MSO definable graph language and f is an MSO graph transduction then $f^{-1}(L)$ is effectively MSO definable.*

Definition 2. *An MSO transducer $\tau : HG_{\Sigma,\Gamma} \to HG_{\Sigma',\Gamma'}$ is $\tau = \langle \rho(\mathcal{W}), \delta(x, \mathcal{W}), (\theta_r(x_1, \ldots, x_{N(r)}, \mathcal{W}))_{r \in \mathcal{R}} \rangle$. $\mathcal{W}$ is a set of parameters; $\rho(\mathcal{W})$ is a **precondition** which input graphs must satisfy; $\delta(x, \mathcal{W})$ is a **domain formula** defining the output domain (i.e. nodes); and $\theta_r(x_1, \ldots, x_{N(r)}, \mathcal{W})$ is a **relation formula** defining relationships between the elements in the output domain (i.e. edges).*[6]

The role of parameters here is to allow nondeterminism. Given a graph G and a parameter assignment α from $\mathcal{W}$ to $V_G \cup E_G$ such that $(G, \alpha) \models \rho(\mathcal{W})$, we define the output of the transducer $\tau(G, \alpha) = (D, R)$ such that $D = \{x \in G \mid (G, x, \alpha) \models \delta(x, \mathcal{W})\}$ and $R = \{\theta_r(x_1, \ldots, x_{N(r)}) \mid (G, x_1, \ldots, x_{N(r)}, \alpha) \models \theta_r(x_1, \ldots, x_{N(r)}, \mathcal{W}), r \in \mathcal{R}\}$. Define $\tau(G) = \{\tau(G, \alpha) \mid (G, \alpha) \models \rho(\mathcal{W})\}$ and for a language L, $\tau(L) = \{\tau(G) \mid G \in L\}$.

3 Hyperedge Replacement Grammars

If f is a function and S is a set, $f|_S$ is the restriction of f to domain elements in S. If f, g are functions, $f \circ g$ is their composition.

Definition 3. *Let G be a hypergraph with an edge e of rank k and let H be a hypergraph also of rank k disjoint from G. The **replacement** of e by H is the graph $G' = G[e/H]$. Let $V_{G'} = (V_G \cup V_H) - ext_H$, $E_{G'} = (E_G \cup E_H) - \{e\}$. Let $ext_H = (v_1, \ldots, v_k)$, $att_G(e) = (u_1, \ldots, u_k)$ and let $f : (V_G \cup V_H) \to V_{G'}$ replace v_i by u_i for $i \in [k]$ and be the identity otherwise. The extension of f to $(V_G \cup V_H)^*$ is also denoted f. Let $E = E_G - \{e\}$, then $att_{G'} = att_G|_E \cup (f \circ att_H)$, $lab_{G'} = lab_G|_E \cup lab_H$.*

Example 5. Replacement is shown in Figure 3. We denote the replacement as $G[X/H]$ since the edge is unambiguous given its label.

Definition 4. *A **hyperedge replacement grammar** (HRG) $G = (N, T, P, S)$ consists of disjoint ranked alphabets N and T of nonterminals and*

---

[6]Technically, this is a non-copying MSO transducer. In general, MSO transductions can define multiple copies of each element of the input domain.

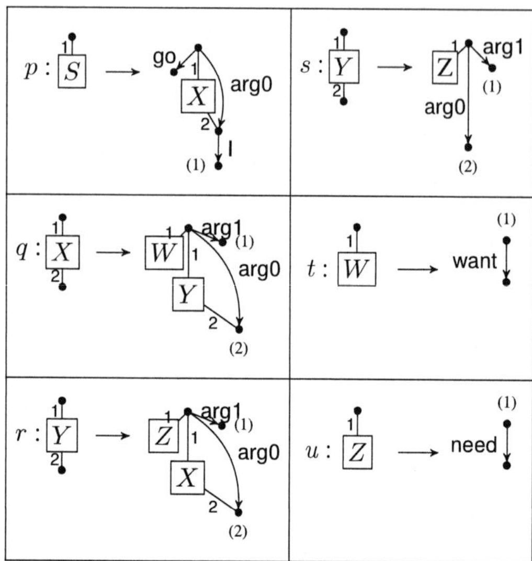

Table 1: Productions of a HRG. The labels $p, q, r, s, t,$ and u label the productions so that we can refer to them in the text. Note that Y can be rewritten either via production r or s.

terminals, a finite set of productions P, and a start symbol $S \in N$. Every production in P is of the form $X \to H$ where $X \in N$ is of rank k and H is a hypergraph of rank k over N and T.

A HRG $\mathcal{G}$ produces graphs in $\text{HG}_{\emptyset, T_\mathcal{G}}$. In each example, we only show terminal edges of rank 2, and depict them as directed edges where the direction is determined by the tentacle labels: tentacle 1 attaches to the source and 2 attaches to the target (Table 1). For each production $p : X \to G$, we use $L(p)$ to refer to its left-hand side (X) and $R(p)$ to refer to its right-hand side (G). An edge is a **terminal edge** if its label is terminal and a **nonterminal edge** if its label is nonterminal. A graph is **terminal** if all of its edges are labeled with terminal symbols. The **terminal subgraph** of a graph is the subgraph induced by its terminal edges. Let $\text{NT}(p) = \{e_1, \ldots, e_n\}$ be an enumeration of the nonterminal edges in $R(p)$, let $|\text{NT}(p)|$ be the number of nonterminal edges in $R(p)$ and let $|\text{NT}(P)| = \max_{p \in P} |\text{NT}(p)|$.

Given a HRG $\mathcal{G}$, we say that graph G **derives** graph G', denoted $G \to G'$, iff there is an edge $e \in E_G$ and a nonterminal $X \in N$ such that $\text{lab}_G(e) = X$ and $G' = G[e/H]$, where $X \to H$ is in P. We extend the idea of derivation to its transitive closure $G \to^* G'$. For every $X \in N$ we also use X to denote the connected graph consisting of a single edge e with $\text{lab}(e) = X$ and nodes $(v_1, \ldots, v_{\text{rank}(X)})$ such that

$\text{att}(e) = (v_1, \ldots, v_{\text{rank}(X)})$, and we define the language $L_X(\mathcal{G}) = \{G \mid X \to^* G, G \text{ is terminal}\}$. The **language of** $\mathcal{G}$ is then $L(\mathcal{G}) = L_S(\mathcal{G})$. We call the family of languages that can be produced by any HRG the **hyperedge replacement languages (HRL)**.

3.1 HRL and MSOT

Since HRGs are context-free, for each HRG $\mathcal{G}$, there is an underlying regular tree grammar $\mathcal{T}_\mathcal{G}$ defining the derivation trees of the graphs in $L(\mathcal{G})$. Each $\mathbf{T} \in \mathcal{T}_\mathcal{G}$ has node labels in $P_\mathcal{G}$ and edge labels in $|\text{NT}(P)|$. If a node has label p and $R(p)$ has n nonterminals $X_1, \ldots, X_n$ then for each $i \in [n]$, there is an i labelled edge from p to a node labelled q where $L(q) = X_i$. The label of the root of $\mathbf{T}$ must be p for some p with $L(p) = S$. Let $\text{VAL} : L(\mathcal{T}_\mathcal{G}) \to L(\mathcal{G})$ be a mapping from derivation trees to graphs so that $G = \text{VAL}(\mathbf{T})$ iff $\mathbf{T}$ is a derivation tree of G. Since HRGs can be ambiguous, this mapping is not injective. (Courcelle, 1991) shows that VAL is an MSO transduction.[7] This does not imply that HRLs are MSOL, since in general MSOL is not closed under MSOT. Hence an MSOT representing the inverse of VAL may not exist for an arbitrary HRG, but we later discuss a subfamily for which it does (§4), allowing us to apply Theorem 1.

To distinguish between elements of a graph and its derivation tree, we denote a grammar by $\mathcal{G}$, graph by G, derivation tree by $\mathbf{T}$, derivation tree node by $\mathbf{v}$, edges and nodes in productions are written with a bar ($\bar{v}$) and nodes and edges in G are unmarked (x).

The transduction VAL preserves the terminal subgraph of every production used in a derivation and fuses nodes from different productions together in the output graph. Node fusion is determined by an equivalence relation $\sim$ generated by a relation $\sim_0$. Let $\text{NT}(p) = (e_1, \ldots, e_n)$ the nonterminal edges of $R(p)$, let $\text{NT}_i(p) = e_i$, and let $\text{ext}_G(i)$ be the ith external node of G.

Definition 5. *Let $\mathcal{G}$ be a HRG and $\mathbf{T}$ be a derivation tree of $\mathcal{G}$, so that $G = \text{VAL}(\mathbf{T})$. Define a binary relation $\sim_0$ on pairs $(\bar{x}, \mathbf{v})$ where $\bar{x}$ is a node in $R(p)$ for some $p \in P$ and $\mathbf{v}$ is a node of $\mathbf{T}$ with label p. Then $(\bar{x}, \mathbf{v}) \sim_0 (\bar{y}, \mathbf{v}')$ iff:*

1. $\mathbf{v}, \mathbf{v}'$ are nodes in $\mathbf{T}$ and $\mathbf{v}'$ is the ith child of $\mathbf{v}$ in $\mathbf{T}$.

---

[7]It can also be viewed in the related framework of interpreted regular tree grammars (Koller and Kuhlmann, 2011).

104

2. $p = lab_{\mathbf{T}}(\mathbf{v})$, $p' = lab_{\mathbf{T}}(\mathbf{v}')$.

3. $\bar{x}$ is the jth node of $NT_i(p)$, $\bar{y} = ext_{R(p')}(j)$.

We define $\sim$ as the reflexive, symmetric, transitive closure of $\sim_0$.

The mapping VAL translates derivation trees to graphs in two steps. First, the terminal subgraph of every instance of every production used in the derivation tree is produced in the output. Then, all equivalent nodes under $\sim$ are fused. (Courcelle, 1991) shows that each step is a MSOT; their composition is also a MSOT.

Example 6. Figure 6 illustrates how VAL maps a derivation tree to a graph.

The mapping VAL can be defined in terms of two finer-grained mappings. Let $E_P = \cup_{p \in P} E_{R(p)}$ and $V_P = \cup_{p \in P} V_{R(p)}$. Then $h_e : E_P \times V_{\mathbf{T}} \to E_G$ maps a pair $(\bar{e}, \mathbf{v})$ to its image e in the graph, where $\bar{e}$ is a terminal edge in p and $lab(\mathbf{v}) = p$. This mapping is one-to-one since edges cannot be fused. $h_v : V_P \times V_{\mathbf{T}} \to V_G$ maps a pair $(\bar{x}, \mathbf{v})$ to its image v, where $\bar{x}$ is a node in p and $lab(\mathbf{v}) = p$. It is not one-to-one since nodes can be fused.

Lemma 1. *Let $\mathcal{G}$ be a HRG, and let G be a graph in $L(\mathcal{G})$ with derivation tree $\mathbf{T}$. If $\bar{x}$ and $\bar{x}'$ are nodes such that $h_v(\bar{x}, \mathbf{v}) = h_v(\bar{x}', \mathbf{v}')$ with $\mathbf{v} \neq \mathbf{v}'$ and, if $\bar{x}$ is internal in $R(p)$ for $p = lab_{\mathbf{T}}(\mathbf{v})$, then $\bar{x}'$ is an external node of $R(p')$ for $p' = lab_{\mathbf{T}}(\mathbf{v}')$ and $\mathbf{v}$ is an ancestor of $\mathbf{v}'$ in $\mathbf{T}$.*

Consequently, if $h_v(\bar{x}, \mathbf{v}) = h_v(\bar{x}', \mathbf{v}')$ then $\bar{x}$ and $\bar{x}'$ cannot both be internal.

4 Regular Graph Grammars

A regular graph grammar (RGG; Courcelle 1991) is a restricted form of HRG. To explain the restrictions, we first require some definitions.

Definition 6. *Given a graph G, a **path** in G from a node v to a node v' is a sequence*

$$(v_0, i_1, e_1, j_1, v_1)(v_1, i_2, e_2, j_2, v_2) \ldots (v_{k-1}, i_k, e_k, j_k, v_k)$$

such that $vert(e_r, i_r) = v_{r-1}$ and $vert(e_r, j_r) = v_r$ for each $r \in [k]$, $v_0 = v$, and $v_k = v'$. The length of this path is k.

A path is **terminal** if every edge in the path is terminal. A path is **internal** if each v_i is internal for $1 \leq i \leq k - 1$. The endpoints v_0 and v_k of an internal path can be external.

Definition 7. *A HRG $\mathcal{G}$ is a **Regular Graph Grammar** if each nonterminal in N has rank at least one and for each $p \in P_{\mathcal{G}}$ the following hold:*

(C1) $R(p)$ *has at least one edge. Either it is a single terminal edge, all nodes of which are external, or each of its edges has at least one internal node.*

(C2) Every pair of nodes in $R(p)$ is connected by a terminal and internal path.

RGLs are HRLs by definition; we will prove that they are also MSOLs by constructing the inverse of VAL, a transducer from RGL graphs to their derivation trees. Since the derivation trees are MSO definable, RGLs must also be MSO definable by Theorem 1. The construction requires a unique **anchor** element (a node or edge) for each production in the grammar. Given an input graph, the transducer first guesses—via parameter assignment—the preimage of each edge and the set of elements whose preimages are anchors. It then checks whether the guess satisfies constraints that must be true for every derived graph:

1. It must be possible to partition the graph into a set of edge-disjoint connected subgraphs, each isomorphic to the terminal subgraph of some production.

2. For each node that is in two such subgraphs, the node must be the image of two nodes in the productions that are allowed to be fused under the grammar.

If these constraints are satisfied, the transducer outputs each guessed anchor and an edge between anchors that it identifies to be in a parent-child relationship.

Every valid parameter assignment corresponds to a different output from the transducer, and we will show that all derivation trees for any input graph in the grammar lie in this output set.

Theorem 2. *RGL $\subseteq$ MSOL.*

The proof of each Lemma and Proposition in this section either appears here or in the supplementary materials. The proof of Theorem 2 is provided in §4.2.1.

4.1 Anchors and Parameters

There are two types of productions in RGGs: those with a single terminal edge, all nodes of which are external; and those where each edge has an internal node. We call the former **ext-productions** and the latter **int-productions**. For each int-production, we arbitrarily choose one of its internal nodes to be its anchor. For each ext-production, we choose its single terminal edge to be the anchor. By Lemma 1, this choice ensures

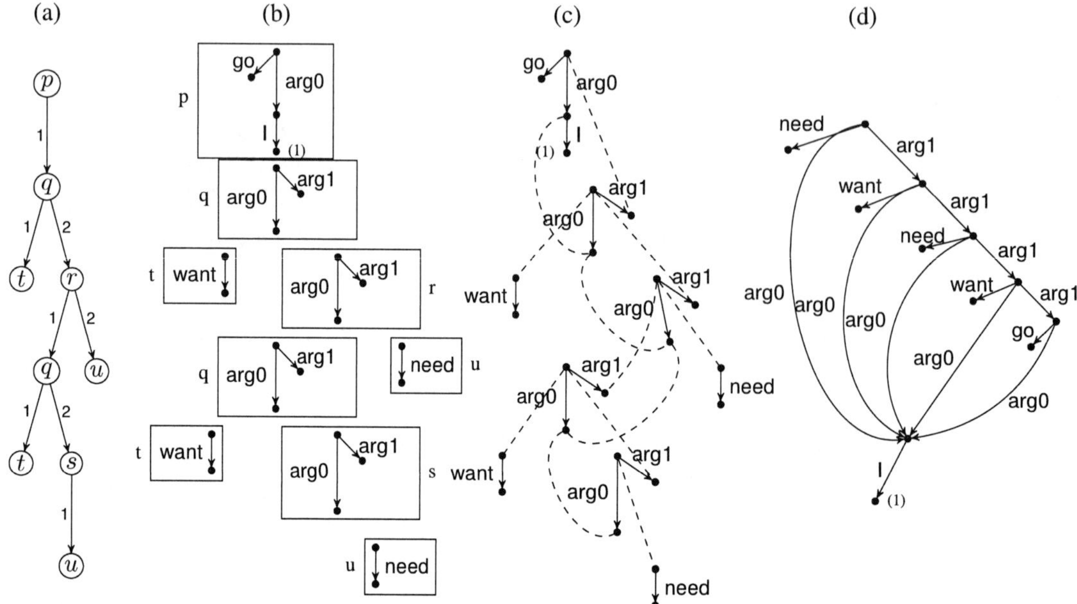

Figure 6: A derivation tree (a), the terminal subgraphs of every copy of every production in the derivation tree (b), the relation $\sim$ illustrated with dashed lines (c) and the resulting graph (d).

that a pair of anchors cannot be fused, so the set of anchors in any derived graph is guaranteed to be in one-to-one correspondence with the nodes of its derivation tree.

We define two sets of parameters: $\mathcal{E}$ and $\mathcal{C}$, where $\mathcal{E}$ guesses preimages of edges, and $\mathcal{C}$ guesses anchors (which may be either nodes or edges). To define $\mathcal{E}$ precisely, we require some notation. Let $\mathcal{G}$ be an RGG, and for each $p \in P$, let $\mathrm{T}(p) = \{\bar{f}_{p,1}, \ldots, \bar{f}_{p,|\mathrm{T}(p)|}\}$ enumerate the terminal edges of $R(p)$ and let $\gamma_{p,j}$ be the label of $\bar{f}_{p,j}$ for each $p \in P$ and $j \in [|\mathrm{T}(p)|]$. Let $|\mathrm{NT}(p)|$ be the number of nonterminal edges in p and let $|\mathrm{NT}(P)| = \max_{p \in P} |\mathrm{NT}(p)|$. Given a node $\mathbf{v}$ in a derivation tree $\mathbf{T}$, we say that $\mathbf{v}$ is an i-child if it is the ith child of some other node in $\mathbf{T}$. By convention, the root node is the only 0-child.

Let G be in $L(\mathcal{G})$ and let $\mathbf{T}$ be a derivation tree of G. For each $i \in [0, |\mathrm{NT}(P)|]$, $p \in P$ and $j \in [|T(p)|]$, we define a parameter $E_{i,p,j}$:

$$E_{i,p,j} = \{e \in E_G \mid h_e(\bar{f}_{p,j}, \mathbf{v}) \text{ and } \mathbf{v} \text{ is an } i\text{-child.}\}$$

Let $\mathcal{E} = \{E_{i,p,j}\}$ for $i \in [0, |\mathrm{NT}(P)|]$, $p \in P$ and $j \in [|T(p)|]$.

For each $i \in [|\mathrm{NT}(P)|]$ and $p \in P$, define

$$C_{i,p} = \{h(\bar{c}_p, \mathbf{v}) \mid p = \mathrm{lab}_{\mathbf{T}}(\mathbf{v}), \mathbf{v} \text{ is an } i\text{-child.}\}$$

Where $h = h_e \cup e_v$ since $\bar{c}_p$ can either be an edge or a node. Let $\mathcal{C} = \cup_{i,p} C_{i,p}$.

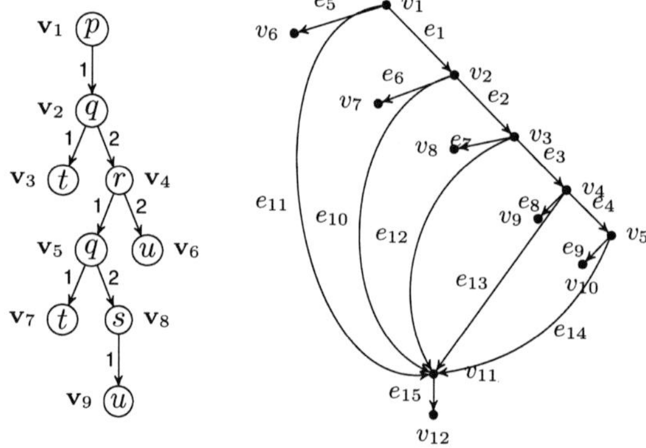

Figure 7: The derivation tree from Figure 6 (a) and the graph from Figure 6 (d) with variable names for the nodes and edges.

Let $\mathcal{W} = \mathcal{E} \cup \mathcal{C}$.

Example 7. Table 2 shows the productions of Table 1 with labels on each node and edge. Figure 7 shows the derivation tree and graph from Figure 6 with variable names added. We use these variable names to refer to specific nodes and edges in the text. For example, $h_v(\bar{c}_s, \mathbf{v}_8) = v_1$, and $h_e(\bar{f}_{u,1}, \mathbf{v}_9) = e_5$.

Example 8. Using the labels in Table 2 and Figure 7, we see that $E_{0,p,1} = \{e_9\}$, $E_{1,q,2} = \{e_{12}, e_{14}\}$, and $v_1 = h(\bar{c}_p, \mathbf{v}_8)$ is an anchor.

106

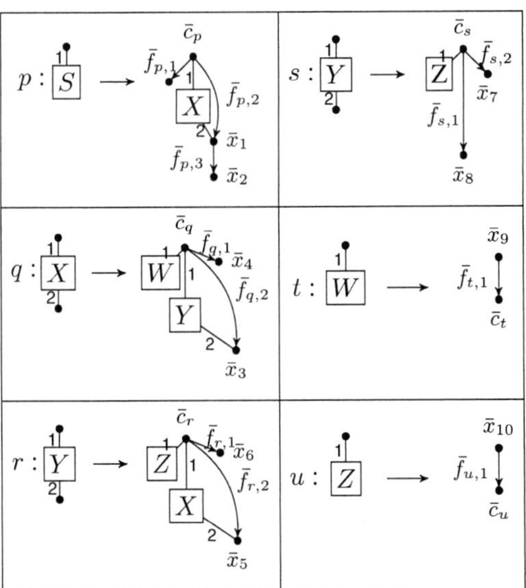

Table 2: The productions from Table 1 with variable names added to each of the nodes and terminal edges. Node variables of the form $\bar{c}_x$ for $x \in \{p, q, r, s, t, u\}$ indicate anchors.

4.1.1 Path Properties of RGLs

The precondition will exploit the properties of RGGs, particularly the properties of paths between nodes. Let $\mathcal{G}$ be an RGG, $G \in L(\mathcal{G})$, and let $\mathbf{T}$ be a derivation tree of G. In the following, we relate paths within individual productions in P (denoted π) to paths in G (denoted λ). For each e in G, we define $\mathrm{o}(e) = (i, p, j)$ iff $e \in E_{i,p,j}$.

For every path λ in G of the form

$$(v, i_1, e_1, j_1, v_1)(v_1, i_2, e_2, j_2, v_2) \ldots (v_{k-1}, i_k, e_k, j_k, v')$$

we define its **trace** as the sequence

$$\mathrm{tr}(\lambda) := (\mathrm{o}(e_1), i_1, j_1)(\mathrm{o}(e_2), i_2, j_2) \ldots (\mathrm{o}(e_k), i_k, j_k).$$

Now let π be a path

$$(\bar{v}, i_1, \bar{e}_1, j_1, \bar{v}_1) \ldots (\bar{v}_{k-1}, i_k, \bar{e}_k, j_k, \bar{v}')$$

in $R(p)$ for some $p \in P$. Let $\mathbf{v} \in V_\mathbf{T}$, $p = lab_\mathbf{T}(\mathbf{v})$. We denote by $h(\pi, \mathbf{v})$ the following path in G:

$$(h(\bar{v}, \mathbf{v}), i_1, h(\bar{e}_1, \mathbf{v}), j_1, h(\bar{v}_1, \mathbf{v})) \ldots$$
$$(h(\bar{v}_{k-1}, \mathbf{v}), i_1, h(\bar{e}_k, \mathbf{v}), j_1, h(\bar{v}', \mathbf{v}))$$

If $\mathbf{v}$ is an i-child of some node in $V_\mathbf{T}$ then $\mathrm{tr}(h(\pi, \mathbf{v}))$ is the sequence

$$((i, p, m_1), i_1, j_1) \ldots ((i, p, m_k), i_k, j_k)$$

where $\bar{e}_j = \bar{f}_{p,m_j}$ for each $j \in [k]$. Note that $\mathrm{tr}(\pi) = \mathrm{tr}(h(\pi, \mathbf{v}))$. The trace is a property that remains constant when a path is projected from a production into a graph. This projection is not one-to-one since a production can be applied several times; a trace appears in the graph once for each application of the corresponding production in a derivation. For $\mathbf{v} \in V_\mathbf{T}$, we write $\pi \in R(lab_\mathbf{T}(\mathbf{v}))$ to denote that π is a path in the production which is the label of $\mathbf{v}$.

Example 9. Let π be the path $(\bar{x}_3, 2, \bar{f}_{q,2}, 1, \bar{c}_q)(\bar{c}_q, 1, \bar{f}_{q,1}, 2, \bar{x}_4)$ in production q in Table 2. $h(\pi, \mathbf{v}_4)$ for $\mathbf{v}_4$ is the path $(v_1 1, 2, e_{13}, 1, v_4)(v_4, 1, e4, 2, v_5)$ in Figure 7, and its trace is $((1, q, 2), 2, 1)((1, q, 1), 1, 2)$.

Lemma 2 (Lemma 5.5 from (Courcelle, 1991)). *Let $\mathcal{G}$ be an RGG, G be a graph in $L(\mathcal{G})$, and $\mathbf{T}$ be a derivation tree of G. Let λ be a path in G of the form $h(\pi, \mathbf{v})$ for some $\mathbf{v} \in V_\mathbf{T}$ and some terminal path $\pi \in R(lab_\mathbf{T}(\mathbf{v}))$. The final node of π may be internal or external but every other node must be internal. If λ' is another path in G with the same trace and the same initial node as λ, then $\lambda' = \lambda$.*

Lemma 2 guarantees a unique trace for every path in a graph that is the projection of a path in a single production. By property C2 of RGGs, this guarantee must hold for at least one path from the anchor node of an int-production to every other node in the production. For ext-productions, all paths are of the form $\pi = (\bar{e}, i, \bar{v}_i)$, where e is the single nonterminal edge; these paths are also guaranteed unique traces.

4.1.2 MSO Formulas for the Precondition

Given an assignment to our parameters, we can use the path property in Lemma 2 to define some useful MSO statements. The first, ANC, relates anchors to the nodes in the graph. Throughout this section, given a derivation tree $\mathbf{T}$, we will refer to $\alpha_\mathbf{T}$ which is the parameter assignment from $\mathcal{W}$ to $V_G \cup E_G$ as defined above.

Lemma 3 (Lemma 5.6 from (Courcelle, 1991)). *Let $\mathcal{G}$ be an RGG, G be a graph in $L(\mathcal{G})$, and $\mathbf{T}$ be a derivation tree of G. For every $p \in P$, every $i \in [0, |NT(P)|]$, and every node $\bar{x} \in R(p)$, one can construct a formula $\mathrm{ANC}_{p,i,\bar{x}}(u, w, \{\mathcal{W}\})$ such that, for every $u \in V_G \cup E_G$, $w \in V_G$:*

$$(G, u, w, \alpha_\mathbf{T}) \models \mathrm{ANC}_{p,i,\bar{x}}(u, w, \{\mathcal{W}\})$$

iff $u = h(\bar{c}_p, \mathbf{v})$ and $w = h_v(\bar{x}, \mathbf{v})$ for some $\mathbf{v} \in V_\mathbf{T}$ which is an i-child and $p = lab_\mathbf{T}(\mathbf{v})$.

We say that node u **anchors** node v if for some p, i and $\bar{x}$, $\mathrm{ANC}_{p,i,\bar{x}}(u, v, \{\mathcal{W}\})$ holds. We use the fact that a node or edge anchors itself to establish its corresponding production.

Example 10. Looking at Table 2 and Figure 7, we can establish that $\text{ANC}_{p,0,\bar{x}_1}(v_5, v_{11}, \{\mathcal{W}\})$ holds and that $v_5 \in C_{0,p}$.

The next MSO formula we construct relates pairs of anchors to each other. Since the anchors define the output domain of the transducer, the formula PAR defines the edges of the output.

Lemma 4 (Lemma 5.7 of (Courcelle, 1991)). *Let $\mathcal{G}$ be an RGG, G be in $L(\mathcal{G})$, $\mathbf{T}$ be a derivation tree of G, and α be the parameter assignment defined with respect to $\mathbf{T}$. One can construct a formula $\text{PAR}_{p,i,p',i'}(u, w, \{\mathcal{W}\})$ such that, for $u, w \in V_G \cup E_G$:*

$$(G, u, w, \alpha) \models \text{PAR}_{p,i,p',i'}(u, w, \{\mathcal{W}\})$$

iff $u = h(\bar{c}_p, \mathbf{v}), w = h(\bar{c}_{p'}, \mathbf{v}')$ for some $\mathbf{v}, \mathbf{v}'$ in $V_{\mathbf{T}}$ where $p = lab_{\mathbf{T}}(\mathbf{v})$, $p' = lab_{\mathbf{T}}(\mathbf{v}')$, $\mathbf{v}$ is an i-child, and $\mathbf{v}'$ is the i'th child of $\mathbf{v}$ in $\mathbf{T}$.

If $\text{PAR}_{p,i,p',i'}(u, u', \{\mathcal{W}\})$ holds, then u will become the parent of u' in the output tree. The proof of this lemma relies on C1 of RGG.

Example 11. From Table 2 and Figure 7, $\text{PAR}_{q,1,s,2}(v_2, v_1, \{\mathcal{W}\})$ holds.

As introduced in §3, we have a binary equivalence relation $\sim$ over pairs of the form $(\bar{x}, \mathbf{v})$ where $\bar{x}$ is a node in a production p and $\mathbf{v}$ is a node in the derivation tree with label p. We use this relation for the precondition of the transducer so that a pair of nodes are only fused if the grammar and derivation tree allows them to be. We project $\sim$ into the graph to construct a relation over anchors such that $\text{FUSE}_{p,i,\bar{x},p',i',\bar{x}'}(u, u', \{\mathcal{W}\})$ holds if and only if $(\bar{x}, \mathbf{v}) \sim (\bar{x}', \mathbf{v}')$, $u = h(\bar{c}_p, \mathbf{v})$, $u' = h(\bar{c}_{p'}, \mathbf{v}')$, and $h(\bar{x}, \mathbf{v}) = h(\bar{x}', \mathbf{v}')$.

Lemma 5. *Let $\mathcal{G}$ be an RGG, G be in $L(\mathcal{G})$, and $\mathbf{T}$ be a derivation tree of G. One can construct a formula $\text{FUSE}_{p,i,\bar{x},p',i',\bar{x}'}(u, u', \{\mathcal{W}\})$ such that, for $u, u' \in V_G \cup E_G$:*

$$(G, u, u', \{\mathcal{W}\}) \models \text{FUSE}_{p,i,\bar{x},p',i',\bar{x}'}(u, u', \{\mathcal{W}\})$$

iff $u = h(\bar{c}_p, \mathbf{v}), u' = h(\bar{c}_{p'}, \mathbf{v}')$ for some $\mathbf{v}, \mathbf{v}'$ in $V_{\mathbf{T}}$ where $p = lab_{\mathbf{T}}(\mathbf{v}), p' = lab_{\mathbf{T}}(\mathbf{v}')$, $\mathbf{v}$ is the ith child of some node, $\mathbf{v}'$ is the i'th child of some node, and $h_v(\bar{x}, \mathbf{v}) = h_v(\bar{x}', \mathbf{v}')$.

Example 12. From Table 2 and Figure 7, we can see that $\text{FUSE}_{p,0,\bar{x}_1,s,2,\bar{x}_8}(v_5, v_1, \{\mathcal{W}\})$ holds since $v_5 = h_v(\bar{c}_p, \mathbf{v}_1)$, $v_1 = h_v(\bar{c}_s, \mathbf{v}_8)$, $v_{11} = h_v(\bar{x}_1, \mathbf{v}_1) = h_v(\bar{x}_8, \mathbf{v}_8)$, $\text{ANC}_{p,0,\bar{x}_1}(v_5, v_6, \{\mathcal{W}\})$ and $\text{ANC}_{s,2,\bar{x}_8}(v_1, v_{11}, \{\mathcal{W}\})$

4.1.3 The Precondition of the Transducer

Let X be in N, then $P_X = \{p \in P | L(p) = X\}$, and an X-derivation tree is a derivation tree with respect to X as the start symbol (in this case, the root will have label in P_X). An S-derivation tree is referred to simply as a derivation tree.

Edge Requirements

(E1) $\alpha(\mathcal{E})$ partitions E_G,

(E2) for all $e \in \alpha(E_{i,p,j})$ e has label $\gamma_{p,j}$

(E3) there is a unique $p \in P_X$ such that $\alpha(E_{0,p,j})$ has exactly one element for each $j \in [|\mathbf{T}(p)|]$ and for every $p' \neq p$, $\alpha(E_{0,p',j})$ is empty for all j.

Recall the MSO statement $\text{PART}(X_1, \ldots, X_n)$ from Equation 1 which expresses that $X_1, \ldots, X_n$ form a partition over the domain. We can also define a partition over a restricted domain Y to be:

$\text{resPART}(Y, X_1, \ldots, X_n):$

$$\forall x \in Y[x \in X_1 \cup \cdots \cup X_n \wedge \neg x \in X_1 \cap X_2$$
$$\wedge \neg x \in X_1 \cap X_3 \wedge \cdots \wedge \neg x \in X_{n-1} \cap X_n]$$

Using this formula, the requirements E1,E2 and E3 are all expressible in MSO as follows:

$$\text{EDGE}_X(\mathcal{W}) : \text{resPART}(E_G, \mathcal{E}) \wedge$$

$$\bigwedge_{i,p,j} \forall e \in E_{i,p,j} \text{lab}_{\gamma_{p,j}}(e) \wedge$$

$$\bigvee_{p \in P_X} [\bigwedge_j \exists! e \in E_{0,p,j} \wedge \bigwedge_{p' \in P, p' \neq p} \bigwedge_j E_{0,p',j} = \emptyset]$$

Let $\text{EDGE}(\mathcal{W}) = \text{EDGE}_S(\mathcal{W})$. In using the symbol $\bigwedge_{i,p,j}$ we are quantifying over $i \in [0, |\text{NT}(P)|]$, $p \in P$, and $j \in [|\mathbf{T}(p)|]$.

Lemma 6. *Let $\mathcal{G}$ be an RGG and let $G \in L(\mathcal{G})$ then for each derivation tree $\mathbf{T}$ of G, $(G, \alpha_{\mathbf{T}}) \models \text{EDGE}(\mathcal{W})$.*

Example 13. For the grammar in Table 2, and derivation tree and graph in Figure 7, we obtain $\mathcal{E} = \{E_{0,p,1} = \{e_9\}, E_{0,p,2} = \{e_{14}\}, E_{0,p,3} = \{e_{15}\}, E_{1,q,1} = \{e_4, e_2\}, E_{1,q,2} = \{e_{13}, e_{11}\}, E_{2,r,1} = \{e_3\}, E_{2,r,2} = \{e_{12}\}, E_{2,s,1} = \{e_1\}, E_{2,s,2} = \{e_{10}\}, E_{1,t,1} = \{e_6, e_8\}, E_{2,u,1} = \{e_7\}, E_{1,u,1} = \{e_5\}\}$. This clearly forms a partition of the edges, and we can easily check that the rest of the requirements of EDGE also hold.

Decomposition into Subgraphs This constraint partitions the graph into a set of connected subgraphs, each of which is isomorphic to the terminal subgraph of the right-hand side of some production. The requirements are:

(S1) Every node in G is attached to some edge,

(S2) for each anchor $u \in C_{i,p}$ we can identify a unique edge $e \in E_{i,p,j}$ for each $j \in |T(p)|$ such that u anchors all of the endpoints of e,

(S3) for each edge $e \in E_{i,p,j}$ we can identify a unique anchor $u \in C_{i,p}$ such that u anchors all of the endpoints of e.

$\text{SUBGRAPH}_{i,p,j}(\mathcal{W}) :$

$(\forall v \exists e \vee_j \text{vert}(e,j) = v) \wedge$

$\left(\forall c \in C_{i,p} \exists! e \in E_{i,p,j} \right.$

$\exists v_1 \text{ANC}_{p,i,\bar{x}_1}(c, v_1, \{\mathcal{W}\}) \wedge \cdots \wedge$

$\exists v_{j_k} \text{ANC}_{p,i,\bar{x}_{j_k}}(c, v_{j_k}, \{\mathcal{W}\}) \wedge$

$\left. edg(e, v_1, \ldots, v_{j_k}) \right) \wedge$

$\left(\forall e \in E_{i,p,j} \exists! c \in C_{i,p} \right.$

$\exists v_1 \text{ANC}_{p,i,\bar{x}_1}(c, v_1, \{\mathcal{W}\}) \wedge \cdots \wedge$

$\exists v_{j_k} \text{ANC}_{p,i,\bar{x}_{j_k}}(c, v_{j_k}, \{\mathcal{W}\}) \wedge$

$\left. edg(e, v_1, \ldots, v_{j_k}) \right)$

Define $\text{SUBGRAPH}(\mathcal{W}) : \bigwedge_{i,p,j} \text{SUBGRAPH}_{i,p,j}$.

Lemma 7. *Let $\mathcal{G}$ be an RGG and let $G \in L(\mathcal{G})$ then for each derivation tree $\mathbf{T}$ of G, $(G, \alpha_{\mathbf{T}}) \models \text{SUBGRAPH}(\mathcal{W})$.*

Example 14. For the graph in Figure 7, we look at $\text{SUBGRAPH}_{1,q,1}$. We need to look at $C_{1,q} = \{v_4, v_2\}$, and $E_{1,q,1} = \{e_4, e_2\}$. Looking at the graph, we can see that each of $\text{ANC}_{q,1,\bar{c}_q}(v_4, v_4, \{\mathcal{W}\})$, $\text{ANC}_{q,1,\bar{x}_4}(v_4, v_5, \{\mathcal{W}\})$, $\text{ANC}_{q,1,\bar{c}_q}(v_2, v_2, \{\mathcal{W}\})$, $\text{ANC}_{q,1,\bar{x}_4}(v_2, v_3, \{\mathcal{W}\})$ hold. The label 'arg1' is on e_2, e_4 and $f_{q,1}$ so we can quickly verify that this shows that the graph satisfies $\text{SUBGRAPH}_{1,q,1}$.

Subgraph Composition

We require that for a graph G with derivation tree $\mathbf{T}$, $(G, \alpha_{\mathbf{T}}) \models \text{ANC}_{p,i,\bar{x}}(u, v, \{\mathcal{W}\})$ for $u \in C_{i,p}$ and $(G, \alpha_{\mathbf{T}}) \models \text{ANC}_{p',i',\bar{x}'}(u', v, \{\mathcal{W}\})$ for $u' \in C_{i',p'}$ if and only if $(G, \alpha_{\mathbf{T}}) \models \text{FUSE}_{p,i,\bar{x},p',i',\bar{x}'}(u, u', \{\mathcal{W}\})$ holds. This part of the precondition ensures that two different ways of looking at how nodes can be fused agree with one another. The first is if a node can be anchored by two different anchors then this node must be the image of two nodes from different production applications. The second is that we have FUSE which is the equivalence relation generated by a relation over the neighbouring nodes in the derivation tree.

$\text{SHARE}(\mathcal{W}) :$

$\bigwedge_{i,p,\bar{x},i',p',\bar{x}'} \forall c_1 \in C_{i,p} \forall c_2 \in C_{i',p'}$

$\left(\text{ANC}_{p,i,\bar{x}}(c_1, v, \{\mathcal{W}\}) \wedge \right.$

$\text{ANC}_{p',i',\bar{x}'}(c_2, v, \{\mathcal{W}\})$

$\left. \leftrightarrow \text{FUSE}_{i,p,\bar{x},i',p',\bar{x}'}(u, u', \{\mathcal{W}\}) \right)$

Lemma 8. *Let $\mathcal{G}$ be an RGG and let $G \in L(\mathcal{G})$ then for each derivation tree $\mathbf{T}$ of G, there exists $\alpha_{\mathbf{T}}$ such that $(G, \alpha_{\mathbf{T}}) \models \text{SHARE}(\mathcal{W})$.*

Example 15. Looking at Table 2 and Figure 7, $\text{FUSE}_{p,0,\bar{x}_1,s,2,\bar{x}_8}(v_5, v_1, \{\mathcal{W}\})$ holds and $\text{ANC}_{p,0,\bar{x}_1}(v_5, v_6, \{\mathcal{W}\})$ and $\text{ANC}_{s,2,\bar{x}_8}(v_1, v_6, \{\mathcal{W}\})$ both also hold.

The proof of each of the above lemmas is available in the supplementary materials. In each of these proofs, we prove by induction on the size of $\mathbf{T}$ that $(G, \alpha_{\mathbf{T}}) \models R(\mathcal{W})$ for $R \in \{\text{EDGE}, \text{SUBGRAPH}, \text{SHARE}\}$. In each induction, we use the equations (defined below) which express $\alpha_{\mathbf{T}}$ in terms of the parameter assignments of sub-trees of $\mathbf{T}$.

Let $G \in L_X(\mathcal{G})$ and $q : X \to H$ such that H has nonterminals $Y_1, \ldots, Y_n$ and $G = H[Y_1/H_1]\ldots[Y_n/H_n]$. Then $H_\eta \in L_{Y_\eta}(\mathcal{G})$ for each $\eta \in [n]$. Let $\mathbf{T}_\eta$ be a derivation tree for H_η and let $\alpha_{\mathbf{T}_\eta}$ be the assignment of $\mathcal{W}$ to the nodes and edges in H_η. Then we can define $\alpha_{\mathbf{T}}(\mathcal{E})$ in terms of the set of $\alpha_{\mathbf{T}_\eta}(\mathcal{E})$s:

$$\alpha_{\mathbf{T}} : \begin{cases} e \in E_{i,p,j} & \text{if } e \in E_{H_\eta}, \alpha_{\mathbf{T}_\eta} : e \in E_{i,p,j}, i \neq 0 \\ e \in E_{\eta,p,j} & \text{if } e \in E_{H_\eta}, \alpha_{\mathbf{T}_\eta} : e \in E_{0,p,j} \\ e \in E_{0,q,j} & \text{if } e \in E_H, e = h_e(\bar{f}_{q,j}, \mathbf{v}_0). \end{cases}$$

$$(2)$$

Where $e = h_e(\bar{f}_{q,j}, \mathbf{v}_0)$ means that e can be uniquely identified as corresponding to $\bar{f}_{q,j}$ since H and $R(q)$ are isomorphic and $\mathbf{v}_0$ is the root of $\mathbf{T}$. For the anchor set,

$$\alpha_{\mathbf{T}}(C) = c \cup_{\eta \in [n]} \alpha_{\mathbf{T}_\eta}(C) \qquad (3)$$

where $c = h(\bar{c}_q, \mathbf{v}_0)$.

4.1.4 RGLs Satisfy the Precondition

The precondition of the transducer is the conjunction of each of these formulas,

$$\rho_X(\mathcal{W}) : \text{EDGE}_X(\mathcal{W}) \wedge \text{SHARE}(\mathcal{W}) \wedge \text{SUBGRAPH}(\mathcal{W})$$

Define $\rho(\mathcal{W}) = \rho_S(\mathcal{W})$.

Proposition 1. *Let $\mathcal{G}$ be an RGG and let $G \in L(\mathcal{G})$, then for each derivation tree $\mathbf{T}$ of G,* *there exists a parameter assignment $\alpha_{\mathbf{T}}$ such that $(G, \alpha_{\mathbf{T}}) \models \rho(\mathcal{W})$.*

Proof. We use the parameter assignment $\alpha_{\mathbf{T}}$ which is defined from $\mathbf{T}$ in §4.1. Lemma 6 proves that $(G, \alpha_{\mathbf{T}}) \models \text{EDGE}(\mathcal{W})$. Lemma 7 proves that $(G, \alpha_{\mathbf{T}}) \models \text{SUBGRAPH}(\mathcal{W})$. Lemma 8 proves that $(G, \alpha_{\mathbf{T}}) \models \text{SHARE}(\mathcal{W})$. Therefore, $(G, \alpha_{\mathbf{T}}) \models \rho(\mathcal{W})$. $\square$

4.2 Parsing as Transduction

The transducer is made up of three types of formulas: the precondition, the domain formulas, and the relation formulas. We have established the precondition $\rho(\mathcal{W})$ and next we define the domain and relation formulas. The domain formulas define the nodes of the derivation tree and so we write $\text{node}(x, \{\mathcal{W}\})$. The relation formulas define which output node is the ith child of another output node, written $\text{child}_i(x, y, \{\mathcal{W}\})$, and the labels of the output nodes, written $\text{lab}_p(x, \{\mathcal{W}\})$.

The domain of the output for a parameter assignment α is $D_{\mathbf{T}}$ where:
$$D_{\mathbf{T}}(\alpha) : \{x \mid (G, x, \alpha) \models \text{node}(x, \{\mathcal{W}\})\}$$
and $\text{node}(x, \{\mathcal{W}\}) : x \in \mathcal{C}$.

The relation formula $\text{child}_r(x, y, \{\mathcal{W}\})$ defines the edges of the output of the transducer. We use the formula $\text{PAR}_{p,i,p',i'}(u, u', \{\mathcal{W}\})$ from Lemma 4, this encodes that the derivation tree node corresponding to u' is the i'th child of the node corresponding to u (which itself is the ith child of some other node).
$$\text{child}_{i'}(x, y, \{\mathcal{W}\}) : \bigvee_{i,p,p'} (\text{PAR}_{p,i,p',i'}(x, y, \{\mathcal{W}\})$$
We also need to assign labels to the tree nodes which can be done via the unary relation:
$$\text{lab}_p(x, \{\mathcal{W}\}) : \bigvee_i x \in C_{i,p}$$

Example 16. Figure 8 shows the output of the transducer when it takes Figure 7 as input with α defined as in the previous examples. The domain formulas specify the existence of the 9 nodes and the relation formulas specify the edges between the nodes, labelled by PAR formulas, and the labels of the nodes, according to the $C_{i,p}$ sets.

We have now defined each part of the transducer τ from graphs to their derivation trees. Let $\mathcal{G}$ be an RGG, and let $X \in N$. Then the corresponding

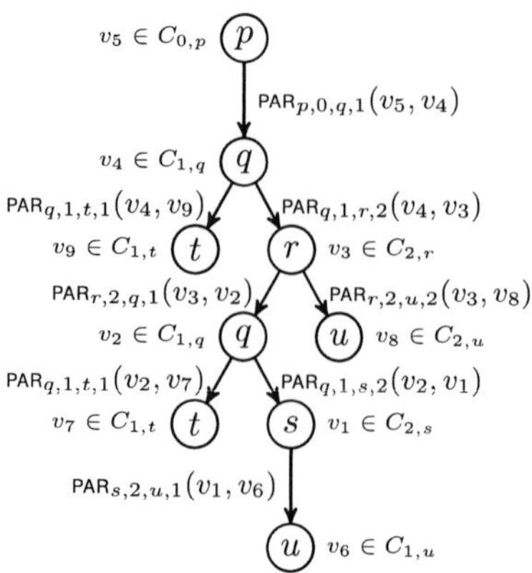

Figure 8: The output of the transducer, variable names are based on those of Figure 7. The PAR formulas are there to explain why the edge exists and the $v \in C_{i,p}$ formulas are there to show where the node labels come from.

transducer τ_X is
$$\langle \rho_X(\{\mathcal{W}\}), \text{node}(x, \{\mathcal{W}\}),$$
$$(\text{lab}_p(x, \{\mathcal{W}\}))_{p \in P},$$
$$(\text{edge}_r(x, y, \{\mathcal{W}\}))_{r \in [|\text{NT}(P)|]}\rangle.$$

For start symbol S of $\mathcal{G}$, let $\tau = \tau_S$. Let G be a graph in $L(\mathcal{G})$, and let α be a parameter assignment such that $(G, \alpha) \models \rho(\{\mathcal{W}\})$. Then the output of the transducer with respect to α is
$$\tau(G, \alpha) = (V_H, \text{lab}_H, (\text{child}_H^i)_{i \in [0, |\text{NT}(P)|]})$$
where $V_H = D_{\mathbf{T}}(\alpha) = \{x \mid (G, x, \alpha) \models \text{node}(x, \{\mathcal{W}\})\}$, $\text{lab}_H : V_H \to P$ such that $\text{lab}_H(x) = p$ if $x \in \alpha(C_{i,p})$ for some i, and $\text{child}_i^H : V_H \to V_H$ such that $\text{child}_i^H(x, y)$ if $(G, x, y, \alpha) \models \text{PAR}_{p,i,p',r}(x, y, \{\mathcal{W}\})$.

4.2.1 Transducer Output and Derivation Trees

We will show that for each $G \in L(\mathcal{G})$ if $\mathbf{T}$ is a derivation tree of G then $\mathbf{T} \in \tau(G)$. We will also show that for each $\mathbf{T} \in \tau(G)$, if it is a derivation tree in $\mathcal{T}_\mathcal{G}$ then it is a derivation tree of G.

Proposition 2. *Let $\mathcal{G}$ be an RGG and τ be the corresponding transducer. Let $G \in L(\mathcal{G})$ and $\mathbf{T}$ be a derivation tree of G. Then $\mathbf{T} \in \tau(G)$.*

By Proposition 2, we know that for each G, $\{\mathbf{T} | \text{val}(\mathbf{T}) = G\} \subseteq \tau(G)$.

Proposition 3. *Let $\mathcal{G}$ be an RGG and $G \in L(\mathcal{G})$. Let α be a parameter assignment such that $(G, \alpha) \models \rho(\mathcal{W})$. Then if $\mathbf{T} = \tau(G, \alpha)$ is in $\mathcal{T}_\mathcal{G}$ then $\text{VAL}(\mathbf{T}) = G$.*

Theorem 2. *RGL $\subseteq$ MSOL.*

Proof. Let $\mathcal{G}$ be an RGG and τ be the corresponding transducer. By Propositions 2 and 3, for each $G \in L(\mathcal{G})$, $\tau(G)$ is a set which contains all of the derivation trees of G and possibly other elements none of which are derivation trees of any $G' \in L(\mathcal{G})$ where $G' \neq G$. Therefore, for each $G \in L(\mathcal{G})$,

$$\tau(G) \cap \mathcal{T}_{\mathcal{G}} = \{\mathbf{T} \in \mathcal{T}_{\mathcal{G}} \mid \text{VAL}(\mathbf{T}) = G\}.$$

Therefore,

$$\tau(L(\mathcal{G})) \cap \mathcal{T}_{\mathcal{G}} = \{\mathbf{T} \in \mathcal{T}_{\mathcal{G}} \mid \text{VAL}(\mathbf{T}) = G, G \in L(\mathcal{G})\}.$$

And since $\{\mathbf{T} \in \mathcal{T}_{\mathcal{G}} \mid \text{VAL}(\mathbf{T}) = G, G \in L(\mathcal{G})\} = \mathcal{T}_{\mathcal{G}}$,

$$\tau^{-1}(\tau(L(\mathcal{G})) \cap \mathcal{T}_{\mathcal{G}}) = \tau^{-1}(\mathcal{T}_{\mathcal{G}}).$$

$$\tau^{-1}(\tau(L(\mathcal{G})) \cap \mathcal{T}_{\mathcal{G}}) = \{G \in L(\mathcal{G}) \mid \tau(G) \cap \mathcal{T}_{\mathcal{G}} \neq \emptyset)\} = L(\mathcal{G}).$$ Therefore,

$$L(\mathcal{G}) = \tau^{-1}(\mathcal{T}_{\mathcal{G}})$$

and so by Theorem 1 and the fact that $\mathcal{T}_{\mathcal{G}}$ is MSO definable, $L(\mathcal{G})$ is MSO definable. $\square$

5 Conclusions and Discussion

Property C1 of RGGs is used repeatedly in the proof that RGL is in MSOL. This property implies connectedness of the terminal subgraph, a property that both Tree-like Grammars (Matheja et al., 2015) and Restricted DAG Grammars (Björklund et al., 2016) share, although both of these formalisms allow nodes that are connected only to nonterminals, which is forbidden in RGG. We suspect that all three families of languages are incomparable. That these restricted forms of HRG all share the property of connectedness suggests that it may be an important property. In particular, we plan to investigate whether connectedness of terminal subgraphs implies that an HRL is in MSOL.

Languages which contain graphs of the form shown in Figure 9 are MSOL but not in RGL or TLG; hence both RGL and TLG are proper sub-families of SCFL. Languages of this form can be produced by RDG, whose relationship to SCFL is unknown. To produce graphs like this, we must allow productions containing nonterminals that are not incident to any internal node. We would need to allow this only in certain circumstances however, as we could easily produce a language of graphs that look like the graph in Figure 9 with equal numbers of a-labelled and b-labelled edges; such languages are not MSO-definable. On a technical level, allowing such extensions would mean that PAR no longer holds. (Courcelle, 1991) dis-

cusses this problem and introduces an alternative representation of derivation trees called **reduced trees** which enable some cases of this type to be defined in MSOL. This point requires further investigation.

Another possible extension would be to consider alternative forms of Lemma 2. Every MSO formula in the transducer depends on this lemma. We could potentially extend RGG if we can define other cases in which a path could be defined in terms of its trace and initial vertex. We intend to investigate such cases in future work.

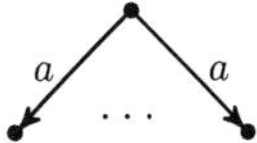

Figure 9: A graph where every edge is labelled a and has the same tail but each edge has a unique head.

Acknowledgments

This work was supported in part by the EPSRC Centre for Doctoral Training in Data Science, funded by the UK Engineering and Physical Sciences Research Council (grant EP/L016427/1) and the University of Edinburgh; and in part by a Google faculty research award (to AL). We thank Clara Vania, Sameer Bansal, Ida Szubert, Federico Fancellu, Antonis Anastasopoulos, Marco Damonte, and the anonymous reviews for helpful discussion of this work and comments on previous drafts of the paper.

References

Omri Abend and Ari Rappoport. 2013. Universal conceptual cognitive annotation (ucca). In *ACL (1)*. The Association for Computational Linguistics, pages 228–238. http://dblp.uni-trier.de/db/conf/acl/acl2013-1.html#AbendR13.

Cyril Allauzen, William Byrne, Adria de Gispert, Gonzalo Iglesias, and Michael Riley. 2014. Pushdown automata in statistical machine translation. *Computational Linguistics* .

Laura Banarescu, Claire Bonial, Shu Cai, Madalina Georgescu, Kira Griffitt, Ulf Hermjakob, Kevin Knight, Philipp Koehn, Martha Palmer, and Nathan Schneider. 2013. Abstract meaning representation for sembanking. In *Proceedings of the 7th Linguistic Annotation Workshop and Interoperability with Discourse*. Association for Computational Linguistics, Sofia, Bulgaria, pages 178–186. http://www.aclweb.org/anthology/W13-2322.

Daniel Bauer and Owen Rambow. 2016. Hyperedge replacement and nonprojective dependency structures. In *Proceedings of the 12th International Workshop on Tree Adjoining Grammars and Related Formalisms (TAG+12), June 29 - July 1, 2016, Heinrich Heine University, Düsseldorf, Germany*. pages 103–111. http://aclweb.org/anthology/W/W16/W16-3311.pdf.

Henrik Björklund, Frank Drewes, and Petter Ericson. 2016. *Between a Rock and a Hard Place – Uniform Parsing for Hyperedge Replacement DAG Grammars*, Springer International Publishing, Cham, pages 521–532. https://doi.org/10.1007/978-3-319-30000-9_40.

Mikolaj Bojanczyk and Michal Pilipczuk. 2016. Definability equals recognizability for graphs of bounded treewidth. In *Proceedings of the 31st Annual ACM/IEEE Symposium on Logic in Computer Science*. ACM, New York, NY, USA, LICS '16, pages 407–416. https://doi.org/10.1145/2933575.2934508.

Taylor L. Booth and Richard A. Thompson. 1973. Applying probability measures to abstract languages. *IEEE Transactions on Computers* 22(5):442–450. https://doi.org/http://doi.ieeecomputersociety.org/10.1109/C.1973.223746.

Julius Richard Büchi. 1960. On a decision method in restricted second-order arithmetic. *Proceedings Logic, Methodology and Philosophy of Sciences* .

Julius Richard Büchi and Calvin Elgot. 1958. Decision problems of weak second order arithmetic and finite automata, part i. *Notices of the American Mathematical Society* page 5;834.

David Chiang, Jacob Andreas, Daniel Bauer, Karl Moritz Hermann, Bevan Jones, and Kevin Knight. 2013. Parsing graphs with hyperedge replacement grammars. In *Proceedings of the 51st Annual Meeting of the Association for Computational Linguistics (Volume 1: Long Papers)*. Association for Computational Linguistics, Sofia, Bulgaria, pages 924–932. http://www.aclweb.org/anthology/P13-1091.

Bruno Courcelle. 1990. The monadic second-order logic of graphs i. recognizable sets of finite graphs. *Information and Computation* pages 12–75.

Bruno Courcelle. 1991. The monadic second-order logic of graphs V: on closing the gap between definability and recognizability. *Theoretical Computer Science* 80(2):153–202. https://doi.org/10.1016/0304-3975(91)90387-H.

Bruno Courcelle and Joost Engelfriet. 2011. *Graph Structure and Monadic Second-Order Logic, a Language Theoretic Approach*. Cambridge University Press.

Frank Drewes, Hans-Jörg Kreowski, and Annegret Habel. 1997. Hyperedge replacement graph grammars. In Grzegorz Rozenberg, editor, *Handbook of Graph Grammars and Computing by Graph Transformation*, World Scientific, pages 95–162.

Manfred Droste and Paul Gastin. 2005. *Weighted Automata and Weighted Logics*, Springer Berlin Heidelberg, Berlin, Heidelberg, pages 513–525. https://doi.org/10.1007/11523468_42.

Dan Flickinger, Yi Zhang, and Valia Kordoni. 2012. Deepbank : a dynamically annotated treebank of the wall street journal. In *Proceedings of the Eleventh International Workshop on Treebanks and Linguistic Theories (TLT11)*. Lisbon, pages 85–96. HU.

Jan Hajič, Eva Hajičová, Jarmila Panevov, Petr Sgall, Ondřej Bojar, Silvie Cinková, Eva Fučíková, Marie Mikulová, Petr Pajas, Jan Popelka, Jiří Semecký, Jana Šindlerová, Jan Štěpánek, Josef Toman, Zdeňka Urešová, and Zdeněk Žabokrtský. 2012. Announcing prague czech-english dependency treebank 2.0. In Nicoletta Calzolari (Conference Chair), Khalid Choukri, Thierry Declerck, Mehmet Uur Doan, Bente Maegaard, Joseph Mariani, Asuncion Moreno, Jan Odijk, and Stelios Piperidis, editors, *Proceedings of the Eight International Conference on Language Resources and Evaluation (LREC'12)*. European Language Resources Association (ELRA), Istanbul, Turkey.

John E. Hopcroft and Jeffrey D. Ullman. 1979. *Introduction to automata theory, languages and computation*. Addison-Wesley.

Bevan Jones, Jacob Andreas, Daniel Bauer, Karl Moritz Hermann, and Kevin Knight. 2012. Semantics-based machine translation with hyperedge replacement grammars. In *Proceedings of COLING*.

Tsutomu Kamimura and Giora Slutzki. 1981. Parallel and two-way automata on directed ordered acyclic graphs. *Information and Control* 49(1):10–51.

Alexander Koller and Marco Kuhlmann. 2011. A generalized view on parsing and translation. In *Proceedings of the 12th International Conference on Parsing Technologies*. Association for Computational Linguistics, Dublin, Ireland, pages 2–13. http://www.aclweb.org/anthology/W11-2902.

Christoph Matheja, Christina Jansen, and Thomas Noll. 2015. *Tree-Like Grammars and Separation Logic*, Springer International Publishing, Cham, pages 90–108. https://doi.org/10.1007/978-3-319-26529-2_6.

Mehryar Mohri, Fernando C. N. Pereira, and Michael Riley. 2008. Speech recognition with weighted finite-state transducers. In Larry Rabiner and Fred Juang, editors, *Handbook on Speech Processing and Speech Communication, Part E: Speech recognition*, Springer.

Xiaochang Peng, Linfeng Song, and Daniel Gildea. 2015. A synchronous hyperedge replacement grammar based approach for AMR parsing. In *Proceedings of the 19th Conference on Computational Natural Language Learning, CoNLL 2015, Beijing, China, July 30-31, 2015*. pages 32–41. http://aclweb.org/anthology/K/K15/K15-1004.pdf.

Daniel Quernheim and Kevin Knight. 2012. Towards probabilistic acceptors and transducers for feature structures. In *Proceedings of the Sixth Workshop on Syntax, Semantics and Structure in Statistical Translation*. Association for Computational Linguistics, Stroudsburg, PA, USA, SSST-6 '12, pages 76–85. http://dl.acm.org/citation.cfm?id=2392936.2392948.

Wolfgang Thomas. 1991. *Automata, Languages and Programming: 18th International Colloquium Madrid, Spain, July 8–12, 1991 Proceedings*, Springer Berlin Heidelberg, Berlin, Heidelberg, chapter On logics, tilings, and automata, pages 441–454. https://doi.org/10.1007/3-540-54233-7-154.

Boris Trakhtenbrot. 1961. Finite automata and logic of monadic predicates. *Doklady Akademii Nauk SSSR* pages 140:326–329.

Graph Transductions and Typological Gaps in Morphological Paradigms

Thomas Graf
Stony Brook University
Department of Linguistics
Stony Brook, NY 11794-4376
`mail@thomasgraf.net`

Abstract

Several typological gaps have attracted a lot of interest in the linguistic literature recently. These concern the Person Case Constraint and the absence of ABA patterns in adjectival gradation, pronoun suppletion, case syncretism, and singular noun allomorphy, among others. This paper is the first to provide a unified explanation of all these phenomena, and it does so via weakly non-inverting graph-transductions. A pattern P is absent from the typology whenever such transductions cannot produce the graph corresponding to P from some fixed underlying base graph. I show that weakly non-inverting graph-transductions are particularly simple from a computational perspective, and consequently all these typological gaps follow from general simplicity desiderata.

1   Introduction

One peculiar property of natural language is that its typology rarely cover the full range of logically possible options. The Person Case Constraint (PCC), for instance, blocks certain combinations of direct objects (DOs) and indirect objects (IOs) based on their person specification.

(1) a. * Roger *me*   *leur*   a
         Roger 1SG.ACC 3PL.DAT has
         présenté.
         shown

     b. Roger *le*   *leur*   a
         Roger 3SG.ACC 3PL.DAT has
         présenté.
         shown

     'Roger has shown me/him to them.'

Modulo cases where both DO and IO have the same person, there are 64 conceivable PCC variants yet only 4 are attested.

A similar case of limited variation is the *ABA generalization, which was first stated by Bobaljik (2012) with respect to adjectival gradation. While many adjectives have regular comparative and superlative forms (*smart, smarter, smartest*), some adjectives display stem suppletion (*good, better, best*). Bobaljik (2012) claims that there are no languages where the comparative is suppletive while the superlative is regular (*good, better, goodest*) — in other words, there are no ABA patterns. Since then the *ABA generalization has been observed in a large number of morphological paradigms, and many proposals have been put forward to explain the absence of ABA patterns.

However, cases of limited complexity like the PCC and the *ABA generalization have not received much attention from mathematical linguists. One reason may be that these restrictions on natural languages do not seem to line up with the usual notions of generative capacity, computational complexity, learnability or minimal description length. The PCC, for instance, is utterly unremarkable from a formal perspective: the constrained elements are string adjacent clitics, and the sets of permitted and blocked configurations are both finite. As a result, every PCC variant is strictly 2-local over strings (McNaughton and Papert, 1971), making it even less complex than simple phonological processes such as intervocalic voicing and locally bounded vowel harmony (Heinz, 2015). Since different PCCs only vary in which one of six IO-DO combinations they allow, there are no quantifiable consequences for learnability, either. The tools of mathematical linguistics are geared towards vertical variation — hierarchies of expressivity and complexity — whereas phenomena like the PCC and the *ABA generalization pertain to horizontal variation, i.e. limitations that seem arbitrary and pointless from a computational perspective.

Proceedings of the 15th Meeting on the Mathematics of Language, pages 114–126,
London, UK, July 13–14, 2017. ©2017 Association for Computational Linguistics

I argue in this paper that mathematical linguistics does in fact have a lot to say about such cases of horizontal variation. Not only does a mathematically informed perspective allow for a level of abstraction where the PCC and the *ABA generalization can be given a unified explanation, it even allows us to derive the limits of variation from computational considerations. Contrary to initial appearances, then, horizontal variation is indeed interwoven with vertical variation upon closer inspection.

More concretely, I show that both the PCC and the *ABA generalization can be decomposed into two components: a base hierarchy that is represented by a graph, and a graph transduction that produces a language-specific ordering from the base hierarchy. The restrictions on cross-linguistic variation arise from limitations on how the graph transduction may change the ordering relations in the base hierarchy. These limitations, in turn, guarantee that the transductions belong to an especially weak class of mappings. Viewed from the perspective of strings, they are input strictly 1-local relations (Chandlee, 2014).

The paper is laid out as follows. The few required basics of graph theory are summarized and exemplified in Sec. 2 in an effort to accommodate readers from various backgrounds. I then discuss Graf's (2014) algebraic account of the PCC, which forms the basis of my graph-theoretic analysis. Said analysis is subsequently extended to a number of phenomena in Sec. 4. All of them are instances of the *ABA generalization or at least closely related to it: adjectival gradation, pronoun allomorphy, case syncretism, and noun stem allomorphy, With the full analysis in place, I then turn to the computational investigation (Sec. 5) and address some methodological concerns about the viability of studying horizontal variation across natural languages (Sec. 6).

2 Preliminaries

Even though the paper presupposes only minimal familiarity with graph theory, I include a slightly more accessible explanation of the basic concepts due to the interdisciplinary subject matter, which might attract readers without the expected mathematical background. The reader can safely skip this section if they are not puzzled by terms like *weakly connected graph* and *graph transduction*.

A *directed graph* $G := \langle V, E \rangle$ consists of a set V of *vertices* and a set $E \subseteq V \times V$ of *edges* that connect these vertices. Both V and E may be empty, so there is no requirement for a graph to contain any vertices or that any of its vertices are connected by edges. We say that vertex v is *immediately reachable* from vertex u iff there is an edge from u to v (i.e. $\langle u, v \rangle \in E$). In the special case where $u = v$ the edge is called a *loop*. Furthermore, u' is *reachable* from u iff there are vertices $v_1, \ldots, v_n$ such that v_1 is immediately reachable from u, and v_{i+1} is immediately reachable from v_i $(1 \leq i < n)$, and u' is immediately reachable from v_n. Reachability thus holds iff $\langle u, v \rangle \in E^+$, where E^+ is the transitive closure of E. In this case we also write $u \lhd v$. If a vertex is reachable from itself, the graph contains a *cycle*.

As an example, consider the following directed graph G:

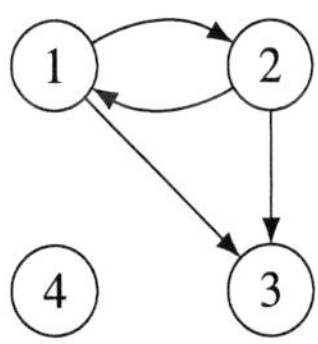

Its set of vertices is $\{1, 2, 3, 4\}$, and the set of edges is $\{\langle 1, 2 \rangle, \langle 2, 1 \rangle, \langle 1, 3 \rangle, \langle 2, 3 \rangle\}$. Therefore 2 is immediately reachable from 1, and the other way round. We also see that 3 is immediately reachable from 1 and 2, but no vertex is immediately reachable from 3. Moreover, 1 is reachable from itself even though it is not immediately reachable from itself. This is the case because 2 is immediately reachable from 1 and from there we can immediately reach 1. Formally, we have $\langle 1, 2 \rangle \in E$ and $\langle 2, 1 \rangle \in E$, which implies $\langle 1, 1 \rangle \in E^+$. This also entails that G contains a cycle even though there is no loop $\langle 1, 1 \rangle \in E$.

A graph is *undirected* iff its edge relation is symmetric: for all $u, v \in V$, $\langle u, v \rangle \in E$ iff $\langle v, u \rangle \in E$. In an undirected graph, u is (immediately) reachable from v iff v is (immediately) reachable from u. An undirected graph is *connected* iff every node is reachable from every other node: $\forall u, v \in V, \langle u, v \rangle \in E^+$. A directed graph is

- *connected* iff for all $u, v \in V$ it holds that $\langle u, v \rangle \in E^+$ or $\langle v, u \rangle \in E^+$,

- *weakly connected* iff adding $\langle v, u \rangle$ to E for every $\langle u, v \rangle \in E$ yields an undirected graph that is connected.

The example graph G above is not connected because I) 1 and 2 are not reachable from 3, and II) no node can reach 4 or be reached from 4. Due to II G is not weakly connected either. If G were weakly connected, then we could turn it into a connected undirected graph by adding the symmetric counterpart of every existing edge. But this only grows the edge relation E of G from $\{\langle 1, 2\rangle, \langle 2, 1\rangle, \langle 1, 3\rangle, \langle 2, 3\rangle\}$ to $\{\langle 1, 2\rangle, \langle 2, 1\rangle, \langle 1, 3\rangle, \langle 3, 1\rangle, \langle 2, 3\rangle, \langle 3, 2\rangle\}$. The resulting graph is still not connected because there is no edge from or to 4. But if 4 were to be removed from the set of vertices, the graph would indeed be weakly connected (but not connected).

A *graph transduction* τ is any computable binary relation between graphs. In this paper, however, I only consider transductions that do not change the set of vertices. Given a graph G, $\tau(G) := \{G' \mid \langle G, G'\rangle \in \tau\}$. In order to distinguish reachability in G from reachability in some $g \in \tau(G)$ I sometimes use the symbol ◄ instead of ◁.

Graph transductions generalize string transductions and tree transductions from strings and trees to arbitrary graphs. String transductions are closely related to phonological and morphological rewrite rules (Johnson, 1972; Kaplan and Kay, 1994; Mohri, 1997; Chandlee, 2014, 2016). Tree transductions are the formal counterpart to syntactic transformations, as is explicitly mentioned in Rounds (1970), one of the earliest papers on tree transducers; others include Engelfriet (1975) and Baker (1978; 1979). For modern surveys see Knight (2007) and Maletti (2010). For the purposes of this paper, the technical aspects of graph transductions are of little concern. The only relevant point is that just like string and tree transductions, graph transductions differ in their computational requirements so that some transductions are easier to compute than others. For a more formal perspective on graph transductions, the reader is referred to Courcelle (1992) and Courcelle and Engelfriet (2012).

3 Person Case Constraint

The vantage point for this project is the algebraic analysis of the Person Case Constraint (PCC) in Graf (2014). Once the analysis is recast in graph-theoretic terms, it is easily extended to the *ABA generalization in Sec. 4. From a didactic perspective this order of topics is slightly lopsided be-cause Graf's (2014) PCC treatment is more complex than the morphological paradigms I extend it to. But it is still fairly simple, and mastering the complex case first will greatly speed up the discussion of the simpler phenomena later on.

As I mentioned in the introduction, the PCC renders the well-formedness of DO-IO-combinations contingent on their person specifications. Four PCCs are attested in the literature (Walkow, 2012). Using 1, 2, and 3 as shorthands for first, second, and third person, respectively, they are defined as follows:

S(trong)-PCC DO must be 3. (Bonet, 1994)

U(ltrastrong)-PCC DO is less prominent than IO, where 3 is less prominent than 2, and 2 is less prominent than 1. (Nevins, 2007)

W(eak)-PCC 3IO combines only with 3DO. (Bonet, 1994)

M(e first)-PCC If IO is 2 or 3, then DO is not 1. (Nevins, 2007)

Note that cases where IO and DO have the same person feature are frequently treated separately in the literature, so I will not consider them here either.

Graf (2014) provides a mathematical account of the PCC that gradually moves from presemilattices as a purely descriptive device to a more theoretical proposal that can be recast in graph-theoretic terms. Rather than reiterate this gradual development, I immediately skip ahead to the three essential components of the final account.

1. All variants of the PCC are subsumed under the **G(eneralized)-PCC**, which states that IO must not be (strictly) less prominent than DO (IO $\not<$ DO). This constraint will produce exactly the four attested PCC variants if combined with the directed graphs in Fig. 1, where m is more prominent than n ($n < m$) iff n is reachable from m.

2. The independently motivated person hierarchy $3 < 2 < 1$ of Zwicky (1977) is posited as a universal base ordering for person. From our perspective, Zwicky's person hierarchy is identical to the graph for the U-PCC.

3. The four PCC-specific prominence rankings in Fig. 1 are obtained from Zwicky's hierarchy by a graph transduction τ that adds or

removes edges while preserving three essential properties of the base structure. In the following, $\lhd$ and $\blacktriangleleft$ denote the transitive closure of the edge relations in the input and output graph, respectively, and all graphs are assumed to contain no loops.

Weak connectedness The output graph produced by τ must be weakly connected.

Weak maximality If there is no y such that $y \lhd x$, then $z \blacktriangleleft x$ only if we also have $x \blacktriangleleft z$.

Strong minimality If there is no y such that $x \lhd y$, then there is no z such that $x \blacktriangleleft z$.

Then $x < y$ iff $y \blacktriangleleft x$.

The reader is invited to verify for themselves that the four graphs in Fig. 1, and only those, can be obtained from Zwicky's person hierarchy without violating any of the three constraints above.

As an example of how this account enforces a specific PCC consider the S-PCC, which only allows IO-DO combinations if DO is 3. Hence the only allowed combinations are IO1-DO3 and IO2-DO3. The G-PCC requires IO $\not< $ DO, and the graph for the S-PCC establishes $2 < 1$, $1 < 2$, $3 < 1$, and $3 < 1$. So any instance where DO is 2 or 1 necessarily results in a violation of the G-PCC: for 1, IO cannot be 2 or 3, and for 2, IO cannot be 1 or 3. With a third person DO, on the other hand, IO can freely vary between 1 and 2. Consequently, the only allowed combinations are indeed those where DO is 3.

This graph-based account is remarkably simple in comparison to syntactic proposals, which not only have to capture the typological variation but must also provide a syntactic encoding for both the G-PCC and the person hierarchy (Anagnostopoulou, 2005; Adger and Harbour, 2007; Nevins, 2007). The specificity of linguistic proposals has certain advantages, as I discuss at the end of Sec. 5, but it also comes with its fair share of problems that the graph-theoretic view avoids. In particular, abstracting away from the details of syntactic implementation provides a greater degree of flexibility and makes the account more accommodating to new data. For example, recent results from Slovenian (Stegovec, 2016) suggest that there are inverted variants of the PCC where the G-PCC is DO $\not< $ IO instead of IO $\not< $ DO. Most syntactic accounts are entirely built around the idea that all instances of the PCC involve an IO $\not< $ DO

asymmetry, and thus they must now be rethought from the ground up or reinterpret the Slovenian data. The graph-based proposal, by contrast, ends up even less complex because the very specific G-PCC has now been reduced to a general ban against prominence mismatches: $x \not< y$, with languages differing in how they instantiate x and y as IO and DO.

For the purposes of this paper, however, the more interesting aspect of the graph-theoretic view is how it captures typological variation in the PCC: languages all start out with the same base hierarchy but may modify it as long as the distinguished roles of the top and bottom positions are not completely destroyed. In fact, weak maximality and strong minimality are instances of more general order preservation properties.

Strongly non-inverting If $x \lhd y$, then it is not the case that $y \blacktriangleleft x$.

Weakly non-inverting If $x \lhd y$, then $y \blacktriangleleft x$ only if $x \blacktriangleleft y$.

The transductions that produce the PCC graphs in Fig. 1 are weakly non-inverting but go a little bit beyond that because they are all strongly non-inverting with respect to 3.

If weak maximality and strong minimality were completely replaced by the property of being weakly non-inverting, this would allow for several new graphs. However, the prominence ranking $<$ is defined in terms of reachability rather than immediate reachability, and all the new graphs turn out to define the same reachability relations as one of the two graphs depicted in Fig. 2. One is a variant of U-PCC where we also have $1 < 2$ and $2 < 3$, the other one a version of the M-PCC with $2 < 3$ and $3 < 2$. As noted in Graf (2014) the former is actually attested in Cairene Arabic as a ban against all DO-IO combinations (Shlonsky, 1997). Although this phenomenon may not be a genuine PCC, we may classify it as the I(ndiscriminate)-PCC. The second PCC variant changes the *Me first*-PCC into a *Me second*-PCC. This M2-PCC is still unattested. Whether the mathematically more pleasing notion of being weakly non-inverting fully captures the PCC thus has to remain an open question.

Even though not all weakly non-inverting graph transductions may be suitable for the PCC, it is certainly the case given our current data that

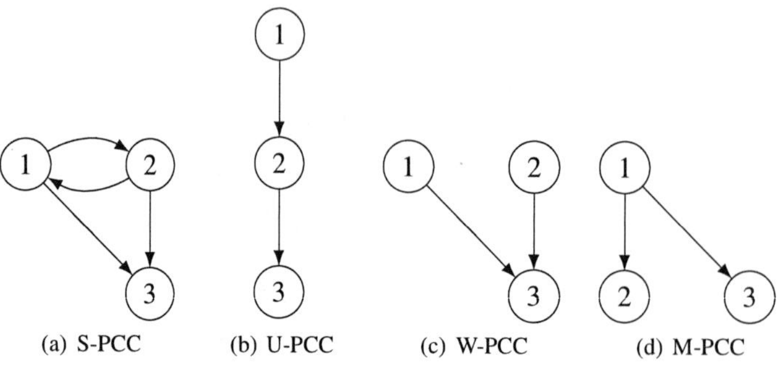

(a) S-PCC   (b) U-PCC   (c) W-PCC   (d) M-PCC

Figure 1: Graphs for the four variants of the PCC

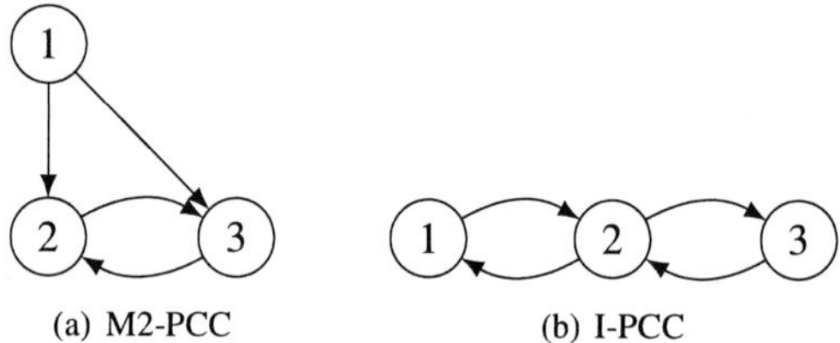

(a) M2-PCC   (b) I-PCC

Figure 2: Replacing weak maximality and strong minimality by the weakly non-inverting property allows for another two PCC-graphs.

all PCC graph transductions are weakly non-inverting. In the next section, I argue that many aspects of morphology are also closely tied to weakly non-inverting graph transductions. In particular, the restriction to weakly non-inverting graph transductions is sufficient to derive the ban against ABA patterns that has attracted a great amount of attention since Bobaljik (2012).

4 Deriving the *ABA Generalization

4.1 Stem Suppletion in Adjectival Gradation

The *ABA generalization refers to a particular gap in various morphological paradigms. Given a morphological subsystem where one may posit an underlying hierarchy $x < y < z$, z cannot pattern with x to the exclusion of y. At the beginning of this paper I already presented an example from suppletion in adjectival gradation, analyzed at great depth in Bobaljik (2012). Bobaljik points out that if a language allows for stem suppletion in either comparatives or superlatives, it must allow for both. Data illustrating this generalization is given in Tab. 1. If one follows the convention to list the three forms in the order *positive, comparative, superlative* and uses letters to indicate which forms use the same stem, one can decompose the

gap into two constraints, *AAB and *ABA.

In Bobaljik (2012), these constraints are explained via structural assumptions. Bobaljik decomposes adjectival forms into a tree template such that comparatives contain the positive base form as a subtree and are in turn themselves subtrees of the corresponding superlative forms. Then *AAB and *ABA follow from specific assumptions about the rewrite rules (tree-to-string transducers in computational terms) that map these tree structures to the output string. Bobaljik and Sauerland (2017) provide a less stipulative explanation grounded in the combinatorics of feature systems, which is closer to my graph-theoretic proposal (although they ultimately reject this content-agnostic solution in favor of the structural account). Both works, however, agree that *ABA is the more important constraint of the two — *AAB seems to be specific to adjectival suppletion whereas *ABA holds for many morphological paradigms (more on that in the next subsections).

The increased importance of *ABA relative to *AAB is noteworthy because the former is indeed more complex than the latter from the perspective of graph transductions. Suppose that there is a universal underlying hierarchy of the form $H_U :=$ *positive $<$ comparative $<$ superlative*, which we may identify with the U-PCC graph and thus abbreviate as $1 < 2 < 3$ (I stipulate that $x < y$ iff $x \blacktriangleleft y$ rather than $y \blacktriangleleft x$ in order to stay close to linguistic intuitions, but this is immaterial for the actual account). Assume furthermore that two forms m and n of an adjective involve the same (original or suppletive) stem only if $m < n$ and $n < m$ in the language-specific hierarchy H_L. Applying this idea to the graphs in Fig. 1 and 2 produces four different patterns: AAB, ABC, ABB, and AAA (see Tab. 2). Cru-

118

Language	Positive	Comparative	Superlative	Pattern
English	smart	smart-er	smart-est	AAA
English	good	bett-er	be-st	ABB
Finnish	hyvä	pare-mpi	parha-in	ABB
Latin	bon-us	mel-ior	opt-imus	ABC
Welsh	da	gwell	gor-au	ABC
unattested	good	bett-er	good-est	*ABA
unattested	good	good-er	be-st	*AAB

Table 1: Examples of attested suppletion patterns from Smith et al. (2016)

PCC graph	Suppletion pattern
S-PCC	AAB
U-PCC	ABC
W-PCC	ABC
M-PCC	ABC
M2-PCC	ABB
I-PCC	AAA

Table 2: Adjectival gradation patterns defined by the six PCC graphs

cially, ABA is not among these graphs. So the ABA pattern cannot be produced from the $1 < 2 < 3$ base order assuming that the graph transductions

- are weakly non-inverting, and

- produce weakly connected graphs, and

- do not relabel any nodes, and

- do not delete any nodes.

Most of these assumptions are innocent from a linguistic perspective. Deletion of nodes makes no sense in this case as it would amount to removing the positive, comparative, or superlative form, but we are only interested in languages with all three forms because the *ABA generalization is trivially satisfied otherwise. Relabeling nodes would create an "anything goes" scenario where adjectival gradation hierarchies could even be mapped to person and number with no rhyme or reason. And the output graphs must be weakly connected because hierarchies in natural language never allow for elements that are completely unordered with respect to the other elements in the hierarchy. This leaves only two non-trivial assumptions that do the actual work of blocking ABA patterns: graph transductions must be weakly non-inverting,

and the base hierarchy is $H_U := positive < comparative < superlative$.

Note that positing this hierarchy does not entail that the ordering needs to be reflected in the structure of adjectives as proposed by Bobaljik (2012). Instead, the hierarchy may be taken to reflect the semantics of these constructions or arise from some other unknown factor. For our purposes, it only matters that we have such an underlying base hierarchy, not what its origins may be. And this is not a peculiarity of this approach: even stating the *ABA generalization for purely descriptive purposes presupposes this order. If one instead assumed an order of, say, *comparative < superlative < positive*, then the banned pattern would be BAA instead of ABA. But the latter is equivalent to ABB, which is allowed in many other morphological paradigms that have nothing to do with adjectives. So an implicit commitment to H_U is required whenever one seeks to analyze adjectival stem suppletion as an instance of the general ban against ABA patterns.

As a matter of fact, though, our finding can be strengthened so that it is compatible with a number of underlying hierarchies rather than just H_U. As long as the directed graph we start with is one of the connected PCC graphs, the ABA pattern cannot be produced.

Theorem 1. *Let τ be a non-deleting, weakly non-inverting graph transduction that does not relabel any nodes and only produces connected graphs, and let S be one of the connected graphs in Fig. 1 and 2. Then no $G \in \tau(S)$ allows for the ABA pattern.*

Proof. Recall that by definition two vertices u and v may have the same realization iff $u \triangleleft v$ and $v \triangleleft u$. Therefore the ABA pattern can only be produced by graphs where both $1 \triangleleft 3$ and $3 \triangleleft 2$ hold but for all $x \in \{1, 3\}$, $x \triangleleft 2$ holds only if $2 \triangleleft x$ does not. No such graph can be produced by τ from any of the four choices for S without deleting 2 or relabeling nodes.

In each S it holds that $1 \lhd 2$ and $2 \lhd 3$ (wherefore $1 \lhd 3$). So if $3 \blacktriangleleft 1$ and $1 \blacktriangleleft 3$ in $G \in \tau(S)$, then necessarily $3 \blacktriangleleft 2$:

1. Because $G \in \tau(S)$ is connected, at least one of the following must hold: $1 \blacktriangleleft 2$, $2 \blacktriangleleft 1$, $3 \blacktriangleleft 2$, $2 \blacktriangleleft 3$.

2. If $2 \blacktriangleleft 3$, then $2 \blacktriangleleft 1$ by transitivity.

3. If $2 \blacktriangleleft 1$, then $1 \blacktriangleleft 2$ because τ is weakly non-inverting.

4. If $1 \blacktriangleleft 2$, then $3 \blacktriangleleft 2$ by transitivity.

But $3 \blacktriangleleft 2$ implies $2 \blacktriangleleft 3$ because τ is weakly non-inverting. So either $3 \blacktriangleleft 2$ and $2 \blacktriangleleft 3$ or it does not hold that both $3 \blacktriangleleft 1$ and $1 \blacktriangleleft 3$. $\qquad\square$

The proof reveals that the *ABA generalization is compatible with any universal base hierarchy that specifies at least *positive* $\leq$ *comparative* and *comparative* $\leq$ *superlative*.

While *ABA follows immediately if the graph transductions must be weakly non-inverting, *AAB is much harder to derive. As shown in Tab. 2, AAB patterns are produced by the S-PCC graph. In order to block this graph, one has to disallow $2 \blacktriangleleft 1$. But then the I-PCC graph would be blocked, too. This only leaves the stipulative option of banning $2 \blacktriangleleft 1$ unless $3 \blacktriangleleft 1$. Intuitively, this states that 1 loses its privileged status only if 1, 2, and 3 are all equally prominent. Just as in the case of the PCC, then, we have to slightly strengthen the requirements on the graph transduction to avoid overgeneration. That this strengthening pertained to 3 in the case of the PCC but to 1 in the case of adjectives is not significant since the two are each other's duals. We could have just as well identified H_U with the inverse of the U-PCC graph and obtained a strengthening requirement with respect to 3 this way.

Putting aside these minor details, we can now say with certainty that the PCC and the *ABA generalization are remarkably similar from a graph-theoretic perspective. Both operate within a class of graphs that are obtained from an underlying base order by some weakly non-inverting graph transduction. Each one puts an additional restriction on the transduction, and in each case the restriction is designed to preserve the special status of an element at the top/bottom of the underlying hierarchy.

4.2 Other Morphological Paradigms

As mentioned earlier on, the ban against ABA patterns also holds with respect to other morphological paradigms. Some of those can be explained in exactly the same manner as the ABA ban with adjectives, whereas others require minor modifications.

Pronoun allomorphy The simplest case arises with pronoun allomorphy. Harbour (2015) conducts an extensive survey of pronoun systems and shows that all of them adopt one of four systems with respect to person:

- all persons are the same (AAA),

- first and second person are the same (AAB),

- second and third person are the same (ABB),

- all persons are different (ABC).

Again the ABA pattern is missing, and this fact is expected if graph transductions must be weakly non-inverting and the underlying person hierarchy fixes $3 < 2$ and $2 < 1$, as we already had to assume for the PCC. However, a quick glance at Tab. 2 reveals that an even stronger result holds: AAA, AAB, ABB, and ABC are exactly the patterns that can be generated under our account. Pronominal systems, then, are the first instance where our base assumptions give a full characterization of the morphological paradigm and no extra stipulations are needed.

Case syncretism Caha (2009; 2013) proposes the *Strong Case Contiguity Hypothesis* according to which case syncretism may only target contiguous areas of Blake's Case Hierarchy (Blake, 2001):

$$\text{Nom} > \text{Acc} > \text{Gen} > \text{Dat} > \text{Inst} > \text{others}$$

This means that a language may mark, say, accusative, genitive, dative and instrumental the same, but not accusative and instrumental to the exclusion of dative and genitive. In other words, the Strong Case Contiguity Hypothesis extends the *ABA generalization beyond systems with three-way contrasts.

Using Blake's Case Hierarchy as a baseline, it is possible with our current assumptions to generate graphs that instantiate some ABA patterns.

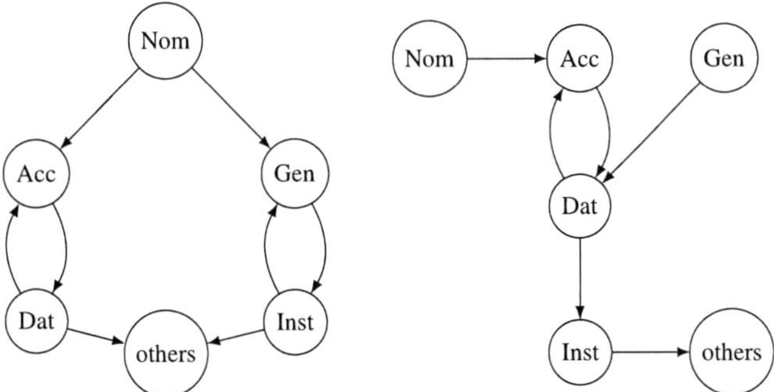

Figure 3: Two graphs that generate ABA patterns for case

Two examples are displayed in Fig. 3. A notable property of these graphs is that they are not connected, even though they are weakly connected. If the graph transductions are limited to producing connected graphs, then no ABA patterns can be generated anymore. So case syncretism may once again not be too different from the PCC, adjectival gradation, or pronoun allomorphy, except that it puts more stringent restrictions on what a valid case hierarchy may look like: no two cases may be unordered with respect to each other.

That said, Harðarson (2016) points out some apparent exceptions to Caha's Strong Case Continuity Hypothesis in Germanic languages, which display accusative-dative syncretism in some case paradigms but not accusative-genitive-dative syncretism. One solution would be to posit a more relaxed version of Blake's hierarchy where genitive and dative are unordered with respect to each other. This allows for all the syncretism patterns of Blake's original hierarchy but also includes accusative-dative syncretism. Whether this is the right way to deal with these exceptions has to remain an open issue for now. Thankfully the typological literature on this topic is very rich (see (Zompí, 2016) and references therein), so a deeper exploration should be possible in the near future.

Noun stem allomorphy Case syncretism has also been studied with respect to the noun stems that are chosen for specific cases. In Latin, for example, the nominative of 'man' is *hom-o*, whereas the accusative is *homin-em*. Nominative and accusative thus are formed with different stems of the same noun. In the following, I only consider the behavior of singular stems because the typology of plural stem allomorphy is still understudied to the best of my knowledge.

McFadden (2017) proposes that all languages obey the *Nominative Stem-Allomorphy Generalization*: if noun stem allomorphy is conditioned by case, it distinguishes the nominative from all other cases. In other words, noun stem allomorphy always displays an AB^n pattern. For a language with three cases, McFadden's generalization permits only AAA and ABB while excluding AAB, ABA, and ABC.

This is an even more restrictive paradigm than the one we encountered for case syncretism. But it can still be explained in terms that fit naturally into the graph-theoretic framework. Note that, as indicated in Tab. 2, AAA and ABB are exactly the patterns generated by the graphs in Fig. 2 — the complement set of our four main PCC graphs from Fig. 1. While at first counterintuitive, this actually makes it possible to describe noun stem allomorphy as the combination of case syncretism with an inverted PCC. First, suppose once more that the graph transduction must produce a connected graph, as we did for case syncretism. Then we only need to enforce two more properties for graph transductions. The first one is weak maximality, which was also part of our account for the PCC. The second is weak non-maximality:

Weak non-maximality If there is a y such that $y \lhd x$, then $x \blacktriangleleft z$ iff $z \blacktriangleleft x$.

When applied to Blake's hierarchy, these two properties ensure that nominative is always a maximal vertex, whereas all other vertices are reachable from each other. This guarantees that only AB^n and A^n patterns are possible.

Interim summary We have looked at five different phenomena where typological variation is much more narrow than one would expect from a computational perspective — the PCC, adjectival gradation, pronoun suppletion, case syncretism, and singular noun stem allomorphy. In all five cases, the typology could be derived from a very natural and independently motivated base hierarchy in combination with certain assumptions about structure preservation. For each language, the base hierarchy is converted into a language-specific hierarchy by some graph transduction τ that must not delete or relabel any nodes, has to produce weakly connected graphs, and, crucially, is weakly non-inverting. In some cases, this already explains the full range of variation, while other paradigms seem to invoke additional restrictions on τ. A succinct overview is given in Tab. 3.

5   Why These Properties?

The previous two sections have established that the range of typological variation across many morphological paradigms is accurately delimited if one assumes that there are universally shared base hierarchies that may only be manipulated in narrowly restricted ways. At the very center of my formal investigation was the requirement that graph transductions be weakly non-inverting. While descriptively adequate, it seems puzzling that natural languages should obey such a particular property. And even if one grants that being weakly non-inverting is advantageous for some reason, why is the requirement not strengthened so that all graph transductions must be strongly non-inverting. If one is good, then the other should be even better. I contend that there are indeed reasons that make weakly non-inverting graph transductions particularly simple from a computational perspective, whereas strongly non-inverting graph transductions do not further improve on this simplicity. Weakly non-inverting graph transductions therefore represent a sweet spot between flexibility and computational simplicity.

Several computational considerations underly this claim. First, the property of being weakly non-inverting enforces a limited amount of order preservation, and order preservation is known to play a central role for other aspects of language, too. Mönnich (2006; 2007) shows that standard Minimalist grammars (Stabler, 1997), a formalization of the Minimalist syntax (Chomsky, 1995), generate tree languages that are the image of regular tree languages under direction preserving MSO transductions. The tree languages of tree adjoining grammars (Joshi, 1985), on the other hand, are the image of regular tree languages under inversely direction preserving MSO transductions (Mönnich, 2006, 2012). Either way order preservation seems to be an important aspect of tree transductions in syntax, so it is not unreasonable that graph transductions in morphology may display similar limitations.

But there are stronger arguments that go beyond mere analogy. If the computational complexity of transductions is severely restricted, they are simply incapable of reversing order and hence are necessarily weakly non-inverting. Unfortunately the current knowledge of very weak graph transductions is not as well-developed as that for string and tree transductions, so I will illustrate my point with string transductions instead.

Note first that all the graphs in this paper are strings or string-like. When a graph is not a string, that is because there are two vertices that either form a cycle or are not immediately reachable from each other. We may use the dedicated symbols - and $|$ for these cases such that u-v means that u and v form a cycle, and $u|v$ denotes that u is not immediately reachable from v, and *vice versa*. With this notation, the four PCC graphs in Fig. 1 correspond to the strings 1-2 3, 1 2 3, 1$|$2 3, and 1 2$|$3, respectively. The notation will bring to light that weakly non-inverting graph transductions invoked in this paper correspond to extremely weak string transductions.

Suppose we have a string transduction τ that can only insert - or $|$ after a symbol. When this transduction is applied to the input string 1 2 3, it produces nine strings:

$$
\begin{array}{lll}
1\,2\,3 & 1\,2\text{-}3 & 1\,2|3 \\
1\text{-}2\,3 & 1\text{-}2\text{-}3 & 1\text{-}2|3 \\
1|2\,3 & 1|2\text{-}3 & 1|2|3
\end{array}
$$

Among those strings, 1$|$2-3, 1-2$|$3 and 1$|$2$|$3 are not weakly connected graphs. The remaining six strings represent exactly the graphs in Fig 1 and Fig 2. This establishes that τ computes a weakly non-inverting graph transduction.

Towards the end of the discussion of case syncretisms I entertained the hypothesis that the base hierarchy might not be totally ordered. In order to emulate such cases, the string transduction τ must also be allowed to delete the symbols - and $|$. Now suppose that our base is 1 2 3$|$4 5, which may be regarded as a truncated version of the partially or-

Phenomenon	Target graph	Additional properties of τ
PCC	weakly connected	weak maximality, strong minimality
Adjectival gradation	weakly connected	$2 \blacktriangleleft 1 \rightarrow 3 \blacktriangleleft 1$
Pronoun allomorphy	weakly connected	none
Case syncretism	connected	none
Noun stem suppletion	connected	weak maximality, weak non-maximality

Table 3: Parameters of each morphological paradigm

dered case hierarchy I proposed. Then τ yields all strings of the form $1\,u\,2\,v\,3\,x\,4\,y\,5$, where u, v, x, and y may each be $|$, -, or the empty string ε. Close inspection of this pattern reveals that τ still encodes a weakly non-inverting graph transduction.

Among string transductions, τ belongs to a very weak class. It is computed by a transducer with a single state (Fig. 4) and can be regarded as input strictly 1-local (ISL-1) in the sense of Chandlee (2014).[1] Transductions that invert the order

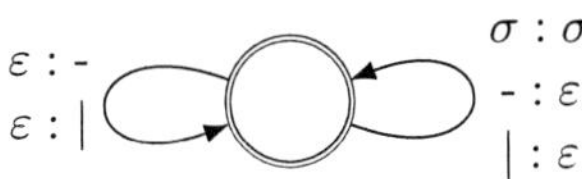

Figure 4: The single state transducer for computing weakly non-inverting graph transductions over string representations.

of symbols in arbitrary strings do not belong to this class. Even switching the order of adjacent symbols cannot be accomplished. This is very clear from the ISL perspective. A transduction is k-ISL iff the output for a given node n depends on the label of n and the labels of the preceding $k-1$ symbols. Chandlee (2014; 2016) proves that k-ISL transductions are only capable of k-bounded metathesis, which means that two symbols in the input string can be switched iff they are separated by at most $k-2$ symbols. This immediately entails that the order of two symbols can be reversed iff $k \geq 2$, wherefore 1-ISL transductions are incapable of reversing order.

It seems, then, that the restriction to weakly non-inverting graph transductions can be derived

from general simplicity desiderata. The recourse to strings is inelegant but unfortunately necessary as long as the class of ISL transductions has not been lifted from strings to graphs. Hopefully this will be rectified in the near future.

This still leaves open the question, though, why subparts of morphology and morphosyntax should impose additional criteria, in particular odd ones like $2 \blacktriangleleft 1 \rightarrow 3 \blacktriangleleft 1$ for adjectival gradation. While it is of course possible that a better computational understanding of graph transductions may eventually offer a satisfying explanation, a more likely scenario is that these properties are "echoes" of mechanisms that operate at a lower level of description. The graph-theoretic approach provides a more unified perspective than alternative proposals in the literature because it deliberately abstracts away from how these graphs and transductions are implemented in the grammar. There is no mention of features, agreement operations, or structural constraints because those vary wildly across domains and would obscure what the phenomena have in common. But these low-level processes might be subject to additional constraints that limit the range of typological variation even more. Abstracting away from them means losing the motivation behind those restrictions.

This highlights that the graph-theoretic view supplements existing approaches in linguistics, rather than replacing them. Its abstract nature makes it a lot easier to state general properties that are shared by all morphological paradigms. But when studying a single phenomenon in depth, the more fine-grained approaches favored by linguists may provide the necessary level of detail to explain aspects that are reduced to *ad hoc* stipulations in the graph-theoretic view.

6 Remarks on Data Reliability

This paper is, in essence, a mathematical exploration of a few particularly prominent typological

---

[1]Note however, that the transduction discussed here is a relation produce as multiple outputs may be produced from a single input, whereas Chandlee only studies transducers that compute functions. This difference, while mathematically important, has no immediate bearing on the overall argument, which hinges only on the fact that ISL transductions can only consider a locally bounded context when rewriting strings.

universals. A common concern in this regard is the reliability of the data on the basis of which these universals are posited. While the generalizations I discussed in Sec. 3 and 4 draw from a wide range of typologically diverse languages, even very extensive surveys such as Smith et al. (2016) include only about 60 languages. Considering that there are an estimated 6000 languages spoken today, this covers only 1% of the potential data. What more, languages frequently display a large amount of variation across their dialects, wherefore the amount of undetected "typological dark matter" may be even larger. One has to wonder, then, how reliable these generalizations are and whether it is even worthwhile to explore them from a formal perspective.

Unsurprisingly, I believe that they are worth exploring and that the arguments that are commonly marshaled against the enterprise do not stand up to closer scrutiny. If claims about substantive universals are unreliable due to the relative scarcity of data, then mathematical linguists should not put much stock into the mild context-sensitivity hypothesis either. After all, there may be unknown languages out there that are not even context-sensitive. One may argue that this is unlikely because the parsing algorithm for such a language could not run in polynomial time, but this holds only if one adopts the competence-performance distinction.

More importantly, it is just as conceivable that variation in the realm of substantive universals is also limited by independent factors — I presented a computational argument along these lines in Sec. 5, but beyond that there may be general principles of human cognition that prefer, say, a base ordering of $1 < 2 < 3$ over $3 < 1 < 2$. Therefore the exploration of substantive universals is methodologically no different from the study of formal universals; if the latter is a viable enterprise, the former is too. Being doubtful about all substantive universals while embracing formal universals cannot be motivated on logical grounds. And refraining from making any claims in the absence of rock solid data is unscientific: science proceeds in the absence of perfect knowledge, and every inductive step necessarily requires a leap of faith regarding the universality of the existing data.

That said, there is of course a bigger risk of overfitting the data in the area of morphosyntax because the models must characterize much smaller classes. The mildly context-sensitive hypothesis leaves a lot more room for cross-linguistic variation, and if it were to be disproved the problem would be solved by adding an additional mechanism to push weak generative capacity to the required level. Designing a model around four attested variants of the PCC or the *ABA generalization increases the risk that a single data point will render the whole model unsalvageable.

This has happened before: all Minimalist accounts of the PCC assumed a strict asymmetry with DO more prominent than IO, and consequently they may now need to be redesigned from the ground up if some languages do indeed display a mirror-PCC with IO more prominent than DO (Stegovec, 2016). However, the root of the problem is not that these proposals treated the data that was known at that point as an upper bound on the range of variation, but rather that their parameters were too tightly intertwined to allow for easy modification in the future.

The graph-transduction perspective in this paper, on the other hand, is similar to other mathematical approaches in that it displays a great amount of malleability to accommodate a shifting empirical landscape. Suppose for the sake of argument that an ABA pattern exists in some languages for some morphological or morphosyntactic domain. That would disprove the *ABA generalization, but it would not change the fact that ABA patterns are much rarer than any of the alternatives. From a formal perspective, this is easy to accommodate by moving to weighted graph transductions that penalize reversal and thus make ABA patterns more costly. The main insights about the importance of being weakly non-inverting stay the same, but they are extended from the Boolean domain to a weighted one.

The move towards a quantitative perspective is prudent anyways because it generalizes claims about the possibility of certain paradigms to claims about their relative frequency, which can be tested even with a non-exhaustive data set. For example, Tab. 2 lists three different graphs that produce ABC patterns, whereas only one graph each gives rise to AAB, ABB, and AAA. It seems unlikely that ABC is typologically three times more common than AAA, but a more sophisticated analysis may be able to derive better quantitative predictions. At any rate the approach presented in this paper has the requisite flexibility to be viable even with limited data, and in particular to avoid irreparable damage due to overfitting.

Conclusion

The account proposed in this paper derives typological gaps from two components: a fixed underlying hierarchy shared across all languages (a person hierarchy, case hierarchy, and so on), and heavily restricted graph transductions that generate the language-specific graph(s) from said hierarchy. The most important restriction is that the transductions be weakly non-inverting. Not only does this property severely limit their ability to alter the underlying hierarchy, it also reduces their complexity tremendously. Applying concepts from the theory of subregular string transductions, we may view these transductions as input strictly 1-local, which is the weakest non-trivial class of transductions. Overall, then, the graph-theoretic view sheds new light on these typological gaps and demonstrates the virtues of a mathematical approach that abstracts away from matters of implementation.

Of course a lot of work remains to be done. The literature on typological generalizations is enormous, and only a few could be touched on here. It will be particularly important to extend this approach to phenomena where multiple hierarchies are combined, e.g. number and person in pronoun hierarchies. Some other phenomena such as resolved agreement have more of a group-theoretic flavor. Resolved agreement refers to cases where an adjective agrees with multiple co-ordinated noun phrases. In Icelandic, for example, the adjective displays masculine agreement if all noun phrases are masculine, feminine if all noun phrases are feminine, and neuter in all other cases. It is still unclear whether the graph-theoretic perspective can be fruitfully expanded to such phenomena or whether algebraic techniques might provide a better fit. Irrespective of the final answer, there is no doubt that the abstraction and flexibility of mathematical approaches will be a great aid in the study of typological gaps.

References

David Adger and Daniel Harbour. 2007. The syntax and syncretisms of the person case constraint. *Syntax* 10:2–37.

Elena Anagnostopoulou. 2005. Strong and weak person restrictions: A feature checking analysis. In Lorie Heggie and Francisco Ordóñez, editors, *Clitics and Affix Combinations: Theoretical Perspectives*, John Benjamins, Amsterdam, pages 199–235.

Brenda S. Baker. 1978. Tree transducers and tree languages. *Information and Control* 37:241–266.

Brenda S. Baker. 1979. Composition of top-down and bottom-up tree transductions. *Information and Control* 41:186–213.

Barry J. Blake. 2001. *Case*. Cambridge University Press, Cambridge.

Jonathan D. Bobaljik. 2012. *Universals in Comparative Morphology: Suppletion, Superlatives, and the Structure of Words*. MIT Press, Cambridge, MA.

Jonathan D. Bobaljik and Uli Sauerland. 2017. *ABA and the combinatorics of morphological features. Ms., University of Connecticut and Leibniz-Zentrum für Allgemeine Sprachwissenschaft (ZAS).

Eulàlia Bonet. 1994. The Person-Case Constraint: A morphological approach. In *The Morphology-Syntax Connection*. Number 22 in MIT Working Papers in Linguistics, pages 33–52.

Pavel Caha. 2009. *The Nanosyntax of Case*. Ph.D. thesis, University of Tromsø.

Pavel Caha. 2013. Explaining the structure of case paradigms by the mechanisms of Nanosyntax: The Classical Armenian nominal declension. *Natural Language and Linguistic Theory* 31:1015–1066.

Jane Chandlee. 2014. *Strictly Local Phonological Processes*. Ph.D. thesis, University of Delaware.

Jane Chandlee. 2016. Computational locality in morphological maps. Ms., Haverford College.

Noam Chomsky. 1995. *The Minimalist Program*. MIT Press, Cambridge, MA.

Bruno Courcelle. 1992. The monadic second-order logic of graphs VII: Graphs as relational structures. *Theoretical Computer Science* 80:153–202.

Bruno Courcelle and Joost Engelfriet. 2012. *Graph Structure and Monadic Second-Order Logic: A Language-Theoretic Approach*. Cambridge University Press, Cambridge, UK.

Joost Engelfriet. 1975. Bottom-up and top-down tree transformations — a comparison. *Mathematical Systems Theory* 9:198–231. https://doi.org/10.1007/BF01704020.

Thomas Graf. 2014. Feature geometry and the Person Case Constraint: An algebraic link. In *Proceedings of CLS 50*. To appear.

Daniel Harbour. 2015. Poor pronoun systems and what they teach us. *Nordlyd* 41(1). https://doi.org/10.7557/12.3314.

Gísli Rúnar Harðarson. 2016. A case for a Weak Case Contiguity Hypothesis — a reply to Caha. *Natural Language and Linguistic Theory* 34:1329–1343.

Jeffrey Heinz. 2015. The computational nature of phonological generalizations. Ms., University of Delaware.

C. Douglas Johnson. 1972. *Formal Aspects of Phonological Description*. Mouton, The Hague.

Aravind Joshi. 1985. Tree-adjoining grammars: How much context sensitivity is required to provide reasonable structural descriptions? In David Dowty, Lauri Karttunen, and Arnold Zwicky, editors, *Natural Language Parsing*, Cambridge University Press, Cambridge, pages 206–250.

Ronald M. Kaplan and Martin Kay. 1994. Regular models of phonological rule systems. *Computational Linguistics* 20(3):331–378. http://www.aclweb.org/anthology/J94-3001.pdf.

Kevin Knight. 2007. Capturing practical natural language transformations. *Machine Translation* 21:121–133.

Andreas Maletti. 2010. Survey: Tree transducers in machine translation. In Henning Bordihn, Rudolf Freund, Thomas Hinze, Markus Holzer, Martin Kutrib, and Friedrich Otto, editors, *Proceedings of the 2nd International Workshop on Non-Classical Models of Automata and Applications*. Österreichische Computer Gesellschaft, volume 263 of `books@ocg.at`, pages 11–32.

Thomas McFadden. 2017. *ABA in stem-allomorphy and the emptiness of the nominative. Ms., Leibniz-Zentrum für Allgemeine Sprachwissenschaft (ZAS).

Robert McNaughton and Seymour Papert. 1971. *Counter-Free Automata*. MIT Press, Cambridge, MA.

Mehryar Mohri. 1997. Finite-state transducers in language and speech processing. *Computational Linguistics* 23:269–311.

Uwe Mönnich. 2006. Grammar morphisms. Ms. University of Tübingen.

Uwe Mönnich. 2007. Minimalist syntax, multiple regular tree grammars and direction preserving tree transductions. In James Rogers and Stephan Kepser, editors, *Model Theoretic Syntax at 10*. pages 83–87.

Uwe Mönnich. 2012. A logical characterization of extended TAGs. In *Proceedings of the 11th International Workshop on Tree Adjoining Grammars and Related Formalisms (TAG+11)*. Paris, France, pages 37–45. http://www.aclweb.org/anthology-new/W/W12/W12-4605.

Andrew Nevins. 2007. The representation of third person and its consequences for person-case effects. *Natural Language and Linguistic Theory* 25:273–313.

William C. Rounds. 1970. Mappings on grammars and trees. *Mathematical Systems Theory* 4:257–287.

Ur Shlonsky. 1997. *Clause Structure and Word Order in Hebrew and Arabic*. Oxford University Press, Oxford.

Peter W. Smith, Beata Moskal, Ting Xu, Jungmin Kang, and Jonathan D. Bobaljik. 2016. Case and number suppletion in pronouns. Ms., University of Connecticut.

Edward P. Stabler. 1997. Derivational Minimalism. In Christian Retoré, editor, *Logical Aspects of Computational Linguistics*, Springer, Berlin, volume 1328 of *Lecture Notes in Computer Science*, pages 68–95. https://doi.org/10.1007/BFb0052152.

Adrian Stegovec. 2016. A Person Case Constraint without Case. Ms., University of Connecticut.

Martin Walkow. 2012. *Goals, Big and Small*. Ph.D. thesis, University of Massachusetts Amherst.

Stanislao Zompí. 2016. *Case Decomposition Meets Dependent-Case Theories. A Study of the Syntax-Morphology Interface*. Master's thesis, University of Pisa.

Arnold Zwicky. 1977. Hierarchies of person. In *Chicago Linguistic Society*. volume 13, pages 714–733.

Introducing Structure into Neural Network-Based Semantic Models

Stephen Clark
University of Cambridge and DeepMind
`stephen.clark@cl.cam.ac.uk`

In this talk I will describe two attempts at introducing syntactic structure into semantic models using neural network architectures. The first study focuses on a particular grammatical construction, namely relative clauses, and centers around the design of a new dataset for testing compositional distributional models. The dataset is called REL-PRON, and consists of pairs of terms and properties, such as

> telescope : device that astronomer uses.

The idea is that a good compositional model will produce a vector representation of the property which is close to the vector for the term.

The second study focuses on deriving semantic vectors for phrases and whole sentences, which are then used in two tasks: sentence entailment, and a dictionary definition-term matching task. A feature of the proposed solution is that the syntactic structure for a sentence is induced in an unsupervised fashion, trained end-to-end in order to optimize for the final task.

I will finish with some thoughts on whether and how insights from traditional models of syntax and semantics can contribute to semantic models based on neural networks.

Proceedings of the 15th Meeting on the Mathematics of Language, page 127,
London, UK, July 13–14, 2017. ©2017 Association for Computational Linguistics

Count-Invariance Including Exponentials

Stepan Kuznetsov
Steklov Mathematical Institute
(Moscow), RAS
sk@mi.ras.ru

Glyn Morrill
Universitat Politècnica
de Catalunya
morrill@cs.upc.edu

Oriol Valentín
Universitat Politècnica
de Catalunya
oriol.valentin@gmail.com

Abstract

We define infinitary count-invariance for categorial logic, extending count-invariance for multiplicatives (van Benthem, 1991) and additives and bracket modalities (Valentín et al., 2013) to include exponentials. This provides an effective tool for pruning proof search in categorial parsing/theorem-proving.

1 Introduction

In logical grammar, which dates back to (Ajdukiewicz, 1935), grammar is *reduced* to logic: an expression is grammatical if and only if an associated logical statement is a theorem of a calculus.

1.1 Sharing

In standard logic information does not have multiplicity. Thus where + is the notion of addition of information and $\leq$ is the notion of inclusion we have $x+x \leq x$ and $x \leq x+x$; both these two properties together amount to idempotency: $x+x = x$. These properties are expressed by the rules of inference of Contraction and Expansion:

$$(1) \quad a. \quad \frac{\Delta(A, A) \Rightarrow B}{\Delta(A) \Rightarrow B} \text{ Contraction}$$

$$b. \quad \frac{\Delta(A) \Rightarrow B}{\Delta(A, A) \Rightarrow B} \text{ Expansion}$$

Linguistic resources do not freely have these properties: grammaticality is not generally preserved under addition or removal of copies of words or expressions. However, there are some constructions manifesting something similiar. Parasitic gaps involve a kind of Contraction. Parasitic gaps cannot occur anywhere, thus:

(2) *the slave that$_i$ John sold e_i to e_i

Rather, we assume here that as the term 'parasitic' suggests, a parasitic gap must fall within an island. Extraction from weak islands can become fully acceptable when accompanied by a cobound non-island extraction:

(3) a. man that$_i$ [the friends of e_i] admire e_i
     b. paper that$_i$ John filed e_i [without
         reading e_i]
     c. paper that$_i$ [the editor of e_i] filed e_i
         [without reading e_i]

And iterated coordination allows a kind of Expansion:

(4) John likes, Mary dislikes and Bill loves London.

That is, in logical grammar a *controlled* use of idempotency, or sharing, is motivated. Girard (1987) introduced exponentials for such control. Versions of the exponentials have been used to treat (parasitic) gaps and iterated coordination and iterated "respectively" in categorial grammar (Morrill, 2017), (Morrill and Valentín, 2015a, 2016b).

1.2 Count-Invariance

Count-invariance for multiplicatives in (sub)linear logic is introduced in van Benthem (1991). This involves simply checking the number of positive and negative occurrences of each atom in a sequent. Thus where $\#(\Sigma)$ is a count of the sequent Σ we have:

(5) $\vdash \Sigma \Longrightarrow \#(\Sigma) = 0$

I.e. the numbers of positive and negative occurrences of each atom must exactly balance for the sequent to be a theorem. This provides a necessary, but of course not sufficient, criterion for theoremhood, and can be checked rapidly. It can be

Proceedings of the 15th Meeting on the Mathematics of Language, pages 128–139,
London, UK, July 13–14, 2017. ©2017 Association for Computational Linguistics

used as a filter in proof search: if backward chaining proof search generates a goal which does not satisfy the count-invariant, the goal can be discarded. This notion of count for multiplicatives was included in the categorial parser/theorem-prover CatLog (Morrill, 2012).

In Valentín et al. (2013) the idea is extended to additives (and bracket modalities). Instead of a single count for each atom of a sequent Σ we have a minimum count $\#_{\min}(\Sigma)$ and a maximum count $\#_{\max}(\Sigma)$ and for a sequent to be a theorem it must satisfy two inequations:

$$(6) \quad \vdash \Sigma \Longrightarrow \#_{\min}(\Sigma) \leq 0 \leq \#_{\max}(\Sigma)$$

I.e. the count functions $\#_{\min}$ and $\#_{\max}$ define an interval which must include the point of balance 0; for the multiplicatives, $\#_{\min} = \#_{\max} = \#$ and (6) reduces to the special case (5). This generalised notion of count is included in the categorial parser/theorem-prover CatLog2.

The structure of the continuation of the paper is as follows. In Section 2 we present the infinitary count algebra which we employ, we define the fragment of categorial logic for which we illustrate count invariance, and we define the the (infinitary) count functions for this fragment. In Section 3 we state and prove our count-invariance theorem. In Section 4 we evaluate the introduction of exponential count invariance experimentally in relation to CatLog parsing/theorem-proving.

2 Infinitary Count Algebra

We consider terms built over constants 0, 1, $\perp$ ($-\infty$: minus infinity), and $\top$ ($+\infty$: plus infinity) by binary operations of plus (+), minus (−), minimum (min) and maximum (max), and the infinitary step functions X and Y as follows where i and j are integers (* indicates undefined):

+	j	$\perp$	$\top$
i	$i+j$	$\perp$	$\top$
$\perp$	$\perp$	$\perp$	$*$
$\top$	$\top$	$*$	$\top$

−	j	$\perp$	$\top$
i	$i-j$	$\top$	$\perp$
$\perp$	$\perp$	$*$	$\perp$
$\top$	$\top$	$\top$	$*$

min	j	$\perp$	$\top$
i	$\frac{|i+j|-|i-j|}{2}$	$\perp$	i
$\perp$	$\perp$	$\perp$	$\perp$
$\top$	j	$\perp$	$\top$

max	j	$\perp$	$\top$
i	$\frac{|i+j|+|i-j|}{2}$	i	$\top$
$\perp$	j	$\perp$	$\top$
$\top$	$\top$	$\top$	$\top$

$$X(i) = \begin{cases} \top & \text{if } i > 0 \\ i & \text{if } i \leq 0 \end{cases}$$

$$Y(i) = \begin{cases} i & \text{if } i \geq 0 \\ \perp & \text{if } i < 0 \end{cases}$$

(7) **Proposition.** 0. $\perp < i < \top$; 1. for $a, b < \top$, $a + b < \top$; 2. for $a, b > \perp$, $a + b > \perp$; 3. for $a > \perp \,\&\, b < \top$, $b - a < \top$; 4. for $b > \perp \,\&\, a < \top$, $b - a > \perp$; 5. for $a, b > \perp$, $\min(a, b) > \perp \,\&\, \max(a, b) > \perp$; 6. for $a, b < \top$, $\min(a, b) < \top \,\&\, \max(a, b) < \top$; 7. for $a > \perp$, $X(a) > \perp$; 8. for $a < \top$, $Y(a) < \top$.

2.1 The count functions

The count function, or count functions, are functions from types and sequents into values in the count algebra such that if sequents are provable their images under the count functions fall within a certain range. It follows that if their images do *not* fall within the required range then the sequents are *not* provable; we give examples after defining the count functions, in the next subsection. This provides an efficient filter on parsing/theorem-proving, as we show in the last section.

Let us assume primitive types $\mathcal{P}$. For $Q \in \mathcal{P} \cup \{[]\}$, $m \in \{\min, \max\}$ and $\overline{\min} = \max$ and $\overline{\max} = \min$ we define

$$\#_{m,Q}(\Gamma \Rightarrow A) = \#^{\circ}_{m,Q}(A) - \#^{\bullet}_{\overline{m},Q}(\Gamma)$$

where $\#^{\circ}$ and $\#^{\bullet}$ are as below. We define the enrichment **LAb!$_{\mathbf{b}}$?** of the Lambek calculus (Lambek, 1958) with types **Tp** as follows:

$$
\begin{aligned}
\mathbf{Tp} \quad ::= \quad & \mathcal{P} \,| \\
& \mathbf{Tp \backslash Tp \,|\, Tp/Tp \,|\, Tp \bullet Tp} \,| \\
& \mathbf{Tp \& Tp \,|\, Tp \oplus Tp} \,| \\
& \mathbf{[\,]^{-1}Tp \,|\, \langle\rangle Tp} \,| \\
& \mathbf{!Tp \,|\, ?Tp}
\end{aligned}
$$

Where $P \in \mathcal{P}$, $p \in \{\bullet, \circ\}$, and $\overline{\bullet} = \circ$ and $\overline{\circ} = \bullet$ we define the count functions:

$$
\begin{aligned}
\#^{p}_{m,Q}(P) &= \begin{array}{ll} 1 & \text{if } Q = P \\ 0 & \text{if } Q \neq P \end{array} \\
\#^{p}_{m,Q}(A\backslash C) &= \#^{p}_{m,Q}(C) - \#^{\overline{p}}_{\overline{m},Q}(A) \\
\#^{p}_{m,Q}(C/B) &= \#^{p}_{m,Q}(C) - \#^{\overline{p}}_{\overline{m},Q}(B) \\
\#^{p}_{m,Q}(A\bullet B) &= \#^{p}_{m,Q}(A) + \#^{p}_{m,Q}(B) \\
\#^{\circ}_{m,Q}(A\&B) &= \overline{m}(\#^{\circ}_{m,Q}(A),\#^{\circ}_{m,Q}(B)) \\
\#^{\bullet}_{m,Q}(A\&B) &= m(\#^{\bullet}_{m,Q}(A),\#^{\bullet}_{m,Q}(B)) \\
\#^{\circ}_{m,Q}(A\oplus B) &= m(\#^{\circ}_{m,Q}(A),\#^{\circ}_{m,Q}(B)) \\
\#^{\bullet}_{m,Q}(A\oplus B) &= \overline{m}(\#^{\bullet}_{m,Q}(A),\#^{\bullet}_{m,Q}(B)) \\
\#^{p}_{m,P}([\ ]^{-1}A) &= \#^{p}_{m,P}(A) \\
\#^{p}_{m,[]}([\ ]^{-1}A) &= \#^{p}_{m,[]}(A) - 1 \\
\#^{p}_{m,P}(\langle\rangle A) &= \#^{p}_{m,P}(A) \\
\#^{p}_{m,[]}(\langle\rangle A) &= \#^{p}_{m,[]}(A) + 1 \\
\#^{\circ}_{m,Q}(!A) &= \#^{\circ}_{m,Q}(A) \\
\#^{\bullet}_{\min,Q}(!A) &= Y(\#^{\bullet}_{\min,Q}(A)) \\
\#^{\bullet}_{\max,P}(!A) &= X(\#^{\bullet}_{\max,P}(A)) \\
\#^{\bullet}_{\max,[]}(!A) &= \top \\
\#^{\circ}_{\min,Q}(?A) &= Y(\#^{\circ}_{\min,Q}(A)) \\
\#^{\circ}_{\max,Q}(?A) &= X(\#^{\circ}_{\max,Q}(A)) \\
\#^{\bullet}_{m,Q}(?A) &= \#^{\bullet}_{m,Q}(A)
\end{aligned}
$$

(8) **Lemma.** $\#^{p}_{m}(A)$ is defined and $\bot <$ $\#_{\max}(A)$ & $\#_{\min}(A) < \top$.

Proof. By induction as in Figure 1; justifications refer to the Proposition (7). ∎

To present sequents we define *configurations* **Config** and *tree terms* **TreeTerm** in terms of types **Tp** as follows, where Λ is the empty string:

$$
\begin{aligned}
\textbf{Config} \quad &::= \quad \Lambda \mid \textbf{TreeTerm}, \textbf{Config} \\
\textbf{TreeTerm} \quad &::= \quad \textbf{Tp} \mid [\textbf{Config}]
\end{aligned}
$$

The rules for **LAb!$_{\mathbf{b}}$?** are shown in Figure 2. Note that $!C$ is of a generalised form necessary to prove Cut-elimination in the presence of $!R$. Note also that $?L$ is an infinitary rule; it is not used in linguistic applications. We include it here for the sake of showing technical completeness of the count invariance. For tree terms and configurations, counts are:

$$
\begin{aligned}
\#^{\bullet}_{m,Q}(\Gamma,\Delta) &= \#^{\bullet}_{m,Q}(\Gamma) + \#^{\bullet}_{m,Q}(\Delta) \\
\#^{\bullet}_{m,P}([\Gamma]) &= \#\bullet_{m,P}(\Gamma) \\
\#^{\bullet}_{m,[]}([\Gamma]) &= \#^{\bullet}_{m,[]}(\Gamma) + 1 \\
\#^{\bullet}_{m,Q}(\Lambda) &= 0
\end{aligned}
$$

Lemma 8 extends to configurations.

2.2 Examples

Relativisation including medial and parasitic extraction is obtained by assigning a relative pronoun a type $(CN\backslash CN)/(!N\backslash S)$ whereby the body of a relative clause is analysed as $!N\backslash S$. By way of example of count-invariance, we show how it discards $N, N\backslash S \Rightarrow !N\backslash S$ corresponding to the ungrammaticality of a relative clause without a gap: *paper that John walks. We have the max N-count: $\#_{\max,N}(N, N\backslash S \Rightarrow !N\backslash S) =$ $\#^{\circ}_{\max,N}(!N\backslash S) - \#^{\bullet}_{\min,N}(N, N\backslash S) = \#^{\circ}_{\max,N}(S) - \#^{\bullet}_{\min,N}(!N) - \#^{\bullet}_{\min,N}(N) - \#^{\bullet}_{\min,N}(N\backslash S) = 0 - Y(\#^{\bullet}_{\min,N}(N)) - 1 - \#^{\bullet}_{\min,N}(S) + \#^{\circ}_{\min,N}(N) = -Y(1) - 1 - 0 + 1 = -1 - 1 + 1 = -1 \not\geq 0$ which means that the count-invariance is not satisfied.

Iterated sentential coordination is obtained by assigning a coordinator the type $(?S\backslash S)/S$. By way of a second example we show how count-invariance discards $N, N, N\backslash S \Rightarrow ?S$ corresponding to the ungrammaticality of unequilibrated coordination: *John Mary walks and Suzy talks. Max N-count is: $\#_{\max,N}(N, N, N\backslash S \Rightarrow ?S) = \#^{\circ}_{\max,N}(?S) - \#\bullet_{\min,N}(N, N, N\backslash S) = X(\#^{\circ}_{\max,N}(S)) - \#^{\bullet}_{\min,N}(N) - \#^{\bullet}_{\min,N}(N) - \#^{\bullet}_{\min,N}(N\backslash S) = X(0) - 1 - 1 - \#^{\bullet}_{\min,N}(S) + \#^{\bullet}_{\max,N}(N) = 0 - 2 - 0 + 1 = -1 \not\geq 0$ which means that the count-invariance is not satisfied.

3 Theorem and Proof

Our main theorem is:

(9) **Theorem.**

$$
\vdash \Gamma \Rightarrow A \Longrightarrow \forall Q \in \mathcal{P} \cup \{[]\},
$$
$$
\#_{\min,Q}(\Gamma \Rightarrow A) \leq 0 \leq \#_{\max,Q}(\Gamma \Rightarrow A)
$$
where as we have said,
$$
\#_{m,Q}(\Gamma \Rightarrow A) = \#^{\circ}_{m,Q}(A) - \#^{\bullet}_{\overline{m},Q}(\Gamma).
$$

Proof. The proof is by induction on the length of derivations. For the base case $P \Rightarrow P$ we have $\#_{m,Q}(P \Rightarrow P) = \#^{\circ}_{m,Q}(P) - \#^{\bullet}_{m,Q}(P) = 0$. The inductive cases are as follows, where we use:

- $a + b = b + a$

- $a + (b + c) = (a + b) + c$

- $a - (b+c) = (a-b)-c$ (including the undefined case)

- $(a + b) - c = (a - c) + b$

$$\#^p_{m,Q}(P) \;=\; \begin{cases} 1 & \text{if } Q = P \\ 0 & \text{if } Q \neq P \end{cases} \qquad\qquad \bot < 0, 1 < \top$$

$$\#^p_{\max,Q}(A\backslash C) \;=\; \#^p_{\max,Q}(C) - \#^{\overline{p}}_{\min,Q}(A) \qquad 4$$

$$\#^p_{\min,Q}(A\backslash C) \;=\; \#^p_{\min,Q}(C) - \#^{\overline{p}}_{\max,Q}(A) \qquad 3$$

$$\#^p_{\max,Q}(C/B) \;=\; \#^p_{\max,Q}(C) - \#^{\overline{p}}_{\min,Q}(B) \qquad 4$$

$$\#^p_{\min,Q}(C/B) \;=\; \#^p_{\min,Q}(C) - \#^{\overline{p}}_{\max,Q}(B) \qquad 3$$

$$\#^p_{\max,Q}(A\bullet B) \;=\; \#^p_{\max,Q}(A) + \#^p_{\max,Q}(B) \qquad 2$$

$$\#^p_{\min,Q}(A\bullet B) \;=\; \#p_{\min,Q}(A) + \#p_{\min,Q}(B) \qquad 1$$

$$\#^\circ_{\min,Q}(A\&B) \;=\; \max(\#^\circ_{\min,Q}(A), \#^\circ_{\min,Q}(B)) \qquad 6$$

$$\#^\circ_{\max,Q}(A\&B) \;=\; \min(\#^\circ_{\max,Q}(A), \#^\circ_{\max,Q}(B)) \qquad 5$$

$$\#^\bullet_{\min,Q}(A\&B) \;=\; \min(\#^\bullet_{\min,Q}(A), \#^\bullet_{\min,Q}(B)) \qquad 6$$

$$\#^\bullet_{\max,Q}(A\&B) \;=\; \max(\#^\bullet_{\max,Q}(A), \#^\bullet_{\max,Q}(B)) \qquad 5$$

$$\#^\circ_{\min,Q}(A\oplus B) \;=\; \min(\#^\circ_{\min,Q}(A), \#^\circ_{\min,Q}(B)) \qquad 6$$

$$\#^\circ_{\max,Q}(A\oplus B) \;=\; \max(\#^\circ_{\max,Q}(A), \#^\circ_{\max,Q}(B)) \qquad 5$$

$$\#^\bullet_{\min,Q}(A\oplus B) \;=\; \max(\#^\bullet_{\min,Q}(A), \#^\bullet_{\min,Q}(B)) \qquad 6$$

$$\#^\bullet_{\max,Q}(A\oplus B) \;=\; \min(\#^\bullet_{\max,Q}(A), \#^\bullet_{\max,Q}(B)) \qquad 5$$

$$\#^p_{\max,P}([\,]^{-1}A) \;=\; \#^p_{\max,P}(A) > \bot$$

$$\#^p_{\min,P}([\,]^{-1}A) \;=\; \#^p_{\min,P}(A) < \top$$

$$\#^p_{\max,[]}([\,]^{-1}A) \;=\; \#^p_{\max,[]}(A) - 1 > \bot$$

$$\#^p_{\min,[]}([\,]^{-1}A) \;=\; \#^p_{\min,[]}(A) - 1 < \top$$

$$\#^p_{\max,P}(\langle\rangle A) \;=\; \#^p_{\max,P}(A) > \bot$$

$$\#^p_{\min,P}(\langle\rangle A) \;=\; \#^p_{\min,P}(A) < \bot$$

$$\#^p_{\max,[]}(\langle\rangle A) \;=\; \#^p_{\max,[]}(A) + 1 > \bot$$

$$\#^p_{\min,[]}(\langle\rangle A) \;=\; \#^p_{\min,[]}(A) + 1 < \top$$

$$\#^\bullet_{\max,P}(!A) \;=\; X(\#^\bullet_{\max,P}(A)) \qquad 7$$

$$\#^\bullet_{\min,Q}(!A) \;=\; Y(\#^\bullet_{\min,Q}(A)) \qquad 8$$

$$\#^\bullet_{\max,[]}(!A) \;=\; \top > \bot$$

$$\#^\circ_{\max,Q}(!A) \;=\; \#^\circ_{\max,Q}(A) > \bot$$

$$\#^\circ_{\min,Q}(!A) \;=\; \#^\circ_{\min,Q}(A) < \top$$

$$\#^\circ_{\max,Q}(?A) \;=\; X(\#^\circ_{\max,Q}(A)) \qquad 7$$

$$\#^\circ_{\min,Q}(?A) \;=\; Y(\#^\circ_{\min,Q}(A)) \qquad 8$$

$$\#^\bullet_{\min,Q}(?A) \;=\; \#^\bullet_{\min,Q}(A) < \top$$

$$\#^\bullet_{\max,Q}(?A) \;=\; \#^\bullet_{\max,Q}(A) > \bot$$

Figure 1: Count functions

$$\frac{}{P \Rightarrow P} \; id, P \in \mathcal{P}$$

$$\frac{\Gamma \Rightarrow A \quad \Delta(C) \Rightarrow D}{\Delta(\Gamma, A\backslash C) \Rightarrow D} \; \backslash L \qquad \frac{\Gamma \Rightarrow B \quad \Delta(C) \Rightarrow D}{\Delta(C/B, \Gamma) \Rightarrow D} \; \backslash L$$

$$\frac{A, \Gamma \Rightarrow C}{\Gamma \Rightarrow A\backslash C} \; \backslash R \qquad \frac{\Gamma, B \Rightarrow C}{\Gamma \Rightarrow C/B} \; /R$$

$$\frac{\Delta(A, B) \Rightarrow C}{\Delta(A \bullet B) \Rightarrow C} \; \bullet L \qquad \frac{\Gamma_1 \Rightarrow A \quad \Gamma_2 \Rightarrow B}{\Gamma_1, \Gamma_2 \Rightarrow A \bullet B} \; \bullet R$$

$$\frac{\Delta(A) \Rightarrow D}{\Delta(A \& B) \Rightarrow D} \; \& L_1 \qquad \frac{\Delta(B) \Rightarrow D}{\Delta(A \& B) \Rightarrow D} \; \& L_2 \qquad \frac{\Gamma \Rightarrow A \quad \Gamma \Rightarrow B}{\Gamma \Rightarrow A \& B} \; \& R$$

$$\frac{\Gamma \Rightarrow A}{\Gamma \Rightarrow A \oplus B} \; \oplus R_1 \qquad \frac{\Gamma \Rightarrow B}{\Gamma \Rightarrow A \oplus B} \; \oplus R_2 \qquad \frac{\Delta(A) \Rightarrow D \quad \Delta(B) \Rightarrow D}{\Delta(A \oplus B) \Rightarrow D} \; \oplus L$$

$$\frac{\Gamma(A) \Rightarrow B}{\Gamma([[\,]^{-1}A]) \Rightarrow B} \; [\,]^{-1}L \qquad \frac{[\Gamma] \Rightarrow A}{\Gamma \Rightarrow [\,]^{-1}A} \; [\,]^{-1}R$$

$$\frac{\Gamma([A]) \Rightarrow B}{\Gamma(\langle\rangle A) \Rightarrow B} \; \langle\rangle L \qquad \frac{\Gamma \Rightarrow A}{[\Gamma] \Rightarrow \langle\rangle A} \; \langle\rangle R$$

$$\frac{\Delta(A) \Rightarrow D}{\Delta(!A) \Rightarrow D} \; !L \qquad \frac{!A_1, \ldots, !A_n \Rightarrow A}{!A_1, \ldots, !A_n \Rightarrow !A} \; !R$$

$$\frac{\Delta(\Gamma, !A) \Rightarrow D}{\Delta(!A, \Gamma) \Rightarrow D} \; !P_1 \qquad \frac{\Delta(!A, \Gamma) \Rightarrow D}{\Delta(\Gamma, !A) \Rightarrow D} \; !P_2 \qquad \frac{\Delta(!A_0, \ldots, !A_n, [!A_0, \ldots, !A_n, \Gamma]) \Rightarrow D}{\Delta(!A_0, \ldots, !A_n, \Gamma) \Rightarrow D} \; !C$$

$$\frac{\Delta(A) \Rightarrow D \quad \Delta(A, A) \Rightarrow D \quad \cdots}{\Delta(?A) \Rightarrow D} \; ?L \qquad \frac{\Gamma \Rightarrow A}{\Gamma \Rightarrow ?A} \; ?R$$

$$\frac{\Gamma_1 \Rightarrow C \quad \Gamma_2 \Rightarrow ?C}{\Gamma_1, \Gamma_2 \Rightarrow ?C} \; ?M$$

Figure 2: Rules for the categorial logic fragment **LAb!**$_{\mathbf{b}}$?

- $a - (b - c) = (a - b) + c$

(Where we write $\#(\Delta)$ with Δ a context we should more precisely understand that Δ is a configuration with a hole where the count of a hole is always zero.)

Multiplicatives

- $$\frac{\Gamma \Rightarrow A \qquad \Delta(C) \Rightarrow D}{\Delta(\Gamma, A\backslash C) \Rightarrow D} \ \backslash L$$

For every atom or bracket,

$\#_m(\Delta(\Gamma, A\backslash C) \Rightarrow D) =$
$\#^\circ_m(D) - \#^\bullet_{\overline{m}}(\Delta) - \#^\bullet_{\overline{m}}(\Gamma) - \#^\bullet_{\overline{m}}(A\backslash C) =$
$\#^\circ_m(D) - \#^\bullet_{\overline{m}}(\Delta) - \#^\bullet_{\overline{m}}(\Gamma) - \#^\bullet_{\overline{m}}(C) + \#^\circ_m(A) =$
$\#^\circ_m(A) - \#^\bullet_{\overline{m}}(\Gamma) + \#^\circ_m(D) - \#^\bullet_{\overline{m}}(\Delta) - \#^\bullet_{\overline{m}}(C) =$
$\#_m(\Gamma \Rightarrow A) + \#_m(\Delta(C) \Rightarrow D)$

The induction hypothesis (*i.h.*) tells us that $\#_{\min}(\Gamma \Rightarrow A) \leq 0$ and $\#_{\min}(\Delta(C) \Rightarrow D) \leq 0$. Thus $\#_{\min}(\Delta(\Gamma, A\backslash C) \Rightarrow D) = \#_{\min}(\Gamma \Rightarrow A) + \#_{\min}(\Delta(C) \Rightarrow D) \leq 0$. Similarly, $0 \leq \#_{\max}(\Delta(\Gamma, A\backslash C) \Rightarrow D) = \#_{\max}(\Gamma \Rightarrow A) + \#_{\max}(\Delta(C) \Rightarrow D)$ by *i.h.* Therefore we have:

$\#_{\min}(\Delta(\Gamma, A\backslash C) \Rightarrow D) \leq 0 \leq$
$\#_{\max}(\Delta(\Gamma, A\backslash C) \Rightarrow D)$

- $$\frac{\Gamma \Rightarrow B \qquad \Delta(C) \Rightarrow D}{\Delta(C/B, \Gamma) \Rightarrow D} \ \backslash L$$

Like $\backslash L$.

- $$\frac{A, \Gamma \Rightarrow C}{\Gamma \Rightarrow A\backslash C} \ \backslash R$$

For every atom or bracket,

$\#_m(\Gamma \Rightarrow A\backslash C) =$
$\#^\circ_m(A\backslash C) - \#^\bullet_{\overline{m}}(\Gamma) =$
$\#^\circ_m(C) - \#^\bullet_{\overline{m}}(A) - \#^\bullet_{\overline{m}}(\Gamma) =$
$\#_m(A, \Gamma \Rightarrow C)$

Therefore by *i.h.*,

$\#_{\min}(\Gamma \Rightarrow A\backslash C) \leq 0 \leq \#_{\max}(\Gamma \Rightarrow A\backslash C)$

- $$\frac{\Gamma, B \Rightarrow C}{\Gamma \Rightarrow C/B} \ /R$$

Like $\backslash R$.

- $$\frac{\Delta(A, B) \Rightarrow C}{\Delta(A\bullet B) \Rightarrow C} \ \bullet L$$

For every atom or bracket,

$\#_m(\Delta(A\bullet B) \Rightarrow C) =$
$\#^\circ_m(C) - \#^\bullet_{\overline{m}}(\Delta) - \#^\bullet_{\overline{m}}(A\bullet B) =$
$\#^\circ_m(C) - \#^\bullet_{\overline{m}}(\Delta) - \#^\bullet_{\overline{m}}(A) - \#^\bullet_{\overline{m}}(B) =$
$\#_m(\Delta(A, B) \Rightarrow C)$

Therefore by *i.h.*,

$\#_{\min}(\Delta(A\bullet B) \Rightarrow C) \leq 0 \leq$
$\#_{\max}(\Delta(A\bullet B) \Rightarrow C)$

- $$\frac{\Gamma_1 \Rightarrow A \qquad \Gamma_2 \Rightarrow B}{\Gamma_1, \Gamma_2 \Rightarrow A\bullet B} \ \bullet R$$

For every atom or bracket,

$\#_m(\Gamma_1, \Gamma_2 \Rightarrow A\bullet B) =$
$\#^\circ_m(A\bullet B) - \#^\bullet_{\overline{m}}(\Gamma_1, \Gamma_2) =$
$\#^\circ_m(A) - \#^\bullet_{\overline{m}}(\Gamma_1) + \#^\circ_m(B) - \#^\bullet_{\overline{m}}(\Gamma_2) =$
$\#_m(\Gamma_1 \Rightarrow A) + \#_m(\Gamma_2 \Rightarrow B)$

Therefore by *i.h.*,

$\#_{\min}(\Gamma_1, \Gamma_2 \Rightarrow A\bullet B) \leq 0 \leq$
$\#_{\max}(\Gamma_1, \Gamma_2 \Rightarrow A\bullet B)$

Additives

- $$\frac{\Delta(A) \Rightarrow D}{\Delta(A\&B) \Rightarrow D} \ \&L_1$$

For every atom or bracket,

$\#_{\min}(\Delta(A\&B) \Rightarrow D) =$
$\#^\circ_{\min}(D) - \#^\bullet_{\max}(\Delta) - \#^\bullet_{\max}(A\&B) =$
$\#^\circ_{\min}(D) - \#^\bullet_{\max}(\Delta) -$
$\max(\#^\bullet_{\max}(A), \#^\bullet_{\max}(B)) \leq$
$\#^\circ_{\min}(D) - \#^\bullet_{\max}(\Delta) - \#^\bullet_{\max}(A) =$
$\#_{\min}(\Delta(A) \Rightarrow D) \leq 0 \ i.h.$

And

$\#_{max}(\Delta(A\&B) \Rightarrow D) =$
$\#^{\circ}_{max}(D) - \#^{\bullet}_{min}(\Delta) - \#^{\bullet}_{min}(A\&B) =$
$\#^{\circ}_{max}(D) - \#^{\bullet}_{min}(\Delta)-$
$min(\#^{\bullet}_{min}(A), \#^{\bullet}_{min}(B)) \geq$
$\#^{\circ}_{max}(D) - \#^{\bullet}_{min}(\Delta) - \#^{\bullet}_{min}(A) =$
$\#_{max}(\Delta(A) \Rightarrow D) \geq 0 \; i.h.$

Therefore:

$\#_{min}(\Delta(A\&B) \Rightarrow D) \leq 0 \leq$
$\#_{max}(\Delta(A\&B) \Rightarrow D)$

- $$\dfrac{\Delta(B) \Rightarrow D}{\Delta(A\&B) \Rightarrow D} \; \&L_2$$

  Like $\&L_1$.

- $$\dfrac{\Gamma \Rightarrow A \qquad \Gamma \Rightarrow B}{\Gamma \Rightarrow A\&B} \; \&R$$

$\#_{min}(\Gamma \Rightarrow A\&B) =$
$\#^{\circ}_{min}(A\&B) - \#^{\bullet}_{max}(\Gamma) =$
$max(\#^{\circ}_{min}(A), \#^{\circ}_{min}(B)) - \#^{\bullet}_{max}(\Gamma) =$
$max(\#^{\circ}_{min}(A) - \#^{\bullet}_{max}(\Gamma),$
$\#^{\circ}_{min}(B) - \#^{\bullet}_{max}(\Gamma)) =$
$max(\underbrace{\#_{min}(\Gamma \Rightarrow A)}_{\leq 0 \; i.h.}, \underbrace{\#_{min}(\Gamma \Rightarrow B)}_{\leq 0 \; i.h.})$
$\underbrace{\phantom{max(\#_{min}(\Gamma \Rightarrow A), \#_{min}(\Gamma \Rightarrow B))}}_{\leq 0}$

  And

$\#_{max}(\Gamma \Rightarrow A\&B) =$
$\#^{\circ}_{max}(A\&B) - \#^{\bullet}_{min}(\Gamma) =$
$min(\#^{\circ}_{max}(A), \#^{\circ}_{max}(B)) - \#^{\bullet}_{min}(\Gamma) =$
$min(\#^{\circ}_{max}(A) - \#^{\bullet}_{min}(\Gamma),$
$\#^{\circ}_{max}(B) - \#^{\bullet}_{min}(\Gamma)) =$
$min(\underbrace{\#_{max}(\Gamma \Rightarrow A)}_{0 \leq \; i.h.}, \underbrace{\#_{max}(\Gamma \Rightarrow B)}_{0 \leq \; i.h.})$
$\underbrace{\phantom{min(\#_{max}(\Gamma \Rightarrow A), \#_{max}(\Gamma \Rightarrow B))}}_{0 \leq}$

Therefore:

$\#_{min}(\Gamma \Rightarrow A\&B) \leq 0 \leq \#_{max}(\Gamma \Rightarrow A\&B).$

- $$\dfrac{\Gamma \Rightarrow A}{\Gamma \Rightarrow A\oplus B} \; \oplus R_1$$

$\#_{min}(\Gamma \Rightarrow A\oplus B) =$
$\#^{\circ}_{min}(A\oplus B) - \#^{\bullet}_{max}(\Gamma) =$
$min(\#^{\circ}_{min}(A), \#^{\bullet}_{min}(B)) - \#^{\bullet}_{max}(\Gamma) \leq$
$\#^{\circ}_{min}(A) - \#^{\bullet}_{max}(\Gamma) =$
$\#_{min}(\Gamma \Rightarrow A) \leq 0 \; i.h.$

And

$\#_{max}(\Gamma \Rightarrow A\oplus B) =$
$\#^{\circ}_{max}(A\oplus B) - \#^{\bullet}_{min}(\Gamma) =$
$max(\#^{\bullet}_{max}(A), \#^{\bullet}_{max}(B)) - \#^{\bullet}_{min}(\Gamma) \geq$
$\#^{\bullet}_{max}(A) - \#^{\bullet}_{min}(\Gamma) =$
$\#_{max}(\Gamma \Rightarrow A) \geq 0 \; i.h.$

- $$\dfrac{\Gamma \Rightarrow B}{\Gamma \Rightarrow A\oplus B} \; \oplus R_2$$

Like $\oplus R_1$.

- $$\dfrac{\Delta(A) \Rightarrow D \qquad \Delta(B) \Rightarrow D}{\Delta(A\oplus B) \Rightarrow D} \; \oplus L$$

For every atom or bracket,

$\#_{min}(\Delta(A\oplus B) \Rightarrow D) =$
$\#^{\circ}_{min}(D) - \#^{\bullet}_{max}(\Delta) - \#^{\bullet}_{max}(A\oplus B) =$
$\#^{\circ}_{min}(D) - \#^{\bullet}_{max}(\Delta)-$
$min(\#^{\bullet}_{max}(A), \#^{\bullet}_{max}(B)) =$
$max(\#^{\circ}_{min}(D) - \#^{\bullet}_{max}(\Delta) - \#^{\bullet}_{max}(A),$
$\#^{\circ}_{min}(D) - \#^{\bullet}_{max}(\Delta) - \#^{\bullet}_{max}(B)) =$
$max(\underbrace{\#_{min}(\Delta(A) \Rightarrow D)}_{\leq 0 \; i.h.}, \underbrace{\#_{min}(\Delta(B) \Rightarrow D)}_{\leq 0 \; i.h.})$
$\underbrace{\phantom{max(\#_{min}(\Delta(A) \Rightarrow D), \#_{min}(\Delta(B) \Rightarrow D))}}_{\leq 0}$

$0 \leq \#_{max}(\Delta(A\oplus B) \Rightarrow D)$ similarly

Bracket modalities

- $$\dfrac{\Gamma(A) \Rightarrow B}{\Gamma([[\;]^{-1}A]) \Rightarrow B} \; [\;]^{-1}L$$

For atoms:

$\#_{m,P}(\Gamma([[\;]^{-1}A]) \Rightarrow B) =$
$\#^{\circ}_{m,P}(B) - \#^{\bullet}_{\overline{m},P}(\Gamma([[\;]^{-1}A])) =$
$\#^{\circ}_{m,P}(B) - \#^{\bullet}_{\overline{m},P}(\Gamma) - \#^{\bullet}_{\overline{m},P}([[\;]^{-1}A]) =$
$\#^{\circ}_{m,P}(B) - \#^{\bullet}_{\overline{m},P}(\Gamma) - \#^{\bullet}_{\overline{m},P}([\;]^{-1}A)) =$
$\#^{\circ}_{m,P}(B) - \#^{\bullet}_{\overline{m},P}(\Gamma) - \#^{\bullet}_{\overline{m},P}(A)) =$
$\#^{\circ}_{m,P}(B) - \#^{\bullet}_{\overline{m},P}(\Gamma(A)) =$
$\#_{m,P}(\Gamma(A) \Rightarrow B)$

I.e. the property for the conclusion follows from the induccion hypothesis for the premise since brackets and bracket modalities are transparent to atom count.

For brackets:

$$\#_{m,[]}(\Gamma([[\ ]^{-1}A]) \Rightarrow B) =$$
$$\#^{\circ}_{m,[]}(B) - \#^{\bullet}_{\overline{m},[]}(\Gamma([[\ ]^{-1}A])) =$$
$$\#^{\circ}_{m,[]}(B) - \#^{\bullet}_{\overline{m},[]}(\Gamma) - \#^{\bullet}_{\overline{m},[]}([[\ ]^{-1}A])) =$$
$$\#^{\circ}_{m,[]}(B) - \#^{\bullet}_{\overline{m},[]}(\Gamma) - \#^{\bullet}_{\overline{m},[]}([\ ]^{-1}A) - 1 =$$
$$\#^{\circ}_{m,[]}(B) - \#^{\bullet}_{\overline{m},[]}(\Gamma) - \#^{\bullet}_{\overline{m},[]}(A) + 1 - 1 =$$
$$\#^{\circ}_{m,[]}(B) - \#^{\bullet}_{\overline{m},[]}(\Gamma) - \#^{\bullet}_{\overline{m},[]}(A)) =$$
$$\#^{\circ}_{m,[]}(B) - \#^{\bullet}_{\overline{m},[]}(\Gamma(A)) =$$
$$\#_{m,[]}(\Gamma(A) \Rightarrow B) =$$

Therefore by *i.h.*,

$$\#_{\min}(\Gamma([[\ ]^{-1}A]) \Rightarrow B) \le 0 \le$$
$$\#_{\max}(\Gamma([[\ ]^{-1}A]) \Rightarrow B)$$

- $$\frac{[\Gamma] \Rightarrow A}{\Gamma \Rightarrow [\ ]^{-1}A} \ [\ ]^{-1}R$$

For atoms:

$$\#_{m,P}(\Gamma \Rightarrow [\ ]^{-1}A) =$$
$$\#_{m,P}(\Gamma \Rightarrow A) =$$
$$\#_{m,P}([\Gamma] \Rightarrow A)$$

Since brackets and bracket modalities are transparant to atom count.

For brackets:

$$\#_{m,[]}(\Gamma \Rightarrow [\ ]^{-1}A) =$$
$$\#^{\circ}_{m,[]}([\ ]^{-1}A) - \#^{\bullet}_{\overline{m},[]}(\Gamma) =$$
$$\#^{\circ}_{m,[]}(A) - 1 - \#^{\bullet}_{\overline{m},[]}(\Gamma) =$$
$$\#^{\circ}_{m,[]}(A) - (\#^{\bullet}_{\overline{m},[]}(\Gamma) + 1) =$$
$$\#^{\circ}_{m,[]}(A) - \#^{\bullet}_{\overline{m},[]}([\Gamma]) =$$
$$\#_{m,[]}([\Gamma] \Rightarrow A)$$

Therefore by *i.h.*

$$\#_{\min}(\Gamma \Rightarrow [\ ]^{-1}A) \le 0 \le \#_{\max}(\Gamma \Rightarrow [\ ]^{-1}A)$$

- $$\frac{\Gamma([A]) \Rightarrow B}{\Gamma(\langle\rangle A) \Rightarrow B} \ \langle\rangle L$$

For atoms,

$$\#_{m,P}(\Gamma(\langle\rangle A) \Rightarrow B) = \#_{m,P}(\Gamma([A]) \Rightarrow B)$$

since brackets and bracket modalities are transparent to atom count.

For brackets,

$$\#_{m,[]}(\Gamma(\langle\rangle A) \Rightarrow B) =$$
$$\#^{\circ}_{m,[]}(B) - \#^{\bullet}_{\overline{m},[]}(\Gamma) - \#^{\bullet}_{\overline{m},[]}(\langle\rangle A) =$$
$$\#^{\circ}_{m,[]}(B) - \#^{\bullet}_{\overline{m},[]}(\Gamma) - (\#^{\bullet}_{\overline{m},[]}(A) + 1) =$$
$$\#^{\circ}_{m,[]}(B) - \#^{\bullet}_{\overline{m},[]}(\Gamma) - \#^{\bullet}_{\overline{m},[]}([A]) =$$
$$\#_{m,[]}(\Gamma([A]) \Rightarrow B)$$

Therefore by *i.h.*

$$\#_{\min}(\Gamma(\langle\rangle A) \Rightarrow B) \le 0 \le \#_{\max}(\Gamma(\langle\rangle A) \Rightarrow B)$$

- $$\frac{\Gamma \Rightarrow A}{[\Gamma] \Rightarrow \langle\rangle A} \ \langle\rangle R$$

For atoms,

$$\#_{m,P}([\Gamma] \Rightarrow \langle\rangle A) = \#_{m,P}(\Gamma \Rightarrow A)$$

since brackets and bracket modalities are transparent to atom count.

For brackets,

$$\#_{m,[]}([\Gamma] \Rightarrow \langle\rangle A) =$$
$$\#^{\circ}_{m,[]}(\langle\rangle A) - \#^{\bullet}_{\overline{m},[]}([\Gamma]) =$$
$$\#^{\circ}_{m,[]}(A) + 1 - \#^{\bullet}_{\overline{m},[]}(\Gamma) - 1 =$$
$$\#^{\circ}_{m,[]}(A) - \#^{\bullet}_{\overline{m},[]}(\Gamma) =$$
$$\#_{m,[]}(\Gamma \Rightarrow A)$$

Therefore by *i.h.*:

$$\#_{\min}([\Gamma] \Rightarrow \langle\rangle A) \le 0 \le \#_{\max}([\Gamma] \Rightarrow \langle\rangle A)$$

3.1 Exponentials

- $$\frac{\Delta(A) \Rightarrow D}{\Delta(!A) \Rightarrow D} \ !L$$

For atoms,

$$\#_{\min,P}(\Delta(!A) \Rightarrow D) =$$
$$\#^{\circ}_{\min,P}(D) - \#^{\bullet}_{\max,P}(\Delta) - \#^{\bullet}_{\max,P}(!A) =$$
$$\#^{\circ}_{\min,P}(D) - \#^{\bullet}_{\max,P}(\Delta) - X(\#^{\bullet}_{\max,P}(A)) \le$$
$$\#^{\circ}_{\min,P}(D) - \#^{\bullet}_{\max,P}(\Delta) - \#^{\bullet}_{\max,P}(A) =$$
$$\#_{\min,P}(\Delta(A) \Rightarrow D) \le 0 \ i.h.$$

For brackets,

$$\#_{\min,[]}(\Delta(!A) \Rightarrow D) =$$
$$\#^{\circ}_{\min,[]}(D) - \#^{\bullet}_{\max,[]}(\Delta) - \#^{\bullet}_{\max,[]}(!A) =$$
$$\#^{\circ}_{\min,[]}(D) - \#^{\bullet}_{\max,[]}(\Delta) - \top \leq$$
$$\#^{\circ}_{\min,[]}(D) - \#^{\bullet}_{\max,[]}(\Delta) - \#^{\bullet}_{\max,[]}(A) =$$
$$\#_{\min,[]}(\Delta(A) \Rightarrow D) \leq 0 \; i.h.$$

For atoms and brackets,

$$\#_{\max,Q}(\Delta(!A) \Rightarrow D) =$$
$$\#^{\circ}_{\max,Q}(D) - \#^{\bullet}_{\min,Q}(\Delta) - \#^{\bullet}_{\min,Q}(!A) =$$
$$\#^{\circ}_{\max,Q}(D) - \#^{\bullet}_{\min,Q}(\Delta) - Y(\#^{\bullet}_{\min,Q}(A)) \geq$$
$$\#^{\circ}_{\max,Q}(D) - \#^{\bullet}_{\min,Q}(\Delta) - \#^{\bullet}_{\min,Q}(A) =$$
$$\#_{\max,Q}(\Delta(A) \Rightarrow D) \geq 0 \; i.h.$$

$$\bullet \quad \frac{!A_1, \ldots, !A_n \Rightarrow A}{!A_1, \ldots, !A_n \Rightarrow !A} \; !R$$

For atoms,

$$\#_{m,P}(!A_1, \ldots, !A_n \Rightarrow !A) =$$
$$\#^{\circ}_{m,P}(!A) - \#^{\bullet}_{\overline{m},P}(!A_1, \ldots, !A_n) =$$
$$\#^{\circ}_{m,P}(A) - \#^{\bullet}_{\overline{m},P}(!A_1, \ldots, !A_n) =$$
$$\#_{m,P}(!A_1, \ldots, !A_n \Rightarrow A)$$

For brackets,

$$\#_{m,[]}(!A_1, \ldots, !A_n \Rightarrow !A) =$$
$$\#^{\circ}_{m,[]}(!A) - \#^{\bullet}_{\overline{m},[]}(!A_1, \ldots, !A_n) =$$
$$\#^{\circ}_{m,[]}(A) - \#^{\bullet}_{\overline{m},[]}(!A_1, \ldots, !A_n) =$$
$$\#^{\circ}_{m,[]}(A) - \#^{\bullet}_{\overline{m},[]}(!A_1, \ldots, !A_n) =$$
$$\#_{m,[]}(!A_1, \ldots, !A_n \Rightarrow A) \geq 0 \; i.h.$$

$$\bullet \quad \frac{\Delta(!A_0, \ldots, !A_n, [!A_0, \ldots, !A_n, \Gamma]) \Rightarrow D}{\Delta(!A_0, \ldots, !A_n, \Gamma) \Rightarrow D} \; !C$$

For atoms,

$$\#_{\min}(\Delta(!A_0, \ldots, !A_n, \Gamma) \Rightarrow D) =$$
$$\#^{\circ}_{\min}(D) - \#^{\bullet}_{\max}(\Delta, \Gamma)$$
$$-\#^{\bullet}_{\max}(!A_0) - \cdots - \#^{\bullet}_{\max}(!A_n) =$$
$$\#^{\circ}_{\min}(D) - \#^{\bullet}_{\max}(\Delta, \Gamma)$$
$$-X(\#^{\bullet}_{\max}(A_0)) - \cdots - X(\#^{\bullet}_{\max}(A_n)) \leq$$
$$\#^{\circ}_{\min}(D) - \#^{\bullet}_{\max}(\Delta, [\Gamma])-$$
$$X(\#^{\bullet}_{\max}(A_0)) - \cdots - X(\#^{\bullet}_{\max}(A_n))-$$
$$X(\#^{\bullet}_{\max}(A_0)) - \cdots - X(\#^{\bullet}_{\max}(A_n)) =$$
$$\#_{\min}(\Delta(!A_0, \ldots, !A_n, |$$
$$[!A_0, \ldots, !A_n, \Gamma]) \Rightarrow D) \leq 0$$

For brackets,

$$\#_{\min}(\Delta(!A_0, \ldots, !A_n, \Gamma) \Rightarrow D) =$$
$$\#^{\circ}_{\min}(D) - \#^{\bullet}_{\max}(\Delta, \Gamma)$$
$$-\#^{\bullet}_{\max}(!A_0) - \cdots - \#^{\bullet}_{\max}(!A_n) =$$
$$\#^{\circ}_{\min}(D) - \#^{\bullet}_{\max}(\Delta, \Gamma) - \top - \cdots - \top \leq$$
$$\#^{\circ}_{\min}(D) - \#^{\bullet}_{\max}(\Delta, [\Gamma]) - \top - \cdots - \top-$$
$$\top - \cdots - \top =$$
$$\#_{\min}(\Delta(!A_0, \ldots, !A_n,$$
$$[!A_0, \ldots, !A_n, \Gamma]) \Rightarrow D) \leq 0$$

And for atoms and brackets,

$$\#_{\max}(\Delta(!A_0, \ldots, !A_n, \Gamma) \Rightarrow D) =$$
$$\#^{\circ}_{\max}(D) - \#^{\bullet}_{\min}(\Delta, \Gamma)$$
$$-\#^{\bullet}_{\min}(!A_0) - \cdots - \#^{\bullet}_{\min}(!A_n) =$$
$$\#^{\circ}_{\max}(D) - \#^{\bullet}_{\min}(\Delta, \Gamma)$$
$$-Y(\#^{\bullet}_{\min}(A_0)) - \cdots - Y(\#^{\bullet}_{\min}(A_n)) \geq$$
$$\#^{\circ}_{\max}(D) - \#^{\bullet}_{\min}(\Delta, [\Gamma])-$$
$$Y(\#^{\bullet}_{\min}(A_0)) - \cdots - Y(\#^{\bullet}_{\min}(A_n))-$$
$$Y(\#^{\bullet}_{\min}(A_0)) - \cdots - Y(\#^{\bullet}_{\min}(A_n)) =$$
$$\#_{\max}(\Delta(!A_0, \ldots, !A_n,$$
$$[!A_0, \ldots, !A_n, \Gamma]) \Rightarrow D) \geq 0$$

$$\bullet \quad \frac{\Delta(A) \Rightarrow D \quad \Delta(A, A) \Rightarrow D \ldots}{\Delta(?A) \Rightarrow D} \; ?L$$

For atoms and brackets,

$$\#_{\min}(\Delta(?A) \Rightarrow D) =$$
$$\#^{\circ}_{\min}(D) - \#^{\bullet}_{\max}(\Delta) - \#^{\bullet}_{\max}(?A) =$$
$$\#^{\circ}_{\min}(D) - \#^{\bullet}_{\max}(\Delta) - X(\#^{\bullet}_{\max}(A)) \leq$$
$$\#^{\circ}_{\min}(D) - \#^{\bullet}_{\max}(\Delta) - \#^{\bullet}_{\max}(A) =$$
$$\#_{\min}(\Delta(A) \Rightarrow D) \leq 0 \; i.h.$$

And

$$\#_{\max}(\Delta(?A) \Rightarrow D) =$$
$$\#^{\circ}_{\max}(D) - \#^{\bullet}_{\min}(\Delta) - \#^{\bullet}_{\min}(?A) =$$
$$\#^{\circ}_{\max}(D) - \#^{\bullet}_{\min}(\Delta) - Y(\#^{\bullet}_{\min}(A)) \geq$$
$$\#^{\circ}_{\max}(D) - \#^{\bullet}_{\min}(\Delta) - \#^{\bullet}_{\min}(A) =$$
$$\#_{\max}(\Delta(A) \Rightarrow D) \geq 0 \; i.h.$$

$$\bullet \quad \frac{\Gamma \Rightarrow A}{\Gamma \Rightarrow ?A} \; ?R$$

For atoms and brackets,

$$\#_{\min}(\Gamma \Rightarrow ?A) =$$
$$\#^{\circ}_{\min}(?A) - \#^{\bullet}_{\max}(\Gamma) =$$
$$Y(\#^{\circ}_{\min}(A)) - \#^{\bullet}_{\max}(\Gamma) \leq$$
$$\#^{\circ}_{\min}(A) - \#^{\bullet}_{\max}(\Gamma) =$$
$$\#^{\circ}_{\min}(\Gamma \Rightarrow A)$$

And,

$$\#_{\max}(\Gamma \Rightarrow ?A) =$$
$$\#^{\circ}_{\max}(?A) - \#^{\bullet}_{\min}(\Gamma) =$$
$$X(\#^{\circ}_{\max}(A)) - \#^{\bullet}_{\min}(\Gamma) \geq$$
$$\#^{\circ}_{\max}(A) - \#^{\bullet}_{\min}(\Gamma) =$$
$$\#^{\circ}_{\max}(\Gamma \Rightarrow A)$$

$$\bullet \quad \frac{\Gamma \Rightarrow A \qquad \Delta \Rightarrow ?A}{\Gamma, \Delta \Rightarrow ?A} \; ?M$$

For atoms and brackets,

$$\#_{\min}(\Gamma, \Delta \Rightarrow ?A) =$$
$$\#^{\circ}_{\min}(?A) - \#^{\bullet}_{\max}(\Gamma) - \#^{\bullet}_{\max}(\Delta) =$$
$$Y(\#^{\circ}_{\min}(A)) - \#^{\bullet}_{\max}(\Gamma) - \#^{\bullet}_{\max}(\Delta) \leq$$
$$Y(\#^{\circ}_{\min}(A)) + \#^{\circ}_{\min}(A) - \#^{\bullet}_{\max}(\Gamma) - \#^{\bullet}_{\max}(\Delta) =$$
$$\underbrace{\#^{\circ}_{\min}(A) - \#^{\bullet}_{\max}(\Gamma)}_{\leq\, 0 \; i.h.} + \underbrace{\#^{\circ}_{\min}(?A) - \#^{\bullet}_{\max}(\Delta)}_{\leq\, 0 \; i.h.}$$
$$\underbrace{}_{\leq\, 0}$$

And,

$$\#_{\max}(\Gamma, \Delta \Rightarrow ?A) =$$
$$\#^{\circ}_{\max}(?A) - \#^{\bullet}_{\min}(\Gamma) - \#^{\bullet}_{\min}(\Delta_n) =$$
$$X(\#^{\circ}_{\max}(A)) - \#^{\bullet}_{\min}(\Gamma) - \#^{\bullet}_{\min}(\Delta) \geq$$
$$X(\#^{\circ}_{\max}(A)) + \#^{\circ}_{\max}(A)) - \#^{\bullet}_{\min}(\Gamma) - \#^{\bullet}_{\min}(\Delta) =$$
$$\underbrace{\#^{\circ}_{\max}(A) - \#^{\bullet}_{\min}(\Gamma)}_{\geq\, 0 \; i.h.} + \underbrace{\#^{\circ}_{\max}(?A) - \#^{\bullet}_{\min}(\Delta)}_{\geq\, 0 \; i.h.}$$
$$\underbrace{}_{\geq\, 0}$$

$\blacksquare$

4 Evaluation

By way of evaluation of the exponential count invariance we compared the performance of CatLog2 (version f8.1) which uses only multiplicative and additive count invariance with CatLog version j2 which uses in addition the exponential invariance,[1] both running under XGP Prolog on a MacBook Air. The lexicon was the same in both cases.

We timed individually the exhaustive parsing of the expressions in Figure 3. Thus, for the sentence a:

(10) **[john]+likes+the+man** : $S f$

there is the result of lexical lookup:

(11) $[\blacksquare Nt(s(m)) : j],$
$\quad \Box((\langle\rangle \exists g Nt(s(g)) \backslash S f)/\exists a Na) :$
$\quad {}^{\wedge}\lambda A \lambda B(Pres\,((\check{}like\,A)\,B)),$

---

[1] The engines are otherwise the same.

$\blacksquare \forall n(Nt(n)/CNn) : \iota,$
$\Box CNs(m) : man \;\Rightarrow\; S f$

Note that these types include, in addition to the Lambek connectives, normal modalities for intensionality —$\blacksquare$ for rigid designators and $\Box$ for semantically active intensionality— and first-order quantifiers for features; these connectives are transparent to count-invariance. There is the derivation given in Figure 4, which delivers logical form:

(12) $(Pres\,((\check{}like\,(\iota\,\check{}man))\,j))$

CatLog proceeds by focalised proof search (Morrill and Valentín, 2015b). The focusing discipline considerably reduces redundancy in the sequent proof search space. The focusing discipline comprises alternating phases of invertible rule application and focalised non-invertible rule application. The boxes in the derivations mark the *focused* types in focused rule application, i.e. the active types decomposed by non-invertible rule applications. The focusing constrains proof search but in displaying proofs the boxes are limited to this decorative role.

For the sentence d:

(13) **man+[[that+[john]+likes]]** : $CNs(m)$

there is the lexical lookup:

(14) $\Box CNs(m) : man,$
$\quad [[\blacksquare \forall n([]^{-1}[]^{-1}(CNn \backslash CNn)/\blacksquare((\langle\rangle Nt(n)\sqcap$
$\quad !\blacksquare Nt(n))\backslash S f)) : \lambda A \lambda B \lambda C[(B\,C) \wedge (A\,C)],$
$\quad [\blacksquare Nt(s(m)) : j],$
$\quad \Box(((\langle\rangle \exists g Nt(s(g))\backslash S f)/\exists a Na) :$
$\quad {}^{\wedge}\lambda D \lambda E(Pres\,((\check{}like\,D)\,E))]] \;\Rightarrow\; CNs(m)$

Note that these types include also an additive and an exponential which are subject to the count-invariance presented in this paper. There is the derivation given in Figure 5. This uses stoups for the sequent derivation with exponentials (Girard, 2011), (Morrill, 2017). It delivers the logical form:

(15) $\lambda C[(\check{}man\,C) \wedge (Pres\,((\check{}like\,C)\,j))]$

The resulting times in seconds were as follows:

a. John likes the man.

b. Mary thinks that John likes the man.

c. Suzy believes that Mary thinks that John likes the man.

d. man that John likes

e. man that Mary thinks that John likes

f. man that Suzy believes that Mary thinks that John likes

g. Mary talks and Bill sings.

h. John walks Mary talks and Bill sings.

i. Suzy laughs John walks Mary talks and Bill sings.

j. Bill walks Suzy laughs John walks Mary talks and Bill sings.

k. Suzy talks Bill walks Suzy laughs John walks Mary talks and Bill sings.

l. John sings Suzy talks Bill walks Suzy laughs John walks Mary talks and Bill sings.

Figure 3: Example sentences

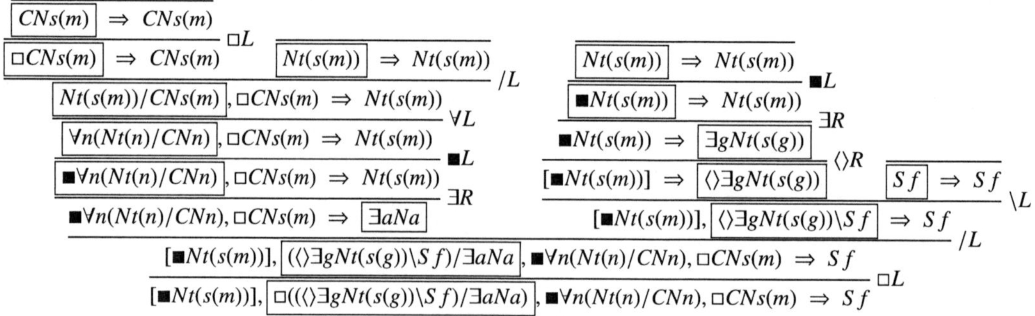

Figure 4: Derivation of example a

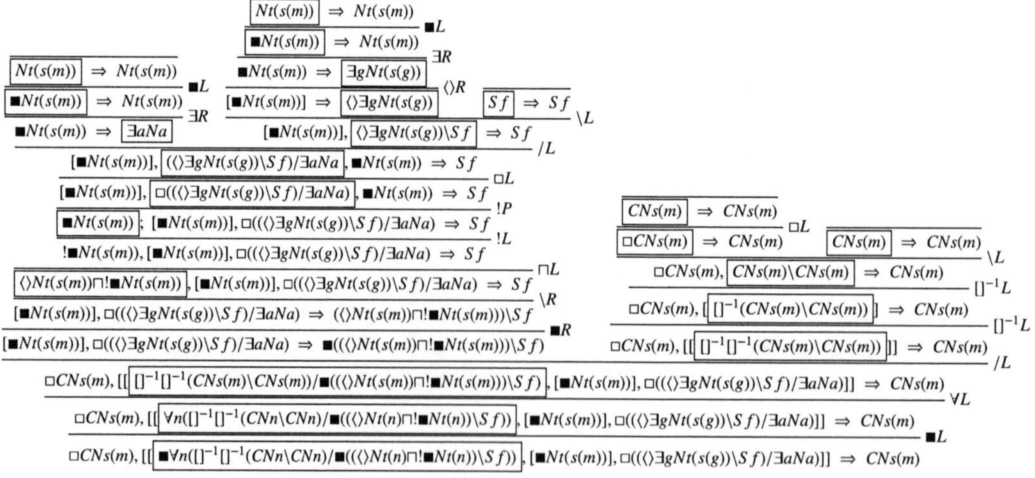

Figure 5: Derivation of example d

(16)

	f8.1 (no exp. inv)	j2 (exp. inv)
a.	1	1
b.	2	2
c.	40	6
d.	2	2
e.	4	4
f.	265	6
g.	1	1
h.	2	1
i.	2	1
j.	2	2
k.	2	3
l.	2	4

We see that for the longer, third, example of a-c there is a speedup. This is mostly in the time taken to discard inappropriate lexical choices (e.g. *that* is lexically ambiguous between a complementiser and a relative pronoun): to show that there are no further analyses. The examples d-f involve the universal exponential in a relative pronoun roughly of the form $(CN\backslash CN)/(((\langle\rangle N\sqcap !N)\backslash S)$; the (semantically inactively) additively conjoined hypothetical subtypes are for subject relativisation and object relativisation respectively (Morrill, 2017). Again we see a considerable speedup with exponential count invariance in the longer third case. The examples g-l involve the existential exponential in a coordinator assignment roughly of the form $(?S\backslash[\ ]^{-1}[\ ]^{-1}S)/S$ to obtain the iteration. Here there is no gain from the exponential type invariance; the overhead causes a slowdown.

For the minicorpus examples of the *Montague Test* (Morrill and Valentín, 2016a), and for the full CatLog2 corpus (Montague minicorpus, typical categorial examples, discontinuity examples, relativisation and coordination examples, and some Scripture) the parsing times in seconds were:

(17)

	f8.1	j2
Montague Test.	37	32
CatLog2 corpus	826	643

We interpret the experiment as showing that the pruning of the search space of count-invariance including exponentials outweighs the overhead that it engenders: it delivers a speedup of around 20%.

Acknowledgements

SK supported by RFBR. GM supported by an ICREA Academia 2012. GM and OV supported by MINECO TIN2014-57226-P (APCOM).

References

K. Ajdukiewicz. 1935. Die syntaktische Konnexität. *Studia Philosophica* 1:1–27. Translated in S. McCall, editor, 1967, *Polish Logic: 1920–1939*, Oxford University Press, Oxford, 207–231.

J.-Y. Girard. 1987. Linear logic. *Theoretical Computer Science* 50:1–102.

J.-Y. Girard. 2011. *The Blind Spot*. European Mathematical Society, Zürich.

J. Lambek. 1958. The mathematics of sentence structure. *American Mathematical Monthly* 65:154–170.

G. Morrill. 2012. CatLog: A Categorial Parser/Theorem-Prover. In *LACL 2012 System Demonstrations*. Nantes, Logical Aspects of Computational Linguistics 2012, pages 13–16.

G. Morrill. 2017. Grammar logicised: relativisation. *Linguistics and Philosophy* 40(2):119–163.

G. Morrill and O. Valentín. 2015a. Computational Coverage of TLG: Nonlinearity. In M. Kanazawa, L.S. Moss, and V. de Paiva, editors, *Proceedings of NLCS'15. Third Workshop on Natural Language and Computer Science*. Kyoto, volume 32 of *EPiC*, pages 51–63.

G. Morrill and O. Valentín. 2015b. Multiplicative-Additive Focusing for Parsing as Deduction. In I. Cervesato and C. Schürmann, editors, *First International Workshop on Focusing, LPAR 2015*. Suva, Fiji, number 197 in EPTCS, pages 29–54.

G. Morrill and O. Valentín. 2016a. Computational coverage of Type Logical Grammar: The Montague Test. In C. Piñón, editor, *Empirical Issues in Syntax and Semantics*, Colloque de Syntaxe et Sémantique à Paris, Paris, volume 11, pages 141–170.

G. Morrill and O. Valentín. 2016b. On the Logic of Expansion in Natural Language. In M. Amblard, P. de Groote, S. Pogodalla, and C. Retoré, editors, *Proceedings of Logical Aspects of Computational Linguistics, LACL'16, Nancy*. Springer, Berlin, volume 10054 of *LNCS, FoLLI Publications on Logic, Language and Information*, pages 228–246.

O. Valentín, D. Serret, and G. Morrill. 2013. A Count Invariant for Lambek Calculus with Additives and Bracket Modalities. In G. Morrill and M.-J. Nederhof, editors, *Proceedings of Formal Grammar 2012 and 2013*. Springer, Berlin, volume 8036 of *Springer LNCS, FoLLI Publications in Logic, Language and Information*, pages 263–276.

J. van Benthem. 1991. *Language in Action: Categories, Lambdas, and Dynamic Logic*. Number 130 in Studies in Logic and the Foundations of Mathematics. North-Holland, Amsterdam.

Conjunctive Categorial Grammars

Stepan Kuznetsov
Steklov Mathematical Institute
Moscow, Russia
`sk@mi.ras.ru`

Alexander Okhotin
St. Petersburg State University
St. Petersburg, Russia
`alexander.okhotin@spbu.ru`

Abstract

Basic categorial grammars are enriched with a conjunction operation, and it is proved that the formalism obtained in this way has the same expressive power as conjunctive grammars, that is, context-free grammars enhanced with conjunction. It is also shown that categorial grammars with conjunction can be naturally embedded into the Lambek calculus with conjunction and disjunction operations. This further implies that a certain NP-complete set can be defined in the Lambek calculus with conjunction.

1 Introduction

This paper establishes a connection between two formal grammar models that emerged within two different theories of syntax.

One theory is the *immediate constituent analysis*, which has its roots in the traditional grammar, and was investigated in the early 20th century by the structural linguists. It reached universal recognition under the name "context-free grammars," introduced in Chomsky's early work. In this paradigm, a grammar assigns certain properties to groups of words, such as "noun phrase" (NP), "verb phrase" (VP) or "sentence" (S). These properties are known as *syntactic categories*, or, in Chomsky's terminology, *nonterminal symbols*. Rules of a grammar, such as S $\to$ NP VP, show how shorter substrings with known properties can be concatenated to form longer strings belonging to a certain category.

In the other theory, which was discovered by Ajdukiewicz (1935), further developed by Bar-Hillel et al. (1960), and is nowadays known as *categorial grammars*, syntactic categories are defined in different way. "Noun phrases" are treated as "the category of phrases *equivalent* to nouns," whereas verb phrases are defined as "the category of phrases, which would form a complete sentence, if a noun, or anything equivalent to a noun, is concatenated on the left," denoted by NOUN\SENTENCE. A categorial grammar explicitly assigns categories to individual words; and then, by definition, a concatenation of any string of type NOUN with any string of type NOUN\SENTENCE forms a complete sentence. The crucial feature of this approach is that the laws that govern reduction of categories, namely, $A\,(A \setminus B)$ to B, and $(B\,/\,A)\,A$ to B, are universal. In contrast, Chomsky's context-free formalism uses different rules for different categories.

A formal connection between these two models was established by Bar-Hillel et al. (1960), who proved them to be equivalent in power: a language is defined by a context-free grammar if and only if it is defined by a basic categorial grammar (assuming languages do not contain the empty string).

More than half a century of research gave birth to many extensions of both basic models. Categorial grammars were the first to get an interesting extension: the *Lambek calculus*, introduced by Lambek (1958), augments the model with additional derivation rules. Later, Pentus (1993) established that this extended model is still equivalent in power to context-free grammars. Pentus' translation yields a context-free grammar of exponential size with respect to the original Lambek grammar. For the special case of *unidirectional* Lambek grammars, which use only one kind of division operators ($\setminus$, $/$), but not both, Kuznetsov (2016), using the ideas of Savateev (2009), presents a polynomial translation into context-free grammars. Other generalizations of categorial grammars include combinatory categorial grammars by Steedman (1996), categorial dependency grammars by Dekhtyar and Dikovsky

140

Proceedings of the 15th Meeting on the Mathematics of Language, pages 140–151,
London, UK, July 13–14, 2017. ©2017 Association for Computational Linguistics

(2008), and others.

Lambek grammars, in their turn, can be generalized further. Modern extensions of the Lambek calculus with new operations, such as those considered by Carpenter (1997), Morrill (2011), and Moot and Retoré (2012), are capable of describing quite sophisticated syntactic phenomena.

From the point of view of modern logic, the Lambek calculus is a substructural logical system, namely, a non-commutative variation of *linear logic*, introduced by Girard (1987), see Abrusci (1990), Yetter (1990). Linear logic offers many logical operations, and some of them can be used in the non-commutative case for extending Lambek grammars.

Morrill (2011) and his collaborators, following and extending Moortgat (1996), consider a system, based on the Lambek calculus, with discontinuous connectives, subexponentials for controlled non-linearity, brackets for controlled non-associativity, and many other operations. The use of negation in categorial grammars was considered by Buszkowski (1996). Kanazawa (1992) investigated the power of Lambek grammars with conjunction and disjunction operations that are "additive operations" in terms of linear logic.

Numerous generalized models have also been introduced in the paradigm of immediate constituent analysis, as extensions of the context-free formalism. One direction is to extend the form of constituents, that is, sentence fragments to which syntactic categories are being assigned in a grammar. The most well-known of these models are the multi-component grammars, introduced by Vijay-Shanker et al. (1987) and by Seki et al. (1991), inspired by an earlier model by Fischer (1968): these grammars define the properties of discontinuous constituents, that is, substrings with a bounded number of gaps. Extensions of another kind augment the model by introducing new logical operators to be used in grammar rules: for instance, *conjunctive grammars*, featuring a conjunction operation, and *Boolean grammars*, further equipped with negation, were introduced by Okhotin (2001, 2004). Earlier, Latta and Wall (1993) argued for the relevance of such operations in linguistic descriptions. The main results on conjunctive grammars indicate that they preserve the practically useful properties of context-free grammars, such as efficient parsing algorithms, while substantially extending their expressive power. The known re-sults on conjunctive grammars are presented in a fairly recent survey by Okhotin (2013).

A few years ago, Kuznetsov (2013) compared the expressive power of Lambek grammars with conjunction, as considered by Kanazawa (1992), with that of conjunctive grammars. It was proved that a large subclass of conjunctive grammars (namely, conjunctive grammars in Greibach normal form) can be simulated in the Lambek calculus with conjunction, but the exact power of the latter remains undetermined.

This paper makes a fresh attempt at introducing conjunction in categorial grammars. The new model extends basic categorial grammars, rather than Lambek grammars, and for that reason it uses categories and rules of a simpler form than in the earlier model by Kanazawa (1992) and Kuznetsov (2013). Yet, it is shown that this model can simulate every conjunctive grammar. A converse simulation is presented as well, which implies the equivalence of the two models.

As compared to the classical equivalence result for context-free grammars and basic categorial grammars, the new result requires a more elaborate construction. One particular difficulty is that the normal form theorems for conjunctive grammars are weaker than those for the context-free grammars: in particular, no analogue of the Greibach normal form is known for conjunctive grammars. For this reason, the simulation of conjunctive grammars by the proposed conjunctive categorial grammars relies on a different normal form by Okhotin and Reitwießner (2010). This leads to a representation of the whole class of conjunctive grammars, in contrast to the result by Kuznetsov (2013), which is valid only for grammars in Greibach normal form.

The second result is that conjunctive categorial grammars, as defined in this paper, can be represented in the Lambek calculus with the conjunction operation, as considered by Kanazawa (1992), and therefore this extension of the Lambek calculus is at least as powerful as are the conjunctive grammars. Furthermore, it is proved that Kanazawa's model (Kanazawa, 1992) can describe an NP-complete language, which conjunctive grammars cannot describe unless $P = NP$.

2 Basic Categorial Grammars and Context-Free Grammars

Let Σ be a finite *alphabet* of the language being defined. Its elements are called *symbols*. In linguistic descriptions, symbols typically represent words of the language. The set of non-empty strings over Σ is denoted by Σ^+. Throughout this paper, any subset of Σ^+ is a *language*, that is, all languages are assumed to be without the empty string.

The models considered in this paper are derived from two classical formal grammar frameworks: basic categorial grammars and context-free grammars.

Basic categorial grammars (BCG) have their roots in the works of Ajdukiewicz (1935). Let Σ be an alphabet. Let $\mathbf{Pr} = \{p, q, r, \dots\}$ be a finite set of *primitive categories*, and let $s \in \mathbf{Pr}$ be a designated *target category* of all syntactically correct sentences.

The set $\mathbf{BCat}$ of *basic categories* is defined as follows.

- Every primitive category is a basic category.

- If $A \in \mathbf{BCat}$ and $p \in \mathbf{Pr}$, then $(p \backslash A) \in \mathbf{BCat}$ and $(A / p) \in \mathbf{BCat}$.

The definition of a basic categorial grammar is given in terms of logical *propositions*, which are expressions of the form $B(v)$, where $B \in \mathbf{BCat}$ and $v \in \Sigma^+$. This proposition states that v is a string of syntactic category B.

The language of propositions is indeed very simple: all propositions are atomic, there are no variables and quantifiers (if the syntactic category B is considered, in the spirit of first-order logic, as a predicate, then its argument, v, is a constant term).

A categorial grammar is regarded as a *logical calculus* for deriving categorial propositions. It includes a finite set of axioms (axiomatic propositions) of the form $A(a)$, where $A \in \mathbf{BCat}$ and $a \in \Sigma$, and the following inference rules.

$$\frac{p(u) \quad (p \backslash A)(v)}{A(uv)} \qquad \frac{(A / p)(u) \quad p(v)}{A(uv)}$$

The string w belongs to the language generated by the BCG if and only if the proposition $s(w)$ is derivable by means of this calculus.

Example 1. *The basic categorial grammar with the following axiomatic propositions and with s as the target category ($\mathbf{Pr} = \{s, p, q\}$) describes the language $\{ba^n ca^n \mid n \geqslant 0\}$.*

$$(s / p)(b), \quad p(c), \quad (p / q)(a), \quad (p \backslash q)(a)$$

Another, more well-known formal grammar framework is the phrase-structure formalism, defined by Chomsky (1956) and later renamed into *context-free grammars* (CFG). In a CFG, there is a fixed finite set of categories N (usually called "non-terminal symbols"), and one of them is designated as the initial symbol $S \in N$. The grammar is defined by a finite set of rules (or "productions") of the form $A \to \beta$, where $A \in N$ and $\beta \in (\Sigma \cup N)^+$.

Even though Chomsky's original definition of context-free grammars was given in terms of string rewriting, it is more convenient—at least in this paper—to present it as a logical derivation similar to the one in categorial grammars. Propositions in the context-free framework are of the form $\beta(u)$, where $\beta \in (\Sigma \cup N)^+$ and $u \in \Sigma^+$. Intuitively, such a proposition means that u can be derived from β using the rules of the CFG. Axioms of the calculus of propositions are of the form $a(a)$, $a \in \Sigma$, and the rules of inference are as follows.

$$\frac{\beta_1(u_1) \quad \beta_2(u_2)}{(\beta_1 \beta_2)(u_1 u_2)}$$

$$\frac{\beta(v)}{A(v)} \qquad \text{for each rule } A \to \beta$$

Again, the string w belongs to the language generated by this grammar if and only if the proposition $S(w)$ is derivable.

Example 2. *The language from Example 1 is described by the following CFG.*

$$S \to bA$$

$$A \to aAa \mid c$$

As usual, "$A \to aAa \mid c$" is a short-hand notation for two rules, $A \to aAa$ and $A \to c$.

There is an important difference between BCGs and CFGs: in BCGs, the linguistic information is stored in the axioms (in other words, it is *lexicalized*), while the inference rules are the same for all BCGs. For CFGs, the situation is opposite: axioms are trivial, and all information is kept in the rules. However, these two formalisms are equivalent in power.

Theorem A. *A language is generated by a BCG if and only if it is generated by a CFG.* (Bar-Hillel et al., 1960)

$$
\frac{
  \begin{array}{ccc}
   & \dfrac{c(c)}{} & \\
  a(a) & B(c) & a(a)
  \end{array}
}{}
$$

Figure 1

3 Conjunction in Grammars

In this section, both grammar formalisms are enriched with a *conjunction* operation. Using conjunction, one can impose multiple syntactic constraints on the same phrase at the same time. The extension of context-free grammars with conjunction, called *conjunctive grammars,* was introduced by Okhotin (2001).

Let Σ be the alphabet, and let N be the set of categories ("non-terminal symbols"), with $S \in N$ representing all well-formed sentences. A conjunctive grammar is defined by a finite set of rules of the form $A \to \beta_1 \& \ldots \& \beta_k$, with $\beta_i \in (\Sigma \cup N)^+$. If k is 1, then this is an ordinary rule $A \to \beta$, as in an ordinary context-free grammar.

Propositions in a conjunctive grammar are of the form $\beta(u)$, where $u \in \Sigma^+$ and $\beta \in (\Sigma \cup N)^+$. Axioms of the calculus of propositions are of the form $a(a)$, where $a \in \Sigma$. The first inference rule is as follows.

$$
\frac{\beta_1(u_1) \quad \beta_2(u_2)}{(\beta_1\beta_2)(u_1 u_2)}
$$

The other inference rule is valid for each grammar rule $A \to \beta_1 \& \ldots \& \beta_k$ and for each string v.

$$
\frac{\beta_1(v) \quad \ldots \quad \beta_k(v)}{A(v)}
$$

The string w belongs to the language generated by the grammar if and only if the proposition $S(w)$ is derivable from the axioms.

Example 3. *The following conjunctive grammar describes the language $\{ba^n ca^n ca^n \mid n \geqslant 1\}$.*

$$
S \to bBcA \,\&\, bAcB
$$
$$
A \to aA \mid a
$$
$$
B \to aBa \mid c
$$

The rules for A and B use no conjunction, and have the same effect as in ordinary context-free grammars. Thus, $bBcA(w)$ is true for all strings

of the form $w = ba^n ca^n ca^i$, with $n \geqslant 0$, $i \geqslant 1$, whereas $bAcB(w)$ holds true for strings of the form $w = ba^i ca^n ca^n$. The conjunction of these two conditions is exactly the condition of membership in the desired language, and the rule for S ensures it by derivations of the following form.

$$
\frac{bBcA(ba^n ca^n ca^n) \quad bAcB(ba^n ca^n ca^n)}{S(ba^n ca^n ca^n)}
$$

A full derivation of the string $w = bacaca$ is given in Figure 1.

The notion of a *conjunctive categorial grammar* is defined by extending basic categorial grammars with the conjunction operation. Let $\mathbf{Pr} = \{p, q, r, \ldots\}$ be the set of primitive categories, $s \in \mathbf{Pr}$ is the target category. The set of *conjuncts,* $\mathbf{Conj}$, is defined as follows:

1. every primitive category is a conjunct;

2. if $p_1, \ldots, p_k \in \mathbf{Pr}$, then $(p_1 \wedge \cdots \wedge p_k) \in \mathbf{Conj}$.

The set of *basic categories with conjunction,* $\mathbf{BCat}_\wedge$, is defined as follows.

1. Every primitive category belongs to $\mathbf{BCat}_\wedge$.

2. If $\mathcal{C} \in \mathbf{Conj}$ and $A \in \mathbf{BCat}_\wedge$, then $(\mathcal{C} \backslash A) \in \mathbf{BCat}_\wedge$ and $(A / \mathcal{C}) \in \mathbf{BCat}_\wedge$.

Categorial propositions are expressions of the form $B(v)$, where $v \in \Sigma^+$ and $B \in \mathbf{BCat}_\wedge \cup \mathbf{Conj}$. A conjunctive categorial grammar is a logical theory deriving categorial propositions. It includes an arbitrary finite set of axioms of the form $A(a)$, with $A \in \mathbf{BCat}_\wedge$ and $a \in \Sigma$, and the following inference rules.

$$
\frac{p_1(v) \quad \ldots \quad p_k(v)}{(p_1 \wedge \cdots \wedge p_k)(v)}
$$

$$
\frac{\mathcal{C}(u) \quad (\mathcal{C} \backslash A)(v)}{A(uv)} \qquad \frac{(A / \mathcal{C})(v) \quad \mathcal{C}(u)}{A(vu)}
$$

143

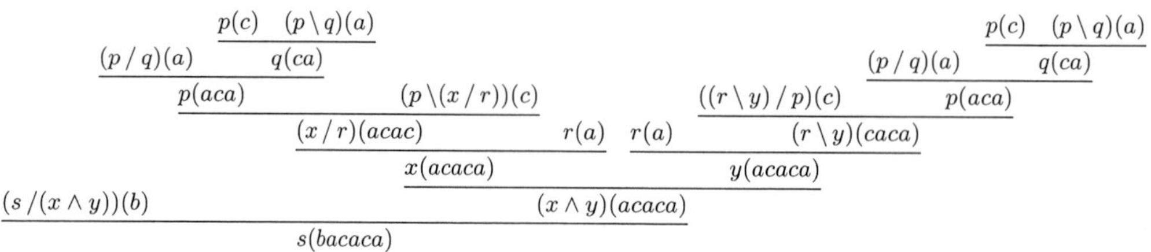

Figure 2

The string w belongs to the language generated by this grammar if and only if the proposition $s(w)$ is derivable.

Example 4. *The categorial conjunctive grammar with the set of primitive categories* $\mathbf{Pr} = \{s, x, y, p, q, r\}$, s *as the target category, and with the following set of axioms describes the language* $\{ba^n ca^n ca^n \mid n \geqslant 1\}$, *the same as in Example 3.*

$$r(a), \quad (r\,/\,r)(a),$$
$$p(c), \quad (p\,/\,q)(a), \quad (p\,\backslash\,q)(a),$$
$$(p\,\backslash(x\,/\,r))(c), \quad ((r\,\backslash\,y)\,/\,p)(c),$$
$$(s\,/(x \wedge y))(b).$$

The category p is defined in the same way as in Example 1. Then, using further categories without conjunction, the propositions $x(a^n ca^n ca^i)$ and $y(a^i ca^n ca^n)$, for all $n \geqslant 0$, $i \geqslant 1$, are derived as in an ordinary categorial grammar. The only strings that satisfy both conditions, x and y, are those of the form $a^n ca^n ca^n$, and these are therefore all strings in the category s, derived as follows.

$$\frac{\dfrac{x(a^n ca^n ca^n) \quad y(a^n ca^n ca^n)}{(s\,/(x \wedge y))(b) \qquad (x \wedge y)(a^n ca^n ca^n)}}{s(ba^n ca^n ca^n)}$$

A complete derivation of the proposition $s(bacaca)$ is presented in Figure 2.

The calculus used in the conjunctive categorial grammar formalism enjoys the following *inverted subformula property* (ISF): if a category of the form $(C \backslash A)$ or $(A\,/\,C)$ appears somewhere in the derivation, then it is a subexpression of some category used in an axiom. (The notion of subexpression on categories is defined in a standard way: each conjunct (in particular, primitive category) is a subexpression of itself, and subexpressions of $(C \backslash A)$ include $C \backslash A$, C, and all subexpressions

of A; symmetrically for $(A\,/\,C)$. To prove the ISF, we trace the rightmost branch of the derivation upwards; finally we reach an axiom that includes the goal category as a subexpression.)

Another useful property is the fact that the rule for $\wedge$ is invertible: if $(p_1 \wedge \ldots \wedge p_k)(v)$ is derivable, then so are $p_1(v), \ldots, p_k(v)$. Indeed, the only way to derive $(p_1 \wedge \ldots \wedge p_k)(v)$ is by applying this rule.

The calculus used in conjunctive categorial grammars also enjoys the following *cut elimination* property.

Lemma 1. *Let $A(u)$ (for some $A \in \mathbf{BCat}_\wedge$ and $u \in \Sigma^+$) be derivable in the given conjunctive categorial grammar. Consider a new conjunctive categorial grammar over an extended alphabet $\Sigma \cup \{b\}$, where $b \notin \Sigma$. The new grammar has all the same axioms as the original grammar, and an additional axiom $A(b)$. Then, if the new grammar derives $B(v_1 b v_2)$, for some $B \in \mathbf{BCat}_\wedge$ and arbitrary, possibly empty, strings v_1, v_2 over Σ, then $B(v_1 u v_2)$ is derivable in the original grammar.*

Proof. Consider the derivation of $B(v_1 b v_2)$ in the extended grammar and substitute u for all occurrences of b. Applications of inference rules remain valid; the same for axioms of the old grammar (they don't include b). The new axiom $A(b)$ becomes $A(u)$, which is derivable in the old grammar by assumption. $\qquad\square$

4 Equivalence of Conjunctive Grammars and Conjunctive Categorial Grammars

The main result of this paper is an extension of Theorem A for grammars with conjunction, stated as follows.

Theorem 1. *A language is generated by a conjunctive grammar if and only if it is generated by a conjunctive categorial grammar.*

144

The proof uses the following two known properties of conjunctive grammars. The first result is their closure under quotient with a single symbol.

Lemma B. *If L is a language over Σ described by a conjunctive grammar, and $a \in \Sigma$ is any symbol, then there exists a conjunctive grammar that describes the language $a^{-1}L = \{w \mid aw \in L\}$.* (Okhotin and Reitwießner, 2010, Thm. 2)

The other result is a normal form theorem. A conjunctive grammar G with the initial symbol S is in the *odd normal form*, if all its rules are of the following form, with $A \in N$, $a \in \Sigma$, $B_i, C_i \in N$, and $a_i \in \Sigma$.

$$A \to a$$
$$A \to B_1 a_1 C_1 \,\&\, \ldots \,\&\, B_k a_k C_k$$
$$S \to aA$$

Rules of the latter kind are allowed only if S is never referenced in any rules.

Theorem C. *Every language described by a conjunctive grammar can be described by a conjunctive grammar in odd normal form.* (Okhotin and Reitwießner, 2010)

Proof of Theorem 1. **The "if" part** is easier. A conjunctive grammar equivalent to a given categorial grammar G has the set N comprised of all categories used in the axioms of G, and of all their subexpressions (categories and conjuncts). The conjunctive rules are now as follows.

$$(p_1 \wedge \ldots \wedge p_k) \to p_1 \,\&\, \ldots \,\&\, p_k,$$
$$A \to C\,(C \setminus A),$$
$$A \to (A\,/\,C)\,C,$$

for all $(p_1 \wedge \ldots \wedge p_k), (C \setminus A), (A\,/\,C) \in N$, and

$$A \to a, \quad \text{if } A(a) \text{ is an axiom in } G.$$

For **the "only if" part** of the proof, the first step is to transform a given conjunctive grammar. Let $\Sigma = \{a_1, \ldots, a_n\}$. For each symbol a_i, by Lemma B, there is a conjunctive grammar G_i that describes the quotient $a_i^{-1}L$. By Theorem C, this grammar can be assumed to be in the odd normal form. It can also be assumed that, for $i \neq j$, the grammars G_i and G_j have disjoint sets of non-terminal symbols. Let S_i be the initial symbol of G_i. Then this grammar is further modified as follows. Every rule

$$A \to B_1 a_1 C_1 \,\&\, \ldots \,\&\, B_k a_k C_k$$

is replaced with $k + 1$ new rules:

$$A \to \widetilde{X}_1 \,\&\, \ldots \,\&\, \widetilde{X}_k \qquad \text{and} \qquad \widetilde{X}_i \to B_i a_i C_i,$$

where $\widetilde{X}_i$ are fresh non-terminals. For the sake of uniformity, rules of the form

$$S_i \to aA$$

are replaced with

$$S_i \to \widetilde{Y} \qquad \text{and} \qquad \widetilde{Y} \to aA,$$

and rules of the form

$$A \to a$$

are replaced with

$$A \to \widetilde{Z} \text{ and } \widetilde{Z} \to a.$$

Finally, a new conjunctive grammar for L is obtained by joining these grammars together, for all i, adding the following extra rules for the new initial symbol $\widetilde{S}$.

$$\widetilde{S} \to a_1 S_1, \qquad \ldots, \qquad \widetilde{S} \to a_n S_n$$

In the resulting grammar, all non-terminals are of two sorts (with and without a tilde), and the rules have the following form.

$$A \to \widetilde{X}_1 \,\&\, \ldots \,\&\, \widetilde{X}_k \qquad \text{(here } k \text{ could be 1)}$$
$$\widetilde{X} \to BaC, \qquad \widetilde{Y} \to aA, \qquad \text{and} \qquad \widetilde{Z} \to a$$

It is then transformed to a conjunctive categorial grammar, with the set of primitive categories $\mathbf{Pr} = \{p_{\widetilde{X}} \mid \widetilde{X} \text{ is a non-terminal decorated with a tilde }\}$, and with the following axioms.

1. For each rule $\widetilde{Z} \to a$, there is an axiom $p_{\widetilde{Z}}(a)$.

2. For each pair of rules $\widetilde{Y} \to aA$ and $A \to \widetilde{X}_1 \,\&\, \ldots \,\&\, \widetilde{X}_k$, the axiom is $\left(p_{\widetilde{Y}}\,/\,(p_{\widetilde{X}_1} \wedge \ldots \wedge p_{\widetilde{X}_k})\right)(a)$.

3. For each triple of rules $\widetilde{X} \to BaC$, $B \to \widetilde{Y}_1 \,\&\, \ldots \,\&\, \widetilde{Y}_k$, and $C \to \widetilde{Z}_1 \,\&\, \ldots \,\&\, \widetilde{Z}_m$, the axiom is

$$\left(((p_{\widetilde{Y}_1} \wedge \ldots \wedge p_{\widetilde{Y}_k}) \setminus p_{\widetilde{X}})\,/\,(p_{\widetilde{Z}_1} \wedge \ldots \wedge p_{\widetilde{Z}_m})\right)(a).$$

The target category is $p_{\widetilde{S}}$.

Claim. *For every non-terminal $\widetilde{X}$ decorated with a tilde, the proposition $p_{\widetilde{X}}(v)$ is derivable in the newly constructed conjunctive categorial grammar if and only if $\widetilde{X}(v)$ is derivable in the original conjunctive grammar.*

Proof. **The "if" part.** The proof proceeds by induction on the derivation size. There are three possible cases.

Case 1. The proposition $\widetilde{X}(v)$ is actually of the form $\widetilde{Z}(a)$ and is derived from $a(a)$ using the rule $\widetilde{Z} \to a$. Then $p_{\widetilde{Z}}(a)$ is an axiom in the conjunctive categorial grammar.

Case 2. The proposition $\widetilde{X}(v)$ is of the form $\widetilde{Y}(av_1)$ and is derived from $a(a)$ and $A(v_1)$ using a rule of the form $\widetilde{Y} \to aA$. Next, a rule of the form $A \to \widetilde{X}_1 \& \ldots \& \widetilde{X}_k$ should be applied for A. Therefore, the propositions $\widetilde{X}_i(v_1)$ are derivable (for all i) in the original conjunctive grammar. Then, by induction hypothesis, $p_{\widetilde{X}_i}(v_1)$ are derivable in the conjunctive categorial grammar, and there is a derivation for $p_{\widetilde{Y}}(av_1)$ shown in Figure 3.

Case 3. The proposition $\widetilde{X}(v)$ is of the form $\widetilde{X}(v_1av_2)$ and is derived from some propositions of the form $B(v_1)$, $a(a)$, and $C(v_2)$, following the rule $\widetilde{X} \to BaC$. Next, for B and C, some rules for the form $B \to \widetilde{Y}_1 \& \ldots \& \widetilde{Y}_k$ and $C \to \widetilde{Z}_1 \& \ldots \& \widetilde{Z}_m$ should be applied. Therefore, the propositions $\widetilde{Y}_i(v_1)$ and $\widetilde{Z}_j(v_2)$ are derivable (for all i, j) in the original conjunctive grammar. Then, by induction hypothesis, $p_{\widetilde{Y}_i}(v_1)$ and $p_{\widetilde{Z}_j}(v_2)$ are derivable in the conjunctive categorial grammar, and there is a derivation for $p_{\widetilde{X}}(v_1av_2)$, as shown in Figure 4.

The "only if" part. This time, it is assumed that $p_{\widetilde{X}}(v)$ is derivable in the newly constructed conjunctive categorial grammar. The proof is by induction on its derivation.

The *axiom case* is trivial: any axiom of the form $p_{\widetilde{Z}}(a)$ is associated with a rule $\widetilde{Z} \to a$ in the original conjunctive grammar, and then $\widetilde{Z}(a)$ is derivable from $a(a)$.

In the *left division* case, $v = v_1w$, and the last step of the derivation is as follows.

$$\frac{\mathcal{C}(v_1) \quad (\mathcal{C} \backslash p_{\widetilde{X}})(w)}{p_{\widetilde{X}}(v_1w)}$$

By the ISF (see above), $(\mathcal{C} \backslash p_{\widetilde{X}})$ is a subexpression of the category in one of the axioms. The only possibility is that $(\mathcal{C} \backslash p_{\widetilde{X}})$ is a subexpression $((p_{\widetilde{Y}_1} \wedge \ldots \wedge p_{\widetilde{Y}_k}) \backslash p_{\widetilde{X}})$ of an axiom

$((p_{\widetilde{Y}_1} \wedge \ldots \wedge p_{\widetilde{Y}_k}) \backslash p_{\widetilde{X}}) / (p_{\widetilde{Z}_1} \wedge \ldots \wedge p_{\widetilde{Z}_m})$. Moreover, again by the ISF, the only way to derive $(\mathcal{C} \backslash p_{\widetilde{X}})(w)$ is to apply the right division rule to the category used in the axiom. This analysis shows that the derivation must end in the way depicted in the earlier Figure 4, where $w = av_2$.

Since the rules for $\wedge$ are invertible (see above), the propositions $p_{\widetilde{Y}_1}(v_1)$, ..., $p_{\widetilde{Y}_k}(v_1)$, $p_{\widetilde{Z}_1}(v_2)$, ..., $p_{\widetilde{Z}_m}(v_2)$ are derivable. By induction hypothesis, $\widetilde{Y}_i(v_1)$ and $\widetilde{Z}_j(v_2)$, for all i, j, are derivable in the original conjunctive grammar. Then, the derivation uses the rules $\widetilde{X} \to BaC$, $B \to \widetilde{Y}_1 \& \ldots \& \widetilde{Y}_k$, and $C \to \widetilde{Z}_1 \& \ldots \& \widetilde{Z}_m$, and is of the following form.

$$\frac{\dfrac{\widetilde{Y}_1(v_1) \ldots \widetilde{Y}_k(v_1)}{B(v_1)} \quad a(a) \quad \dfrac{\widetilde{Z}_1(v_2) \ldots \widetilde{Z}_m(v_2)}{C(v_2)}}{\widetilde{X}(v_1av_2)}$$

The *right division* case is even easier. Here a derivation ends as follows.

$$\frac{(p_{\widetilde{Y}} / \mathcal{C})(w) \quad \mathcal{C}(v_1)}{p_{\widetilde{Y}}(wv_1)}$$

By the ISF, the left premise could be nothing but an axiom of the form $\left(p_{\widetilde{Y}} / (p_{\widetilde{X}_1} \wedge \ldots \wedge p_{\widetilde{X}_k})\right)(a)$ (and $w = a$). Then, $\mathcal{C}(v_1)$ is $(p_{\widetilde{X}_1} \wedge \ldots \wedge p_{\widetilde{X}_k})(v_1)$, and by the invertibility of the $\wedge$ rule, all $p_{\widetilde{X}_i}(v_1)$ are derivable. By the induction hypothesis, $\widetilde{X}_i(v_1)$, for all i, are derivable in the original conjunctive grammar, and there is the following derivation for $\widetilde{Y}(av_1)$, using the rules $\widetilde{Y} \to aA$ and $A \to \widetilde{X}_1 \& \ldots \& \widetilde{X}_k$.

$$\frac{a(a) \quad \dfrac{\widetilde{X}_1(v_1) \quad \ldots \quad \widetilde{X}_k(v_k)}{A(v_1)}}{\widetilde{Y}(av_1)}$$

$\square$

This claim immediately yields the main result, since $\widetilde{S}(w)$ is derivable in the original conjunctive grammar if and only if $p_{\widetilde{S}}(w)$ is derivable in the constructed conjunctive categorial grammar. $\square$

5 Conjunctive Categorial Grammars and Lambek Grammars with Additives

Lambek (1958) suggested a richer logic as a background for categorial grammars, called *the Lambek calculus*. In the Lambek calculus, or **L** for short, syntactic categories built from a set of $\mathbf{Pr} =$

$$\frac{\left(p_{\widetilde{Y}}/(p_{\widetilde{X}_1} \wedge \ldots \wedge p_{\widetilde{X}_k})\right)(a) \quad \dfrac{p_{\widetilde{X}_1}(v_1) \quad \ldots \quad p_{\widetilde{X}_k}(v_1)}{(p_{\widetilde{X}_1} \wedge \ldots \wedge p_{\widetilde{X}_k})(v_1)}}{p_{\widetilde{Y}}(av_1)}$$

Figure 3

$$\frac{(p_{\widetilde{Y}_1} \wedge \ldots \wedge p_{\widetilde{Y}_k})(v_1) \quad \dfrac{\left(((p_{\widetilde{Y}_1} \wedge \ldots \wedge p_{\widetilde{Y}_k}) \backslash p_{\widetilde{X}})/(p_{\widetilde{Z}_1} \wedge \ldots \wedge p_{\widetilde{Z}_m})\right)(a) \quad (p_{\widetilde{Z}_1} \wedge \ldots \wedge p_{\widetilde{Z}_m})(v_2)}{((p_{\widetilde{Y}_1} \wedge \ldots \wedge p_{\widetilde{Y}_k}) \backslash p_{\widetilde{X}})(av_2)}}{p_{\widetilde{X}}(v_1 a v_2)}$$

Figure 4

$\{p_1, p_2, p_3, \ldots\}$ of primitive categories using three binary operations: product ($\cdot$), which means concatenation, left division ($\backslash$), and right division ($/$). The formal recursive definition is as follows.

1. Every primitive category is a category.

2. If A and B are categories, then $(A \cdot B)$, $(A \backslash B)$, and (B / A) are also categories.

The set of all Lambek categories is denoted by **Cat**. As opposed to basic categories, deep nesting of division operations is allowed here, that is denominators are allowed to be non-primitive.

A Lambek categorial grammar consists of a target category $s \in$ **Cat** (usually s is required to be a primitive category) and a finite number of axiomatic propositions of the form $A(a)$, where A is a category and a is a letter of the alphabet.

A string $w = a_1 \ldots a_n$ is considered accepted by the grammar, if, for some categories A_1, ..., A_n, the propositions $A_i(a_i)$ are included in the grammar as axiomatic ones, and the *sequent* $A_1, \ldots, A_n \to s$ is derivable in the Lambek calculus, which consists of the axioms and inference rules listed in Figure 5. In all rules, left-hand sides of the sequents are required to be non-empty.

Note that in Lambek grammars, arrows traditionally point in an opposite direction than in context-free grammars ($\ldots \to s$ vs. $S \to \ldots$).

The following *cut rule* is not officially included in the system, but is admissible (Lambek, 1958).

$$\frac{\Pi \to A \quad \Gamma, A, \Delta \to D}{\Gamma, \Pi, \Delta \to D} \ (\text{cut})$$

As one can easily see, all basic categories, as defined in Section 2, are also Lambek categories: **BCat** $\subset$ **Cat**. Moreover, as noticed by Buszkowski (1985), if a basic categorial grammar is regarded as a Lambek categorial grammar with the same set of axiomatic propositions, then it describes the same language.

Next, the Lambek calculus is extended with the so-called "additive" conjunction and disjunction, as defined by Kanazawa (1992). These new operations correspond to the additive operations in linear logic by Girard (1987). Inference rules for these operations are depicted in Figure 6.

This calculus, denoted by **MALC** ("multiplicative-additive Lambek calculus"), also enjoys cut elimination and the subformula property.

Lambek categories with $\wedge$ and $\vee$ generalize conjunctive categories (and conjuncts):

$$\mathbf{BCat}_\wedge \cup \mathbf{Conj} \subset \mathbf{Cat}_{\wedge, \vee},$$

and every conjunctive categorial grammar can be translated into a Lambek grammar with $\wedge$ and $\vee$. However, one cannot simply take the axiomatic propositions of a conjunctive categorial grammar and use them as axiomatic propositions in the sense of Lambek grammars: this would yield a grammar that is not equivalent to the original one (for instance, the Lambek grammar with the axiomatic propositions from Example 4 does not accept any strings at all). The construction has to be more subtle.

Theorem 2. *Let* $\Sigma = \{a_1, \ldots, a_n\}$ *and consider a conjunctive categorial grammar with the following axiomatic propositions.*

$$\begin{array}{llll} A_{1,1}(a_1), & A_{1,2}(a_1), & \ldots & A_{1,k_1}(a_1), \\ A_{2,1}(a_2), & A_{2,2}(a_2), & \ldots & A_{2,k_2}(a_2), \\ & \vdots & & \\ A_{n,1}(a_n), & A_{n,2}(a_n), & \ldots & A_{n,k_n}(a_n). \end{array}$$

147

$$A \to A$$

$$\frac{\Pi \to A \quad \Gamma, B, \Delta \to D}{\Gamma, \Pi, A \backslash B, \Delta \to D} \ (\backslash \to) \qquad \frac{A, \Pi \to B}{\Pi \to A \backslash B} \ (\to \backslash) \qquad \frac{\Gamma, A, B, \Delta \to D}{\Gamma, A \cdot B, \Delta \to D} \ (\cdot \to)$$

$$\frac{\Pi \to A \quad \Gamma, B, \Delta \to D}{\Gamma, B \,/\, A, \Pi, \Delta \to D} \ (/ \to) \qquad \frac{\Pi, A \to B}{\Pi \to B \,/\, A} \ (\to /) \qquad \frac{\Gamma \to A \quad \Delta \to B}{\Gamma, \Delta \to A \cdot B} \ (\to \cdot)$$

Figure 5: The Lambek Calculus

$$\frac{\Gamma, A_1, \Delta \to D}{\Gamma, A_1 \wedge A_2, \Delta \to D} \ (\wedge \to)_1 \qquad \frac{\Gamma, A_2, \Delta \to D}{\Gamma, A_1 \wedge A_2, \Delta \to D} \ (\wedge \to)_2 \qquad \frac{\Pi \to A_1 \quad \Pi \to A_2}{\Pi \to A_1 \wedge A_2} \ (\to \wedge)$$

$$\frac{\Gamma, A_1, \Delta \to D \quad \Gamma, A_2, \Delta \to D}{\Gamma, A_1 \vee A_2, \Delta \to D} \ (\vee \to) \qquad \frac{\Pi \to A_1}{\Pi \to A_1 \vee A_2} \ (\to \vee)_1 \qquad \frac{\Pi \to A_2}{\Pi \to A_1 \vee A_2} \ (\to \vee)_2$$

Figure 6: Rules for Conjunction and Disjunction

Then the Lambek grammar with atomic propositions $(A_{i,1} \wedge A_{i,2} \wedge \ldots \wedge A_{i,k_i})(a_i)$ *(for* $i = 1, \ldots, n$*) describes the same language as the original conjunctive categorial grammar. (If* $k_i = 1$*, we take just* $A_{i,1}(a_i)$*.)*

Proof. Let $B_i = A_{i,1} \wedge A_{i,2} \wedge \ldots \wedge A_{i,k_i}$. The new Lambek grammar uses axiomatic propositions of the form $B_i(a_i)$, one for each symbol in Σ. It is sufficient to prove the following: for the target category $s \in \mathbf{Pr}$ and for a string $a_{i_1} \ldots a_{i_m}$, the proposition $s(a_{i_1} \ldots a_{i_m})$ is derivable in the conjunctive categorial grammar if and only if the sequent $B_{i_1}, \ldots, B_{i_m} \to s$ is derivable in **MALC**.

The "only if" part. In order to use induction on the length of derivation in the conjunctive categorial grammar, the statement is proved not only for s, but for an arbitrary category $D \in \mathbf{BCat}_\wedge \cup \mathbf{Conj}$.

The proof in the base case is immediate: if $D(a_i)$ is an axiom, then D is one of the $A_{i,j}$ in the conjunction B_i, and the sequent $B_i \to A_{i,j}$ is derivable by several applications of the $(\wedge \to)$ rules.

For the induction step, there are three cases.

Case 1: $D = (p_1 \wedge \ldots \wedge p_k)$. Then, by the induction hypothesis, $B_{i_1}, \ldots, B_{i_m} \to p_j$ is derivable in **MALC** for every j, and $B_{i_1}, \ldots, B_{i_m} \to p_1 \wedge \ldots \wedge p_k$ is derived by the $(\to \wedge)$ rule.

Case 2: $D(a_{i_1} \ldots a_{i_m})$ is derived from $\mathcal{C}(a_{i_1} \ldots a_{i_\ell})$ and $(\mathcal{C} \backslash D)(a_{i_{\ell+1}} \ldots a_{i_m})$ for some $\mathcal{C} \in \mathbf{Conj}$. Then, by the induction hypothesis, the sequents $B_{i_1}, \ldots, B_{i_\ell} \to \mathcal{C}$ and $B_{i_{\ell+1}}, \ldots, B_{i_m} \to \mathcal{C} \backslash D$ are derivable, and then $B_{i_1}, \ldots, B_{i_m} \to D$ can be derived in the following way. First, $B_{i_1}, \ldots, B_{i_\ell}, \mathcal{C} \backslash D \to D$ is derived from $B_{i_1}, \ldots, B_{i_\ell} \to \mathcal{C}$ and $D \to D$, and then it is combined with $B_{i_{\ell+1}}, \ldots, B_{i_m} \to \mathcal{C} \backslash D$ using the cut rule, to get $B_{i_1}, \ldots, B_{i_\ell}, B_{i_{\ell+1}}, \ldots, B_{i_m} \to D$.

Case 3: $D(a_{i_1} \ldots a_{i_m})$ is derived from $(D \,/\, \mathcal{C})(a_{i_1} \ldots a_{i_\ell})$ and $\mathcal{C}(a_{i_{\ell+1}} \ldots a_{i_m})$. The proof is symmetric.

The "if" part. The following more general statement is claimed. *For every* $j = 1, \ldots, m$, *let* B'_{i_j} *be a conjunction of an arbitrary subset of formulae* $A_{i_j,k}$ *used in the conjunction* B_{i_j}; *in other words,* B'_{i_j} *may coincide with* B_{i_j} *or lack some of the conjuncts. Then, for any* $\mathcal{C} \in \mathbf{Conj}$ *(in particular, for* $\mathcal{C} = s \in \mathbf{Pr} \subset \mathbf{Conj}$), *if* $B'_{i_1}, \ldots, B'_{i_m} \to \mathcal{C}$ *is derivable in* **MALC**, *then the proposition* $\mathcal{C}(a_{i_1} \ldots a_{i_m})$ *is derivable in the original conjunctive categorial grammar.*

The claim is proved by induction on the cut-free derivation of the sequent $B'_{i_1}, \ldots, B'_{i_m} \to \mathcal{C}$ in **MALC**.

Case 1. $\mathcal{C} = p_1 \wedge \ldots \wedge p_k$, $k \geq 2$. Since the $(\to \wedge)$ rule in **MALC** is invertible (this follows from the cut elimination), it can be assumed that all $k-1$

applications of this rule were applied immediately.

$$\frac{B'_{i_1}, \ldots, B'_{i_m} \to p_1 \quad \ldots \quad B'_{i_1}, \ldots, B'_{i_m} \to p_k}{B'_{i_1}, \ldots, B'_{i_m} \to p_1 \wedge \ldots \wedge p_k}$$

Then, by the induction hypothesis, all propositions $p_j(a_1 \ldots a_m)$ are derivable in the conjunctive categorial grammar, and from them one can derive $(p_1 \wedge \ldots \wedge p_k)(a_1 \ldots a_m)$.

In all other cases, $\mathcal{C} \in \mathbf{Pr}$.

Case 2: an axiom. Then, $m = 1$, $B'_{i_1} = \mathcal{C}$, and, since all elements of B'_{i_1} should be of the form $A_{i_1,k}$, the proposition $\mathcal{C}(a_1)$ is an axiom of the conjunctive categorial grammar, and therefore derivable.

Case 3: the last rule of the derivation is $(\wedge \to)$. Then, $B'_{i_\ell} = B''_{i_\ell} \wedge A_{i_\ell,k}$:

$$\frac{B'_{i_1}, \ldots, B''_{i_\ell}, \ldots, B'_{i_m} \to \mathcal{C}}{B'_{i_1}, \ldots, B''_{i_\ell} \wedge A_{i_\ell,k}, \ldots, B'_{i_m} \to \mathcal{C}}$$

or

$$\frac{B'_{i_1}, \ldots, A_{i_\ell,k}, \ldots, B'_{i_m} \to \mathcal{C}}{B'_{i_1}, \ldots, B''_{i_\ell} \wedge A_{i_\ell,k}, \ldots, B'_{i_m} \to \mathcal{C}}$$

In both cases the induction hypothesis is applied: since B''_{i_ℓ} or $A_{i_\ell,k}$ can act as B'_{i_ℓ}, the proposition $\mathcal{C}(a_{i_1} \ldots a_{i_\ell} \ldots a_{i_m})$ is derivable in the conjunctive categorial grammar.

Case 4: the last rule is $(\backslash \to)$. In this case, $B'_{i_h} = A_{i_h,k} = \mathcal{C}' \backslash A'$, for some h and for $\mathcal{C}' \in \mathbf{Conj}$ and $A' \in \mathbf{BCat}_\wedge$, and the sequent $B'_{i_1}, \ldots, B'_{i_{\ell-1}}, B'_{i_\ell}, \ldots, B'_{i_{h-1}}, \mathcal{C}' \backslash A', B'_{i_{h+1}}, \ldots, B'_{a_m} \to \mathcal{C}$ is derived from $B'_{i_\ell}, \ldots, B'_{i_{h-1}} \to \mathcal{C}'$ and $B'_{i_1}, \ldots, B'_{i_{\ell-1}}, A', B'_{i_{h+1}}, \ldots, B'_{i_m} \to \mathcal{C}$. By the induction hypothesis, the proposition $\mathcal{C}'(a_{i_\ell} \ldots a_{i_{h-1}})$ can be derived in the conjunctive categorial grammar, and, since $(\mathcal{C}' \backslash A')(a_{i_h})$ is an axiom, the proposition $A'(a_{i_\ell} \ldots a_{i_{h-1}} a_{i_h})$ is also derivable.

Now, the conjunctive categorial grammar is extended by adding a new symbol a_{n+1} to the original alphabet $\Sigma = \{a_1, \ldots, a_n\}$, with a new axiom, $A'(a_{n+1})$. For the new grammar, we have the same B_j for $j = 1, \ldots, n$, and $B_{n+1} = A'$. Since $B'_{i_1}, \ldots, B'_{i_{\ell-1}}, A', B'_{i_{h+1}}, \ldots, B'_{i_m} \to \mathcal{C}$ is derivable in $\mathbf{MALC}$, by the induction hypothesis, the proposition $\mathcal{C}(a_{i_1} \ldots a_{i_{\ell-1}} a_{n+1} a_{i_{h+1}} \ldots a_{i_m})$ is derivable in the extended conjunctive categorial grammar.

By Lemma 1, the desired proposition $\mathcal{C}(a_{i_1} \ldots a_{i_{\ell-1}} a_{i_\ell} \ldots a_{i_{h-1}} a_{i_h} a_{i_{h+1}} \ldots a_{i_m})$,

where the string $u = a_{i_\ell} \ldots a_{i_{h-1}} a_{i_h}$ has been substituted for a fresh symbol $b = a_{n+1}$, can be derived in the original conjunctive categorial grammar.

Case 5: the last rule is $(/ \to)$. Symmetric. $\square$

This embedding immediately implies that every language generated by a conjunctive grammar can be generated by an $\mathbf{MALC}$-grammar. This supersedes the result by Kuznetsov (2013).

In the classical case without the conjunction, a converse result was shown by Pentus (1993): every language generated by a Lambek grammar is context-free. Whether an analogous property holds for $\mathbf{MALC}$ (that is, whether every $\mathbf{MALC}$-language is generated by a conjunctive grammar) remains an open problem. Establishing any such upper bound on the power of the new model would require proving a non-trivial variant of the famous theorem by Pentus (1993), which would likely be difficult.

However, there is some evidence that $\mathbf{MALC}$ should be strictly more powerful than conjunctive grammars. First, there is a result by Okhotin (2011) that conjunctive grammars can describe a certain P-complete language representing the Circuit Value Problem (CVP) under a suitable encoding. On the other hand, the class of languages generated by $\mathbf{MALC}$-grammars is, by definition, closed under symbol-to-symbol homomorphisms. These two facts are sufficient to develop a $\mathbf{MALC}$ represenation for an NP-complete language, which is the last result of this paper.

Theorem 3. *The family of languages generated by $\mathbf{MALC}$-grammars contains an NP-complete language.*

Sketch of proof. It is not difficult to transform the grammar for the CVP given by Okhotin (2011), so that each CVP instance is represented in the form $u_{k,C}v$, where $u_{k,C} \in \Sigma^*$ is a description of a circuit C with k inputs, while $v \in \{0,1\}^k$ contains the input values, and $0, 1 \notin \Sigma$. The grammar then describes the set of all such strings, on which the circuit evaluates to 1 on the given input values.

$$\mathrm{CVP} = \{u_{k,C}v \mid C(v) = 1\}$$

Let $h \colon \Sigma \cup \{0,1\} \to \Sigma \cup \{?\}$ be a homomorphism that maps both digits to the question mark symbol, leaving all other symbols intact: $h(0) = h(1) = ?$, $h(a) = a$ for all $a \in \Sigma$. This transforms the

149

Circuit Value Problem to the Circuit Satisfiability Problem, which is NP-complete.

$$h(\mathrm{CVP}) = \{u_{k,C}?^k \mid \exists v \in \{0,1\}^k : C(v) = 1\}$$

Since the language CVP is described by a conjunctive grammar, by Theorem 1, it is also described by a conjunctive categorial grammar, and then, by Theorem 2, also by an **MALC**-grammar. Next, as observed by Kanazawa (1992), its symbol-to-symbol homomorphic image $h(\mathrm{CVP})$ must have an **MALC**-grammar as well. $\qquad\square$

On the other hand, every language described by a conjunctive grammar can be parsed in polynomial time—to be exact, in time $O(n^\omega)$, where $\omega < 3$ is the exponent in the complexity of matrix multiplication (Okhotin, 2014). This leads to the following corollary.

Corollary 1. *Under the assumption that* P $\neq$ NP, *conjunctive categorial grammars are strictly weaker in power than* **MALC**.

It would be interesting to establish an unconditional separation of these two classes.

Acknowledgments

Stepan Kuznetsov's research was supported by the Russian Foundation for Basic Research (grants 15-01-09218-a and 14-01-00127-a) and the Presidential Council for Support of Leading Scientific Schools of Russia (grant NŠ-9091.2016.1).

References

V. M. Abrusci. 1990. A comparison between Lambek syntactic calculus and intuitionistic linear propositional logic. *Zeitschrift für mathematische Logik und Grundlagen der Mathematik* 36(1):11–15.

K. Ajdukiewicz. 1935. Die syntaktische Konnexität. *Studia Philosophica* 1:1–27.

Y. Bar-Hillel, C. Gaifman, and E. Shamir. 1960. On categorial and phrase-structure grammars. *Bulletin of the Research Council of Israel, Section F* 9F:1–16.

W. Buszkowski. 1985. The equivalence of unidirectional Lambek categorial grammars and context-free grammars. *Zeitschrift für mathematische Logik und Grundlagen der Mathematik* 31:369–384.

W. Buszkowski. 1996. Categorial grammars with negative information. In H. Wansing, editor, *Negation. A Notion in Focus*, De Gruyter, pages 107–126.

B. Carpenter. 1997. *Type-Logical Semantics*. MIT Press, Cambridge, MA.

N. Chomsky. 1956. Three models for the description of language. *IRE Transactions on Information Theory* 2:113–124.

M. Dekhtyar and A. Dikovsky. 2008. Generalized categorial dependency grammars. In *Pillars of Computer Science (Trakhtenbrot Festschrift)*, Springer, volume 4800 of *LNCS*, pages 230–255.

M. J. Fischer. 1968. Grammars with macro-like productions. In *Proc. 9th Annual Symposium on Switching and Automata Theory (SWAT 1968)*. New York, pages 131–142.

J.-Y. Girard. 1987. Linear logic. *Theoretical Computer Science* 50(1):1–102.

M. Kanazawa. 1992. The lambek calculus enriched with additional connectives. *Journal of Logic, Language, and Information* 1:141–171.

S. Kuznetsov. 2013. Conjunctive grammars in greibach normal form and the lambek calculus with additive connectives. In *Proc. Formal Grammar 2012 and Formal Grammar 2013*. Springer, volume 8036 of *LNCS*, pages 242–249.

S. L. Kuznetsov. 2016. On translating Lambek grammars with one division into context-free grammars. *Proceedings of Steklov Institute of Mathematics* 294:129–138.

J. Lambek. 1958. The mathematics of sentence structure. *American Mathematical Monthly* 65:154–170.

M. Latta and R. Wall. 1993. Intersective context-free languages. In *Lenguajes Naturales y Lenguajes Formales IX*. Barcelona, pages 15–43.

M. Moortgat. 1996. Multimodal linguistic inference. *Journal of Logic, Language, and Information* 5(3, 4):349–385.

R. Moot and C. Retoré. 2012. *The Logic of Categorial Grammars: A Deductive Account of Natural Language Syntax and Semantics*. Springer, Heidelberg.

G. V. Morrill. 2011. *Categorial Grammar: Logical Syntax, Semantics, and Processing*. Oxford University Press.

A. Okhotin. 2001. Conjunctive grammars. *Journal of Automata, Languages, and Combinatorics* 6(4):519–535.

A. Okhotin. 2004. Boolean grammars. *Information and Computation* 194(1):19–48.

A. Okhotin. 2011. A simple p-complete problem and its language-theoretic representations. *Theoretical Computer Science* 412(1–2):68–82.

A. Okhotin. 2013. Conjunctive and boolean grammars: the true general case of the context-free grammars. *Computer Science Review* 9:27–59.

A. Okhotin. 2014. Parsing by matrix multiplication generalized to boolean grammars. *Theoretical Computer Science* 516:101–120.

A. Okhotin and C. Reitwießner. 2010. Conjunctive grammars with restricted disjunction. *Theoretical Computer Science* 411(26–28):2559–2571.

M. Pentus. 1993. Lambek grammars are context-free. In *Proc. 8th Annual IEEE Symposium on Logic in Computer Science (LICS 1993)*. Montreal, pages 430–433.

Yu. V. Savateev. 2009. Recognition of derivability for the lambek calculus with one division. *Moscow University Mathematics Bulletin* 64(2):73–75.

H. Seki, T. Matsumura, M. Fujii, and T. Kasami. 1991. On multiple context-free grammars. *Theoretical Computer Science* 88(2):191–229.

M. Steedman. 1996. *Surface structure and interpretation*. MIT Press.

K. Vijay-Shanker, D. J. Weir, and A. K. Joshi. 1987. Characterizing structural descriptions produced by various grammatical formalisms. In *Proc. 25th Annual Meeting of the Association for Computational Linguistics (ACL 1987)*. pages 104–111.

D. N. Yetter. 1990. Quantales and (noncommutative) linear logic. *Journal of Symbolic Logic* 55(1):41–64.

Author Index

Association for Computational Linguistics
209 N. Eighth Street
Stroudsburg, Pennsylvania 18360

ISBN 978-1-5108-4360-8